AF675304

Additive Manufacturing of Metals and Alloys: Microstructure and Mechanical Properties

Additive Manufacturing of Metals and Alloys: Microstructure and Mechanical Properties

Guest Editors

Federico Mazzucato
Samuel Rey-Mermet

MDPI

Basel • Beijing • Wuhan • Barcelona • Belgrade • Novi Sad • Cluj • Manchester

Guest Editors
Federico Mazzucato
Department of Innovative Technologies
University of Applied Sciences and Arts of Southern Switzerland (SUPSI)
Lugano-Viganello
Switzerland

Samuel Rey-Mermet
School of Engineering
HES-SO Valais-Wallis
Sion
Switzerland

Editorial Office
MDPI AG
Grosspeteranlage 5
4052 Basel, Switzerland

This is a reprint of the Special Issue, published open access by the journal *Materials* (ISSN 1996-1944), freely accessible at: https://www.mdpi.com/journal/materials/special_issues/VJT10B3YB7.

For citation purposes, cite each article independently as indicated on the article page online and as indicated below:

Lastname, A.A.; Lastname, B.B. Article Title. *Journal Name* **Year**, *Volume Number*, Page Range.

ISBN 978-3-7258-7166-7 (Hbk)
ISBN 978-3-7258-7167-4 (PDF)
https://doi.org/10.3390/books978-3-7258-7167-4

Contents

About the Editors

Federico Mazzucato

Federico Mazzucato is a senior lecturer and researcher at the Automation, Robotics and Machines (ARM) lab of the Department of Innovative Technologies, University of Applied Sciences and Arts of Southern Switzerland (SUPSI). His research area focuses on metal additive manufacturing, design for additive manufacturing, AM process optimisation and engineering, and advanced metallic structure design.

Samuel Rey-Mermet

Samuel Rey-Mermet is a full-time professor at the Institute of Systems Engineering, School of Engineering, HES-SO Valais-Wallis. His research area focuses on additive manufacturing, powder metallurgy, shape memory alloys, and functional materials.

materials

MDPI

Article

Effect of Laser Parameters on Fracture Properties of Laser-Repaired Cracks with Micro/NanoMaterial Addition: Multiscale Analysis

Yinyin Li [1], Wei Jiang [2,*] and Meiqiu Li [1]

1 School of Mechanical Engineering, Yangtze University, Jingzhou 434023, China; liyyhbsy@yangtzeu.edu.cn (Y.L.); limq@yangtzeu.edu.cn (M.L.)
2 School of Mechanical Engineering, Dalian University of Technology, Dalian 116024, China
* Correspondence: jiangwei@dlut.edu.cn

Abstract: In laser crack repair processes, laser parameters have significant influence on repair quality. Improper combination of laser process parameters may result in defects—such as porosity, ablation, and coarse grain size—in remelted zones. A trans-scale computational model is established by combining crystal plasticity finite elements and variable-node finite elements. The influence of microstructure characteristics such as grain size and porosity of the repair layer on the cumulative plastic slip (CPS) on the dominant slip system at the meso-scale and the J-integral at the macro-scale is studied to explore the effect of laser process parameters on repair quality. The results show that when the laser power is 1800 W and the heating time is 0.5 s, the grain size and porosity of the repaired specimen are the smallest. The J-integral of the repaired specimen is more than 8% smaller than that of the unrepaired specimen and about 3% smaller than that of the repaired specimen, with a laser power of 2000 W and a heating time of 1 s. Pores increase the CPS of the crystal around the pores, especially when a pore have sharp corners. Selecting appropriate laser process parameters can not only refine grain size but also reduce the volume fraction of pores and thus reduce the J-integral and eventually improve repair quality of repaired specimens. The study investigates the relationship of process parameter–microstructure–repair quality in the laser repair process and provides a method for studying the mechanical behavior of materials at macro and micro scales.

Keywords: laser process parameter; microstructure; fracture properties; crystal plasticity; cumulative plastic slip

1. Introduction

As one of common failure modes in engineering structures, fractures widely exist in aerospace, automobile, ship, and petroleum engineering. The presence of cracks not only reduces the bearing capacity of materials, but may even cause catastrophic disasters, resulting in huge losses of life and economy. Therefore, early-stage crack repair could provide substantial savings for both resources and economy [1].

Laser repair is a feasible method of repairing cracks via metallurgical combination of suitable materials and the workpiece to be repaired with high-energy laser beam [2]. Wang et al. [3] studied the microstructure and wear resistance of titanium alloy components repaired with laser cladding and found that repair coatings show similar microhardness and wear resistance compared with the substrate. Cha et al. [4] proposed a novel system to repair aeroengine airfoil by pulsed laser ablation to remove surface cracks in damaged zone. After repairing, an equivalent performance of high cycle fatigue strength and less microstructural damage can be achieved compared to traditional mechanical material removal methods.

The selection of laser process parameters is a key issue in laser repair of cracks. Laser repair is a dynamic physical metallurgical process involving the interaction between laser, powder, and substrate. During the interaction between laser and materials, the temperature and flow fields in the molten pool affect the convection, solidification process, composition distribution, and microstructure, thus affecting the quality of the laser-repaired layer. Therefore, it is of great significance to choose the appropriate laser process parameters to improve repair quality. The existence of pores is one of the factors that are very sensitive to laser process parameters in the repaired layer. When the packing density of added powders is low, the gas between the powders may dissolve in the molten pool [5]. Under the action of laser, the powder will melt and solidify rapidly, and the gas has no chance to escape from the molten pool. Pores are thus formed in the repaired layer. When excessive input energy is provided, gas bubbles will be generated due to vaporization of low-melting-point ingredients within the alloy. If gas bubbles do not have enough time to rise and escape from the molten pool, regular spherical pores will form. When the input energy is insufficient, the metal powder cannot be melted completely, resulting in irregular pores [6]. Khairallah et al. [7] revealed the formation mechanisms of pore defects in strong dynamic melt flow in laser powder bed fusion by a powder-scale model. Han et al. [8] studied the influence the porosity distribution and mechanical response in laser and metal inert gas hybrid welding of 6082-T6 aluminum alloy. It is found that when the porosity is 3%, the fracture feature of material changes from pure ductile to the mixed mode of ductile and brittle. Cracks initiate on the weld surface when joints have no pores and initiate near pores when pores exist in the joint. Pekok et al. [9] investigated the formation mechanisms and characteristics of porosity in aluminum 2024 alloy through selective laser melting using high and low energy densities and found that low energy density reduces grain size and increases the microhardness of alloys.

The microstructure of the repair layer near the crack tip has undergone significant changes after laser irradiation, which will inevitably affect the mechanical properties of the material. The crystal plastic finite element (CPFE) method has outstanding advantages in solving meso-scale mechanics problems, such as grain size effect, grain interaction, strain gradient effect, crystal high-temperature deformation, texture evolution, recrystallization, etc. [10–15]. The influence of microstructure on the fracture performance of the repaired parts is studied using the CPFE method on the micro scale, which is helpful to improve the laser repair quality. Wang et al. [16] investigated the coupling effects of columnar grains and residual stress on the mechanical behavior of 316L stainless steel formed by selective laser melting through CPFE methods. The simulation results can predict the experimental trends well. Zhang et al. [17] investigated the high-cycle and very-high-cycle fatigue properties of AlSi10Mg alloy prepared via selective laser melting using CFEM method and successfully analyzed the effects of pores and inclusions on the fatigue performance of the material. In the aspect of fracture mechanics, Kirane and Ghosh [18] studied the early crack initiation by a CPFE model. It was found that the load shedding between grains is the main reason for early crack initiation. He et al. [19] employed the CPFE method to predict the slip line distribution as well as the crack initiation location and propagation direction, with preference. The significant stress gradient at the crack tip highlights the crucial role of the microstructure in the vicinity of the crack tip in determining the fracture resistance of the repaired layer. By employing the CPFE method to investigate the influence of the microstructure near the crack tip on the fracture properties of the repaired layer, we can gain valuable insights into the laser repair mechanism of cracks.

Jiang et al. [20] studied the effect of laser process parameters on the repair effect of CT specimens with cracks through experiments; while Li et al. [21,22] established a macro–meso trans-scale model to study the macroscopic and microscopic mechanical behavior of materials. On the basis of these studies, this article analyzes the influence of laser process parameters on the microstructure of materials, with a focus on the influence of laser process parameters on porosity in the cladding layer. Then, a macro–meso trans-scale model is used to study the influence of microstructure such as grain size, pore distribution, and pore shape on the macro- and micromechanical properties of repaired parts. The results offer a theoretical guidance for selecting appropriate process parameters for laser repair of stainless steel materials with cracks.

2. Experiment and Models

2.1. Experiment and Materials

According to ASTM E647-24 standards [23], compact tension (CT) specimens of 304 stainless steel (304SS) with a thickness of 5 mm were processed as the substrate for laser repair of cracks. The dimensions and photograph of the specimen are shown in Figure 1. For a detailed description of the chemical composition of the specimen, refer to Reference [24]. A notch measuring 2 mm in length and 0.2 mm in width was created to simulate a crack using a wire electric discharge machining system. During the laser repair process, the entire 2 mm notch was effectively repaired.

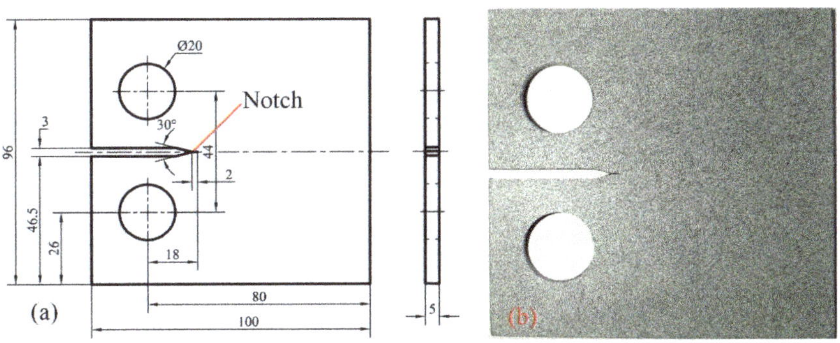

Figure 1. (**a**) The dimensions of CT specimen (unit: mm); (**b**) The photograph of CT specimen.

The micro/nano alloy powder utilized in the laser repair experiment is composed of 3 wt.% nano WC, 1 wt.% nano Al_2O_3, 0.5 wt.% V and 304 stainless steel powder. The particle sizes of 304SS powder, WC, Al_2O_3 and V powder are within the ranges of 30–50 μm, 40–50 nm, 30–50 nm, and 20–25 μm, respectively. All powder materials are provided by Shanghai Xiangtian Nano Materials Co., Ltd., Shanghai, China.

The laser repair experiment was carried out by a CO_2 laser. The mode of the laser is multi-mode, and the light intensity is approximately circular and evenly distributed. The light intensity outside the spot is null, and the heat flux density of circular laser spot is:

$$q(r) = \frac{P}{\pi R^2} \tag{1}$$

where q is the heat flux of internal heat source, r is the distance from the spot center; P represents the laser power; and R is the spot diameter.

With a laser spot diameter D of 3 mm, a laser crack repair experiment with different laser power P and laser heating time t are arranged as follows:

(1) Select laser heating time t as 1.0 s and laser power P as 1600 W, 1800 W, and 2000 W, respectively.
(2) Select laser power P as 1800 W and laser heating time t as 0.5 s, 1.0 s, and 2.0 s, respectively.

After repairing, the repair effect was studied by tensile test with the repaired CT specimen. Real time variation of displacement field of specimen with the load was recorded by a digital image correlation (DIC) system during the tensile test. Figure 2 shows the experimental setup.

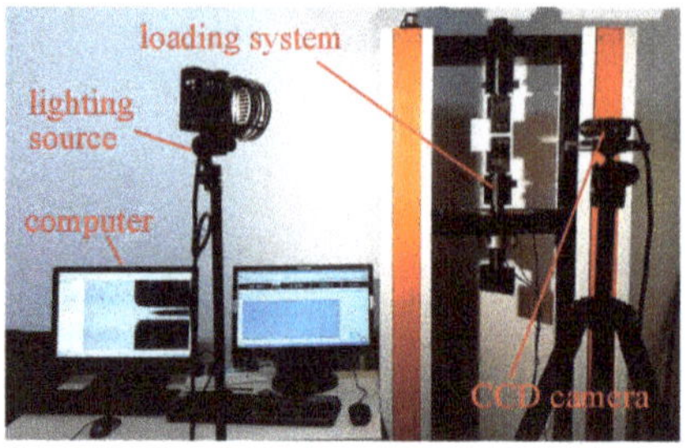

Figure 2. CT test device.

After cutting, grinding, polishing and corrosion, metallographic specimens were prepared. The microstructure of the specimen was analyzed using a scanning electron microscope (SEM) Q45, which was produced by FEI Company, Hillsboro, OR, USA.

2.2. Trans-Scale Calculation Model

Figure 3a illustrates the macro–micro trans-scale model of the laser repaired specimen. The model can be divided into four parts. Part "1" is established using conventional finite elements with an average element size of 0.5 mm, representing the peripheral part of the CT specimen used for applying loads and boundary conditions. The element type is a four-node plane stress element. Part "2" is constructed using Voronoi tessellations and crystal plasticity finite elements with an average element size of 0.98 μm, simulating the stress concentration region at the crack tip. Part "3" represents the variable-node elements used to connect the fine-scale elements and macro-scale elements. Part "4" is composed of four-node plane stress elements with an average size of 1 μm. The addition of these elements serves to enclose the model corresponding to "2" into a regular shape, thereby facilitating the establishment of multi-scale models. Referring to the method proposed by Aghababaei and Joshi [25], the model with pores in the repaired layer is established by removing several elements. Figure 3b indicates a partially enlarged model at the repaired layer, where white holes are used to represent the randomly distributed pores.

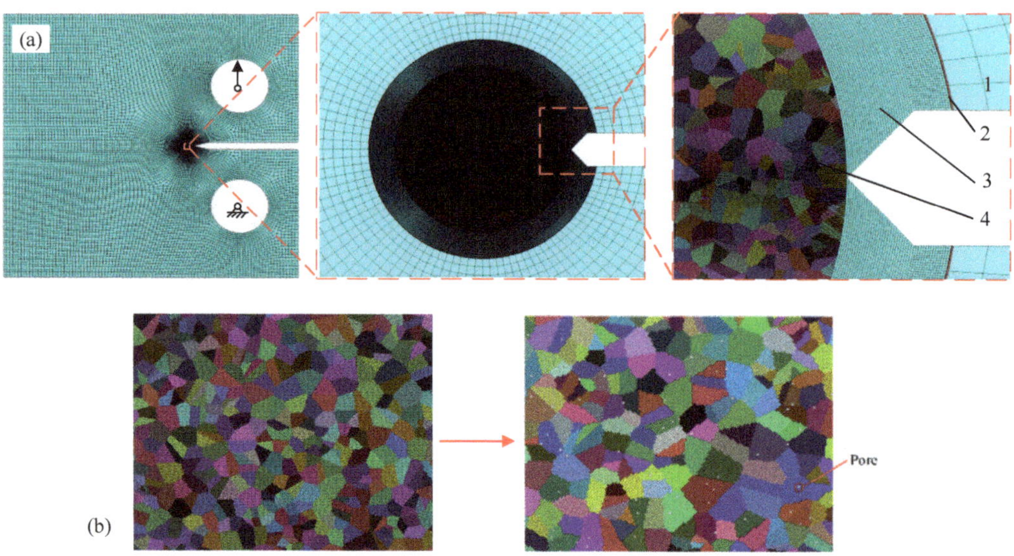

Figure 3. Trans-scale calculation model for CT specimen with pores. (**a**) Overall model, (**b**) schematic diagram illustrating the process of incorporating pores into the model.

To conform to the loading condition in the tensile test, one loading hole is fixed, while the other loading hole is subjected to tensile load. The variable-node elements, i.e., the element indicated by "3", is realized by a 13-node element through the UELMAT subroutine in ABAQUS (v. 2017) [21]. The material model of the repaired zone is realized by crystal plasticity theory through UMAT subroutine in ABAQUS [21]. The orientation of the crystals is assumed to be random. The materials of other regions are exposed by piecewise linear/power hardening materials [22]. The crystal slip hardening modulus adopts the simple form proposed by Peirce et al. [26] and is expressed as:

$$h_{\alpha\alpha} = \mathrm{h}(\gamma) = h_0 \mathrm{sech}^2 \left| \frac{h_0 \gamma}{\tau_s - \tau_0} \right| \tag{2}$$

$$h_{\alpha\beta} = q\mathrm{h}(\gamma)(\alpha \neq \beta) \tag{3}$$

where $h_{\alpha\beta}$ is the slip hardening modulus, which is divided into self-hardening modulus $h_{\alpha\alpha}$ and latent hardening modulus $h_{\alpha\beta}(\alpha \neq \beta)$; γ is the cumulative shear strain; h_0 is the initial hardening modulus; τ_s is the saturation stress; and τ_0 is the initial yield stress. For simplicity, the constant q is set to unity [27]. The initial hardening modulus h_0, the saturation stress τ_s and the initial yield stress τ_0 of 304 stainless steel substrate with grain of 25 μm are obtained through inversion of the tensile stress–strain curve [21], as shown in Table 1.

Table 1. Crystal plasticity constant of 304 stainless steel substrate.

Initial Hardening Modulus h_0	Saturation Stress τ_s	Initial Yield Stress τ_0
464.40 MPa	191.08 MPa	78.86 MPa

3. Results

3.1. Experimental Results

Figure 4 shows the microstructure of five specimens repaired by different laser power and heating time, with the pores in the microstructure indicated by arrows [20]. The laser process parameters used for these specimens and corresponding microstructure characteristics are listed in Table 2. When laser power is small, e.g., 1600 W, the grains are relatively fine, but a large number of pores appeared in the repaired layer, as shown in Figure 4a. When laser power increases to 1800 W, the dendrite began to transform to equiaxed grains, the overall microstructure is compact, and the number of pores in the repaired layer decreased (see Figure 4b). When laser power is too high, e.g., 2000 W, columnar dendrites appear in the repaired layer, the grain size and the number of pores increases (see Figure 4c). The effect of heating time on the microstructure is visible in Figure 4b,d,e. When the laser heating time is 0.5 s, as shown in Figure 4d, the grains in the repaired layer are small and the number of pores is low. When the laser heating time increases to 1.0 s, the grains in the repaired layer grow larger and the number of pores increases (see Figure 4b). When the laser heating time further increases to 2.0 s, the grain size and the number of pores in the repaired layer continue to increase (see Figure 4e). The grain size and porosity of the repaired layer were calculated by ImageJ software (v. 1.51j8) [28], as listed in Table 2.

Table 2. Grain size and porosity of specimens repaired with different laser process parameters.

Specimen	A	B	C	D	E
P-D-T (W-mm-s)	1600-3-1	1800-3-1	2000-3-1	1800-3-0.5	1800-3-2
Grain size (μm)	7.42	7.93	9.33	5.27	8.65
Porosity (%)	1.50	0.50	1.76	0.44	0.63

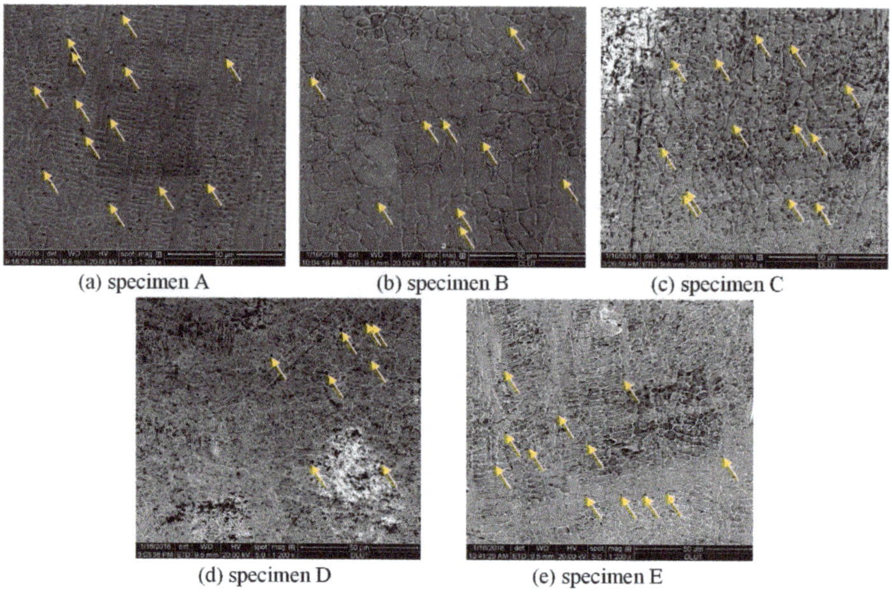

Figure 4. Microstructures at the repaired layer, with pores in the microstructure indicated by yellow arrows [20].

3.2. Numerical Study on the Influence of Grain Size on Repair Quality

Phan et al. [29] applied crystal plasticity theory to study the effect of microstructure on mechanical behavior, and concluded that grain size has no effect on initial hardening modulus h_0 and the saturation stress τ_s but only on critical resolved shear stress τ_0.

Weng et al. [30] provided the relationship between critical resolved shear stress τ_0 and grain size a:

$$\tau_0 = \tau_0^{\infty} + k_0 a^{-1/2} \quad (4)$$

where τ_0^{∞} is the initial critical resolved stress in a free single crystal and k_0 is the Hall–Petch constant.

From macro scale, the Hall–Petch relationship can represent the relationship between material strength σ_{HP} and friction stress σ_0, Hall–Petch coefficient k, and grain size a:

$$\sigma_{HP} = \sigma_0 + k a^{-1/2} \quad (5)$$

For 304 austenitic stainless steel, the friction stress σ_0 and the Hall–Petch coefficient k are 180 MPa and 7.589 MPa $\cdot$ mm$^{0.5}$, respectively [31].

Thus, the critical resolved shear stress τ_0 is inverted with the goal of minimizing the error between the simulated yield stress σ_m and the calculated yield stress σ_{HP} by a CPFE model of tensile specimens. The computed results are presented in Table 3.

Table 3. The calculated critical resolved shear stress τ_0 of 304SS.

Grain Size (μm)	5	5.27	7.42	7.93	8.65	9.33	15	25
Yield stress σ_{HP} (MPa)	287.33	284.55	268.11	265.23	261.60	258.57	241.97	228.00
Critical resolved shear stress τ_0 (MPa)	100.25	99.17	94.08	92.59	91.82	90.51	84.65	78.86

Yield stress σ_{HP} Critical resolved shear stress τ_0 Figure 5 illustrates the J-integral of the specimens with varying grain sizes at repaired zone under a 20 kN load, calculated using the trans-scale calculation method. The grain distribution corresponding to some grain sizes is also shown in the figure. The figure clearly demonstrates that the J-integral increases as the grain size increases. Specifically considering the influence of grain size, specimens A, B, C, D, and E in Table 2 exhibit J-integral values that are 2.79%, 2.62%, 2.19%, 3.77%, and 2.38% lower, respectively, than those of specimens with a grain size of 25 μm (matching the grain size of 304SS substrate). During the laser repair process, the high energy density and rapid solidification rate prevent the grains in the repaired layer from growing sufficiently, leading to smaller grains. The refinement of grains results in an increased grain boundary area per unit volume. Grain boundaries serve as barriers to dislocation movement from one grain to another, leading to initial hardening. Moreover, smaller grain sizes correspond to a greater number of grain boundaries per unit volume and a reduced number of dislocations per unit length that need to be accommodated. Thus, the fewer opportunities there are for microcracks to nucleate and grow, the lower the internal stress of the material. Therefore, fine-grained structure is beneficial to the improvement of crack initiation resistance and the fracture properties of repaired parts.

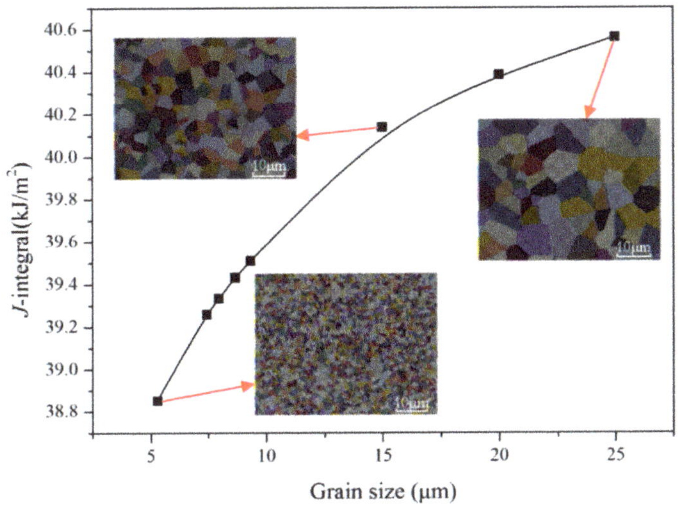

Figure 5. Variation of J-integral of the specimens with different grain sizes at repaired zone.

3.3. *Formatting of Mathematical Components*

The individual CPS is an important parameter for describing crack propagation at the mesoscale. When the CPS on a slip system reaches a critical value, the crack will propagate along the slip system with the largest CPS [32]. The slip system with the largest CPS is called the dominant slip system. The individual CPS is characterized by its definition, which is:

$$\gamma^{(\alpha)} = \int_0^t \left|\dot{\gamma}^{(\alpha)}\right| dt \quad (6)$$

where $\dot{\gamma}^{(\alpha)}$ is the slip rate on α slip system, t represents time.

Figure 6 shows the distribution of pores near the crack tip of Specimen B simulated using a crystal plasticity model. The grain size and pore volume fraction of the repaired layer in the model are 7.93 μm and 0.5%, respectively, and the element size is 0.98 μm. Figure 6a shows the grains and pores near the crack tip; Figure 6b shows the positions of pores and elements on Grain A at the crack tip. Elements 1–9 are located along the direction of the crack front, and the position of their first integral point is indicated by a white "*" in the figure. The orientation of Grain A is (75°, 15°, 15°). The CPS on the dominant slip system with and without pores in Grain A is shown in Figure 6c. It is found that the CPS on dominant slip system with pores is greater than that without pores, and the closer to the position of pores, the greater the difference between them. The results show that the material more easily reaches the crack growth criterion under external load when pores are present—that is, the fracture properties decrease.

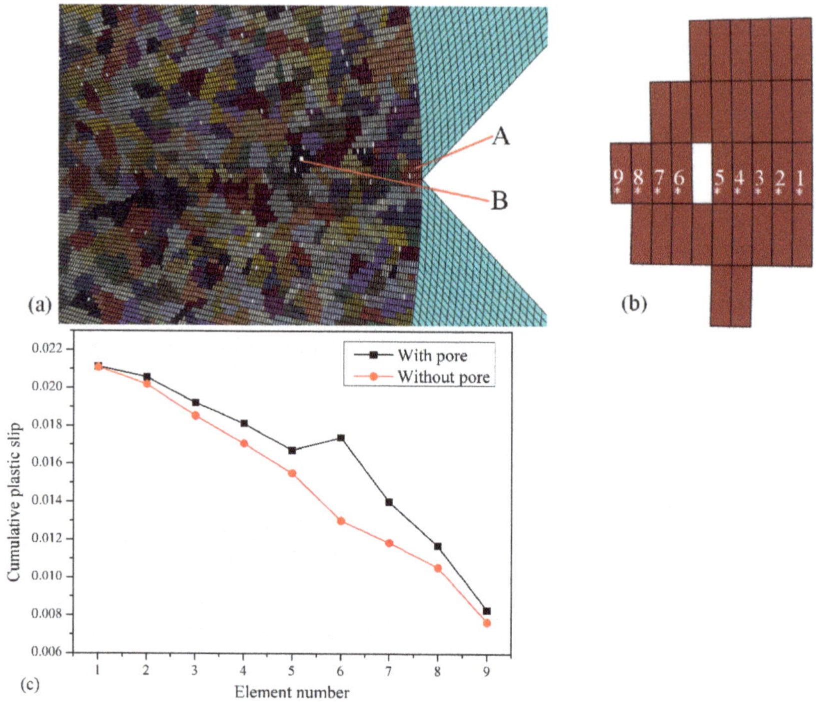

Figure 6. (**a**) The distribution of pores near the crack tip. (**b**) Grain A and element number. (**c**) CPS on dominant slip system near the pore along the front direction of the crack.

Figure 7a shows the CPS on all slip systems of grain B shown in Figure 6, Figure 7b shows the CPS on the dominant slip system of each element. The CPS data is extracted from the integration points marked with red "*" in Figure 7a. It is obvious that when the pore has sharp corners, the CPS is further increased, as shown in elements 1, 5, 6, and 10 in Figure 7b. Additionally, when the edges of pore are flat, the CPS is relatively small, as shown in elements 3 and 8 in Figure 7. Therefore, large pores or pores with sharp corners should be avoided in the repaired layer.

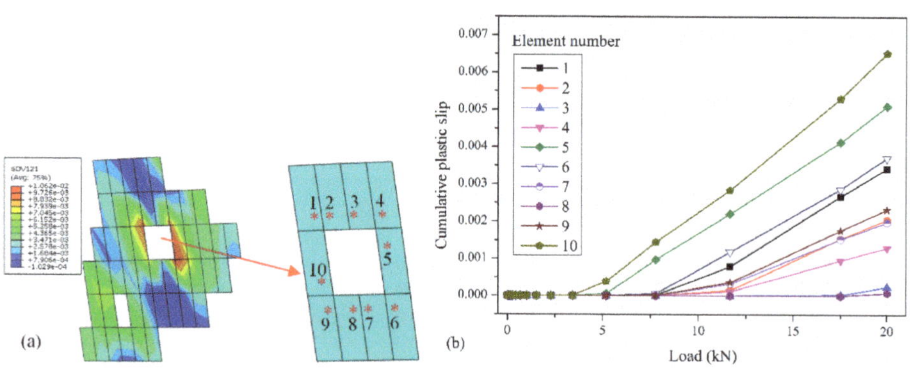

Figure 7. (**a**) Contour of CPS on all slip systems of grain B. (**b**) CPS on dominant slip system of each element near the large pore.

Figure 8 shows the CPS on the dominant slip system near the crack tip when the position of the pore changes. The CPS on the dominant slip system at the integration point indicated by the red arrow in Figure 8a is used to investigate the changes in CPS. It can be seen from Figure 8b that the CPS near the crack tip gradually increases with decreasing distance from the porosity to the crack tip, and the magnitude of the increase is also increasing. In other words, the closer the pores in the repaired layer are to the crack tip, the better the crack repair quality.

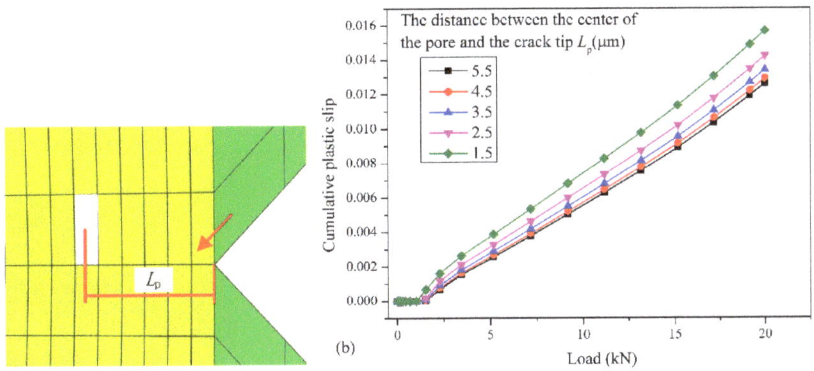

Figure 8. (**a**) Schematic of pore position. (**b**) The CPS on the dominant slip system varies with the position of pore.

3.4. Macro-Scale Analysis

Figure 9 illustrates the variation of J-integral with volume fractions of pores distributed in the repair layer under a 20 kN load. The grain distribution corresponding to some grain sizes is also shown in the figure. The calculations assume a pore radius of 0.98 μm. From the figure, it can be observed that within the calculated range, the J-integral linearly increases with the increase in the volume fraction of pores. When the volume fraction of pores in the repaired layer is 0.26% and 1.3%, the J-integral of the specimens under 20 kN load is increased by 0.2% and 1.1%, respectively, compared with those without pores. The linear relationship between J-integral and volume fraction of pores in the range of study indicates that the fracture properties of repaired parts will be significantly reduced in the presence of pores.

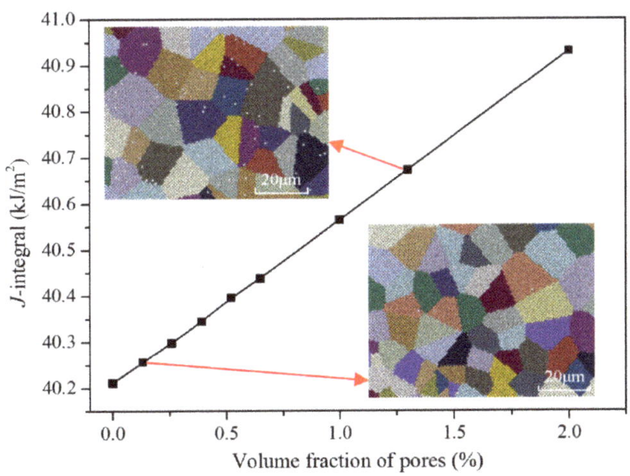

Figure 9. Variation of J-integral with volume fraction of pores under a 20 kN load.

Figure 10 shows the variation of J-integral with the pore size at the load of 20 kN with the assumption that the volume fraction of pores is 1%. When the pore size is large, the J-integral will be affected by the position of pores. The influence of pore size on J-integral is studied by six calculations when the pores are randomly distributed. It can be observed that J-integral increases with the increase of pore size. When pore size reaches 2.4 μm, J-integral is greatly affected by pore position. On the basis of the random distribution of pores, one of pores is fixed at 5.5 μm before the crack tip, and the calculated J-integral is significantly greater than the average value of random distribution. Moreover, the larger the pore size, the greater the effect on J-integral. Therefore, when there are large pores in the repaired layer, especially near the crack tip, the fracture properties of repaired parts will be reduced significantly.

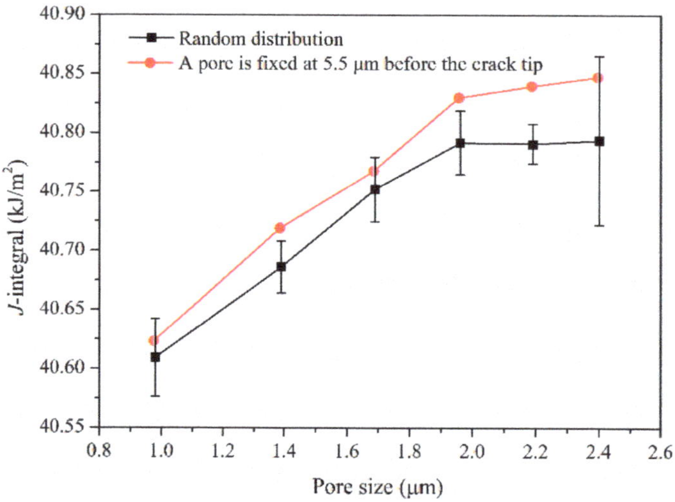

Figure 10. Variation of J-integral with pore size under a 20 kN load.

4. Discussion of Calculation and Experimental Results

Figure 11 shows the J-integral calculated using trans-scale model and the J-integral of different specimens under a load of 20 KN. The laser process parameters corresponding to Specimens A, B, C, D, and E are shown in Figure 4 and Table 2. Specimen F has not been repaired. The microstructure in the trans-scale model is also established based on the microstructure of Specimens A, B, C, D, E, and F. It can be seen that the laser process parameters of Specimen D are relatively suitable, the porosity is low, and the grain size is small. Correspondingly, the J-integral is more than 8% smaller than Specimen F and about 3% smaller than Specimen C. The laser power of Specimens A and C is either too small or too large, causing the porosity and grain size increase and the corresponding J-integral to increase. The heating time of specimens B and E is longer, the energy input for the molten pool is high, the porosity and grain size in the repaired layer increase, so the corresponding J-integral increases. Selecting appropriate laser process parameters can not only refine grain size but also reduce the volume fraction of pores in the repaired layer. Therefore, it is very important to select appropriate laser process parameters to improve repair quality.

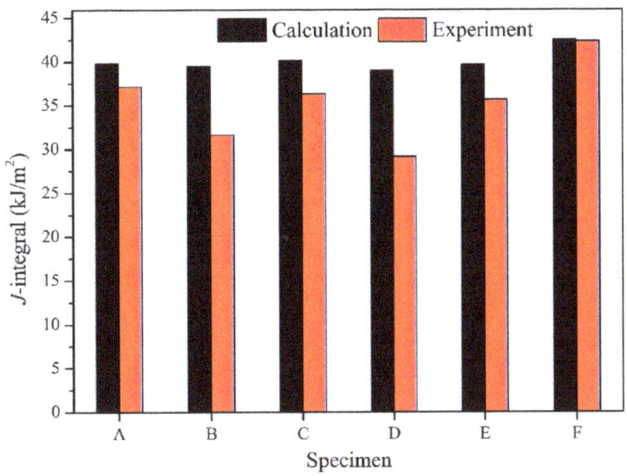

Figure 11. J-integral of repaired specimens obtained from calculation and experiment under a load of 20 kN.

5. Conclusions

In this article, meso–macro-mechanical analysis has been performed to study the influence of laser process parameters on the laser repair quality. The key findings can be summarized as follows:

When the laser power is 1800 W and the heating time is 0.5 s, the porosity of the repaired specimen is low, and the grain size is small. The corresponding J-integral is more than 8% smaller than that of the unrepaired specimen and approximately 3% smaller than that of the repaired specimen subjected to a laser power of 2000 W and a heating time of 1 s. Both excessive and insufficient laser power, as well as prolonged heating time, can result in increased porosity in the repair layer and a reduction in the fracture properties of the repaired components.

The pores in the repair layer enhanced the CPS of the surrounding crystals, particularly for pores with sharp corners, where the CPS exhibited a more pronounced increase. As the distance from the hole to the crack tip decreases, the CPS near the crack tip gradually rises, and the extent of this increase also becomes more significant.

The J-integral of the repaired specimen increases linearly with the pore volume fraction in the repair layer. By selecting appropriate laser processing parameters, the grain size can be refined and the pore volume fraction can be minimized, thereby reducing the J-integral and ultimately improving the repair quality.

The current established model primarily focuses on the influence of microstructural features, such as pores, on the fracture properties of repaired components. However, it has not fully addressed the complex mechanisms associated with the heat-affected zone (HAZ) during the laser repair process. As a critical area in this process, the microstructural and performance changes within the HAZ can significantly impact the overall properties of the repaired components. Therefore, including the HAZ into the research scope is essential for enhancing the properties evaluation and optimization of repair parts, and it is also a vital direction for future research. In addition, this study primarily focused on the microstructural characteristics of pores; however, it is important to note that other types of defects, such as microcracks, may also develop during laser repair processes. These defects can significantly affect the fracture properties of the repaired components. Due to limitations in experimental conditions and technical capabilities, it is currently not possible to fully capture and quantify the characteristics of these defects and their specific effects on material properties. Future research will concentrate on how laser processing parameters and material composition can mitigate the formation of microcracks in the repair layer and will as analyze the influence of microcracks on the fracture properties of the repaired components. By optimizing material composition and laser technology, the occurrence of microcracks in the repair layer can be reduced or eliminated, thereby enhancing the structural integrity and service life of the repaired parts.

Author Contributions: Conceptualization, Y.L., W.J. and M.L.; methodology, Y.L.; software, Y.L.; validation, Y.L.; formal analysis, Y.L.; investigation, Y.L.; resources, M.L.; data curation, Y.L.; writing—original draft preparation, Y.L.; writing—review and editing, W.J. and M.L.; visualization, Y.L.; supervision, W.J.; project administration, M.L.; funding acquisition, Y.L., W.J. and M.L. All authors have read and agreed to the published version of the manuscript.

Funding: This research was funded by the National Natural Science Foundation of China, with grant numbers 51575076 and 52174018; the Natural Science Foundation of Hubei Province, grant number 2024AFB440; the Scientific Research Project of Hubei Provincial Department of Education, grant number Q20231311.

Institutional Review Board Statement: Not applicable.

Informed Consent Statement: Not applicable.

Data Availability Statement: Data are contained within the article.

Conflicts of Interest: The authors declare no competing interests.

References

1. Xu, P.; Lin, C.X.; Zhou, C.Y.; Yi, X.P. Wear and corrosion resistance of laser cladding AISI 304 stainless steel/Al_2O_3 composite coatings. *Surf. Coat. Technol.* **2014**, *238*, 9–14. [CrossRef]
2. Amado, J.M.; Tobar, M.J.; Yáñez, A.; Amigó, V.; Candel, J.J. Crack free tungsten carbide reinforced Ni(Cr) layers obtained by laser cladding. *Phys. Procedia* **2011**, *12*, 338–344. [CrossRef]
3. Wang, W.F.; Wang, M.C.; Jie, Z.; Sun, F.J.; Huang, D.W. Research on the microstructure and wear resistance of titanium alloy structural members repaired by laser cladding. *Opt. Laser Eng.* **2008**, *46*, 810–816. [CrossRef]
4. Cha, D.; Diaz, O.G.; Liao, Z.; Gilbert, D.; Axinte, D.; Kell, J.; Norton, A.; O'Key, M.; Osborne, M.R.; Main, D. Development of a novel system for in-situ repair of aeroengine airfoil via pulsed laser ablation. *J. Manuf. Syst.* **2020**, *55*, 126–131. [CrossRef]
5. Zhang, B.; Li, Y.T.; Bai, Q. Defect formation mechanisms in selective laser melting: A review. *Chin. J. Mech. Eng.* **2017**, *30*, 1476. [CrossRef]
6. Yang, T.; Liu, T.T.; Liao, W.H.; MacDonald, E.; Wei, H.L.; Zhang, C.D.; Chen, X.Y.; Zhang, K. Laser powder bed fusion of AlSi10Mg: Influence of energy intensities on spatter and porosity evolution, microstructure and mechanical properties. *J. Alloy Compd.* **2020**, *849*, 156300. [CrossRef]
7. Khairallah, S.A.; Anderson, A.T.; Rubenchik, A.; King, Y.E. Laser powder-bed fusion additive manufacturing: Physics of complex melt flow and formation mechanisms of pores, spatter, and denudation zones. *Acta Mater.* **2020**, *196*, 30. [CrossRef]
8. Han, X.H.; Yang, Z.B.; Ma, Y.; Shi, C.Y.; Xin, Z.B. Porosity distribution and mechanical response of laser-MIG hybrid butt welded 6082-T6 aluminum alloy joint. *Opt. Laser Technol.* **2020**, *132*, 106511. [CrossRef]
9. Pekok, M.A.; Setchi, R.; Ryan, M.; Han, Q.Q.; Gu, D.D. Effect of process parameters on the microstructure and mechanical properties of AA2024 fabricated using selective laser melting. *Int. J. Adv. Manuf. Technol.* **2021**, *112*, 175–192. [CrossRef]
10. Roters, F.; Eisenlohr, P.; Hantcherli, L.; Tjahjanto, D.D.; Bieler, T.R.; Raabe, D. Overview of constitutive laws, kinematics, homogenization and multiscale methods in crystal plasticity finite-element modeling: Theory, experiments, applications. *Acta Mater.* **2010**, *58*, 1152–1211. [CrossRef]
11. Guo, X.R.; Mao, N.D.; Kong, T.Q.; Zhang, J.; Shen, J.J.; Wang, C.H.; Sun, C.Y.; Li, P.P.; Xiong, Z.P. Strain rate-dependent plastic behavior of TWIP steel investigated by crystal plasticity model. *Mat. Sci. Eng. A* **2024**, *891*, 145986. [CrossRef]
12. Liang, J.B.; Jiao, L.; Yan, P.; Song, Y.F.; Li, S.Y.; Dai, Z.C.; Wang, X.B. A crystal plastic finite element model for the effect of surface integrity on multiaxial fatigue life after multistage machining processes. *Fatigue Fract. Eng. Mater. Struct.* **2024**, *47*, 964–985. [CrossRef]
13. Skamniotis, C.; Grilli, N.; Cocks, A.C.F. Crystal plasticity analysis of fatigue-creep behavior at cooling holes in single crystal Nickel based gas turbine blade components. *Int. J. Plasticity* **2023**, *166*, 103589. [CrossRef]
14. Wang, H.; Lu, C.; Tieu, K. A crystal plasticity FEM investigation of a Cu single crystal processed by accumulative roll-bonding. *J. Mater. Res. Technol.* **2019**, *8*, 5057–5065. [CrossRef]
15. Zhao, J.W.; Jiang, Z.Y.; Wang, Z.H.; Sang, S.B.; Dobrzanski, L.A.; Yang, M.; Ma, X.G.; Wang, Y.Z. An analysis of micro deep drawing of ferritic stainless steel 430 using crystal plasticity finite element method. *J. Mater. Res. Technol.* **2022**, *20*, 2247–2261. [CrossRef]
16. Wang, Z.F.; Jiang, B.B.; Wu, S.L.; Liu, W.L. Anisotropic tension-compression asymmetry in SLM 316L stainless steel. *Int. J. Mech. Sci.* **2023**, *246*, 108139. [CrossRef]

17. Zhang, W.J.; Hu, Y.Y.; Ma, X.F.; Qian, G.A.; Zhang, J.M.; Yang, Z.M.; Berto, F. Very-high-cycle fatigue behavior of AlSi10Mg manufactured by selected laser melting: Crystal plasticity modeling. *Int. J. Fatigue* **2021**, *145*, 106109. [CrossRef]
18. Kirane, K.; Ghosh, S. A cold dwell fatigue crack nucleation criterion for polycrystalline Ti-6242 using grain-level crystal plasticity FE Model. *Int. J. Fatigue* **2008**, *30*, 2127–2139. [CrossRef]
19. He, Z.; Qiu, W.; Fan, Y.N.; Han, Q.N.; Shi, H.J.; Ma, X. Effects of secondary orientation on fatigue crack initiation in a single crystal superalloy. *Fatigue Fract. Eng. Mater. Struct.* **2018**, *41*, 935–948. [CrossRef]
20. Jiang, W.; Li, Y.Y.; Fang, G.L.; Guo, F.L. Effect of laser parameters on microstructure and fracture properties of repaired cracks with micro/nano material addition. *J. Manuf. Sci. Eng.* **2020**, *142*, 054501. [CrossRef]
21. Li, Y.Y.; Jiang, W. Multiscale finite element based trans-scale calculation method for polycrystalline materials. *Mater. Res. Express* **2019**, *6*, 036507. [CrossRef]
22. Jiang, W.; Li, Y.Y. Multiscale analysis of microstructural effects on the fracture properties of laser repaired cracks with micro/nano composite material addition. *Theor. Appl. Fract. Mech.* **2023**, *125*, 103914. [CrossRef]
23. *ASTM E647-24*; Standard Test Method for Measurement of Fatigue Crack Growth Rates. ASTM International: West Conshohocken, PA, USA, 2024.
24. Jiang, W.; Jiang, X.F. Laser repair with addition of nano-WC on microstructure and fracture behavior of 304 stainless steel. *J. Eng. Mater. Technol.* **2017**, *139*, 041002. [CrossRef]
25. Aghababaei, R.; Joshi, S.P. Grain size-inclusion size interaction in metal matrix composites using mechanism-based gradient crystal plasticity. *Int. J. Solids Struct.* **2011**, *48*, 2585–2594. [CrossRef]
26. Peirce, D.; Asaro, R.J.; Needleman, A. Material rate dependence and localized deformation in crystalline solids. *Acta Metall. Mater.* **1983**, *31*, 1951–1976. [CrossRef]
27. Yao, W.Z.; Krill, C.E.; Albinski, B.; Schneider, H.C.; You, J.H. Plastic material parameters and plastic anisotropy of tungsten single crystal: A spherical micro-indentation study. *J. Mater. Sci.* **2014**, *49*, 3705–3715. [CrossRef]
28. Hastie, J.C.; Kartal, M.E.; Carter, L.N.; Attallah, M.M.; Mulvihill, D.M. Classifying shape of internal pores within AlSi10Mg alloy manufactured by laser powder bed fusion using 3D X-ray micro computed tomography: Influence of processing parameters and heat treatment. *Mater. Charact.* **2020**, *163*, 110225. [CrossRef]
29. Phan, V.T.; Nguyen, T.D.; Bui, Q.H.; Dirras, G. Modelling of microstructural effects on the mechanical behavior of ultrafine-grained Nickel using crystal plasticity finite element model. *Int. J. Eng. Sci.* **2015**, *94*, 212–225. [CrossRef]
30. Weng, G.J. A micromechanical theory of grain-size dependence in metal plasticity. *J. Mech. Phys. Solids* **1983**, *31*, 193–203. [CrossRef]
31. Odnobokova, M.; Belyakov, A.; Kaibyshev, R. Development of nanocrystalline 304L stainless steel by large strain cold working. *Metals* **2015**, *5*, 656–668. [CrossRef]
32. Zhang, P.; Zhang, L.; Baxevanakis, K.P.; Zhao, L.G.; Bullough, C. Modelling short crack propagation in a single crystal nickel-based superalloy using crystal plasticity and XFEM. *Int. J. Fatigue* **2020**, *136*, 105594. [CrossRef]

materials

MDPI

Article

Effect of Co Addition on the Microstructure and Mechanical Properties of Sn-11Sb-6Cu Babbitt Alloy

Zhan Cheng [1,2], Meng Wang [1,*], Bo Wang [1,*], Lei Zhang [1], Ting Zhu [3], Ningbo Li [1], Jifa Zhou [1] and Fei Jia [1]

1 Ningbo Intelligent Machine Tool Research Institute Co., Ltd., China National Machinery Institute Group, Ningbo 315700, China; brazecheng@163.com (Z.C.); zhangl@zrime.com.cn (L.Z.); ningbo721403@163.com (N.L.); 15364914458@163.com (J.Z.); 18804017112@163.com (F.J.)
2 College of Mechanical & Electrical Engineering, Nanjing University of Aeronautics and Astronautics, Nanjing 210016, China
3 Shenke Slide Bearing Co., Ltd., Shaoxing 311800, China; skjfkfb@163.com
* Correspondence: wm18838150358@163.com (M.W.); wangbo@camsouth.com.cn (B.W.)

Abstract: A Babbitt alloy SnSb11Cu6 with 0–2.0 wt.% Co was synthesized using the induction melting process. This study examined the effect of cobalt (Co) on the microstructure, tensile properties, compressive properties, Brinell hardness, and wear properties of SnSb11Cu6 using optical microscopy (OM), scanning electron microscopy (SEM), energy-dispersive X-ray spectroscopy (EDS), X-ray diffraction (XRD), a universal tensile testing machine, a Brinell hardness tester, and a wear testing machine. The results indicate that the optimal quantity of Co can enhance the microstructure of the Babbitt alloy and promote microstructure uniformity, with presence of Co_3Sn_2 in the matrix. With the increase in Co content, the tensile and compressive strength of the Babbitt alloy first increased and then decreased, and the Brinell hardness gradually increased with the increase in Co content. The presence of trace Co has a minimal effect on the dry friction coefficient of the Babbitt alloy. When the Co content exceeds 1.5 wt.%, the friction properties of the Babbitt alloy deteriorate significantly. The optimized Babbitt alloy SnSb11Cu6-1.5Co was subsequently fabricated into wires, followed by conducting cold metal transfer (CMT) surfacing experiments. The Co element can promote the growth of interfacial compounds. The microstructure at the interface of the Babbitt alloy/steel is dense, and there is element diffusion between it. The metallurgical bonding is good, and there are serrated compounds relying on the diffusion layer to extend to the direction of the additive layer with serrated compounds extending and growing from the diffusion layer to the additive layer. Overall, Babbitt alloys such as SnSb11Cu6 exhibit improved comprehensive properties when containing 1.5 wt.% Co.

Keywords: Babbitt alloy; Co element; microstructure; interface morphology; strength; wear performance

1. Introduction

Babbitt alloys are extensively utilized in critical bearing components and steel bearing linings, including turbine shafts in nuclear power plants, internal combustion engine spindles, intermediate bearings, and connecting rod bearings [1,2]. This is attributed to its excellent embedment, compliance and anti-seizure properties [3,4]. However, as machinery and equipment continue to exhibit high performance, their usage is progressively extending to high-speed, heavy-load, high-pressure, and other applications [5–7]. This trend unavoidably leads to a reduction in their service life, attributed to their limited load-carrying capacity and inadequate heat resistance [8,9]. Therefore, the development of high-performance Babbitt alloy materials is crucial.

Currently, cold metal transition (CMT) welding holds significant advantages in the production of high-performance sliding bearings [10]. It fulfills the demand for high-performance materials in modern industry. However, the processing difficulty of Babbitt wire is limited by the influence of alloy composition and wire diameter. Among them, the processing difficulty of Babbitt alloy wire with a thin wire diameter containing high antimony and copper components is greatly increased [11]. This seriously limits the development of Babbitt additive manufacturing technology in the field of sliding bearings.

To improve the performance of the Babbitt alloy, researchers have carried out a lot of research on the composition and preparation process of tin-based Babbitt alloys. Zhao [12] conducted a study on the arc deposition process of a Sn Babbitt alloy with 0–2 wt.% of zinc (Zn) on the surface of a steel substrate. The researchers observed that the addition of Zn resulted in a significantly uniform distribution of hard phase SnSb and Cu_6Sn_5, leading to improved hardness, strength, and bonding properties. Nikolay [13] proposed the utilization of multiwalled carbon nanotubes as a modifier to suppress the growth of reinforcing particles and enhance the fatigue resistance of bainite coatings. This was achieved by preparing composite particles of bainite powder and multiwalled carbon nanotubes through planetary milling. Subsequently, these particles were deposited onto the substrate surface using a plasma melting process, leading to an enhancement in frictional properties. Ramadan [14] used stir casting to create a bimetallic bearing material by adding nano alumina to a Babbitt alloy and casting it onto carbon steel. The study found that alumina nanoparticles changed Cu_6Sn_5 shape, enhancing bond strength and wear resistance. It was observed that a portion of silver (Ag) in Sn-based solid solution underwent solidification within the solution, while another portion combined with the Sn atoms to generate Ag_3Sn phase, thereby contributing to precipitation strengthening [15].

The aforementioned study employs a range of methods to achieve doping modification of a Babbitt alloy through the addition of metal elements or micro/nano components. Previous research has demonstrated that the incorporation of cobalt (Co) into Sn alloy can enhance the stability of the microstructure and the mechanical properties of the alloy [16–18]. However, there is no existing report on the effect of trace Co on the characteristics of Sn-based Babbitt alloy.

Based on the above research, SnSb11Cu6 alloys with different Co contents were prepared by induction melting. Then, the modified Babbitt alloy welding wire was prepared. The effects of Co addition on the microstructure, interface morphology, tensile strength, compressive strength, Brinell hardness, and friction and wear properties of SnSb11Cu6 alloys were systematically studied in order to provide a reference for obtaining a Babbitt alloy with excellent performance and promote the popularization and application of CMT welding technology in the preparation of Babbitt alloy bearings.

At present, the practical application of making a Babbitt alloy into welding wire in CMT (cold metal transfer) additive manufacturing technology is mainly reflected in its use as a filling material for manufacturing or repairing worn or damaged bearing components. Currently, ZSnSb8Cu4 is the main Babbitt alloy welding wire successfully marketed for CMT additive manufacturing. ZSnSb11Cu6 and ZSnSb12Cu6Cd1, which improve performance with doping elements such as Ni, Cd, and As, are gradually being introduced into the market.

2. Materials and Methods

This experiment utilized Sn ingots, antimony (Sb) ingots, copper (Cu) particles, and Co blocks, all of which had a purity of 99.9%. Babbitt alloys, SnSb11Cu6 with 0–2.0 wt.% Co, were prepared by adjusting the content of cobalt element. The chemical composition is presented in Table 1.

Table 1. Chemical compositions of Babbitt alloy SnSb11Cu6 with 0–2.0 wt.% Co.

Alloy	Sn	Sb	Cu	Co
1	Bal.	11.11	5.96	0
2	Bal.	11.08	5.99	0.49
3	Bal.	11.14	6.07	1.07
4	Bal.	11.09	6.01	1.48
5	Bal.	11.12	5.98	2.03

The smelting process involves induction heating and nitrogen protection at a temperature of 550 °C for a duration of 60 min. Throughout the process of melting, dried charcoal was utilized as the covering agent. Following the heat preservation process, the temperature was lowered to 480 °C, at which point ammonium chloride and zinc chloride were introduced to purify the molten substance. The high borosilicate glass rod was stirred for 5 min, and the surface scum was removed by using a slag scoop. Subsequently, the ingot was poured into the preheated tensile sample and the extrusion sample mold, and the ingot structure was obtained by air cooling to room temperature. The tensile sample mold is shown in Figure 1.

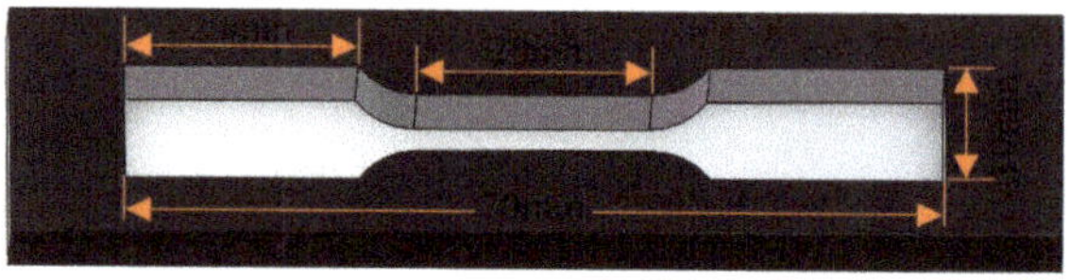

Figure 1. Casting mold of tensile strength sample.

The interface was cut between the ingot and the additive using wire cutting. Following coarse grinding, fine grinding, and subsequent fine polishing, the specimen was subjected to discoloration through corrosion using a 4% nitric acid alcohol solution. A Zeiss optical microscope and Zeiss Evo10 scanning electron microscope equipped with Oxford Ultim 40 energy-dispersive X-ray spectroscopy (EDS) (ZEISS Group, Oberkochen, Germany) were employed for macro and micro structure observation, and analysis of the micro components. The intercepted ingot casting machine underwent processing to create standard tensile samples and cylindrical compression samples. The tensile and compressive strength of the specimens were tested by a universal testing machine (MTS Systems Corporation, Eden Prairie, MN, USA) according to the national standards GB/T 228.1-2010 [19] and GB/T 16748-1997 [20]. The displacement rate of the beam during the experiment was 2 mm/min. The Brinell hardness of the material was assessed in accordance with the national standard GB/T 231.1-2009 [21] for metallic materials. The Brinell hardness test (Laizhou Huayin Test Instrument Co., Ltd., Laizhou, China) was conducted as follows: Part 1: Test methods. The loading time was 60 s and the load was 2452 N. Each sample underwent hardness testing five times, and the average values were calculated. The coefficient of friction was determined using HT-1000 High Temperature Friction and Wear Test (Lanzhou Zhongke Kaihua Technology Development Co., Ltd., Lanzhou, China). The experiment utilized 45 steel balls as the grinding material, with a load of 150 g, a loading time of 20 min, a motor frequency of 100 rpm, and a friction radius of 2 mm.

The ingot underwent a series of processes including cutting, peeling, extrusion, stretching, and reduction to produce an alloy wire with a diameter of 1.60 mm. The arc deposition process test was carried out on a Q235 steel plate with a size of 300 mm × 300 mm × 30 mm by using a CMT welding machine (Austrian Fronius company, Wels, Austria) produced by the Austrian Fronius company and Swiss ABB manipulator. Prior to the experiment, the steel plate underwent pickling to eliminate oil and rust from its surface, and the arc deposition process parameters are detailed in Table 2.

Table 2. Process parameters employed in CMT arc deposition of Sn-based Babbitt alloy.

Stacking Number	Welding Voltage/V	Welding Current/A	Wire Feed Rate/m·min^{-1}	Welding Speed/mm·min^{-1}
1	18.2~18.4	110~120	6.1~6.6	600
2	11.9~12.0	60~70	6.7~7.6	480

3. Experimental Results and Analysis

3.1. Microstructure Analysis

The properties of the alloy were determined by its microstructure. The microstructure and phase composition of the Babbitt Babbittalloy SnSb11Cu6 with 0–2.0 wt.% Co were analyzed. The optical microscopy (OM), scanning electron microscopy (SEM), and X-ray diffraction (XRD) analysis results are depicted in Figures 2–4. The microstructure of the Babbitt alloy, as depicted in Figure 2a, predominantly consists of the α-Sn solid solution matrix, SnSb intermetallic compound phase composed of Sn and Sb, and dispersed needle-like and star-like Cu_6Sn_5 within the soft matrix. The soft phase α-Sn solid solution, known as the concave, exhibits favorable potential, compliance, and resistance to blockage. On the other hand, the hard phases SnSb and Cu_6Sn_5 act as convex elements, creating a small gap between the shaft and the bearing bush. This configuration facilitates the transportation of lubricating medium and contributes to friction reduction.

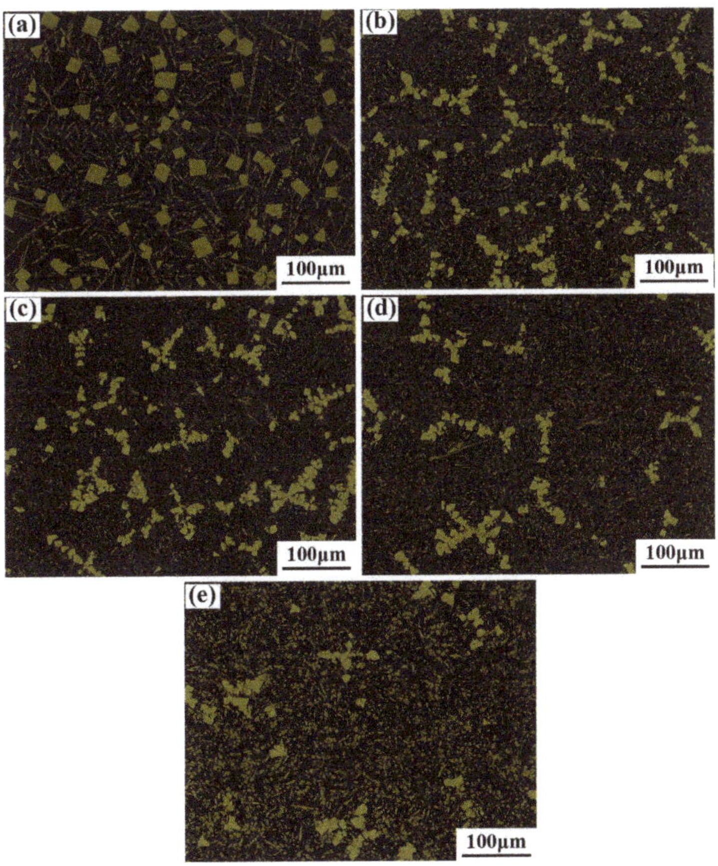

Figure 2. OM microstructure of Babbitt alloy SnSb11Cu6 with different Co contents: (**a**) 0 wt.% Co; (**b**) 0.5 wt.% Co; (**c**) 1.0 wt.% Co; (**d**) 1.5 wt.% Co; (**e**) 2.0 wt.% Co.

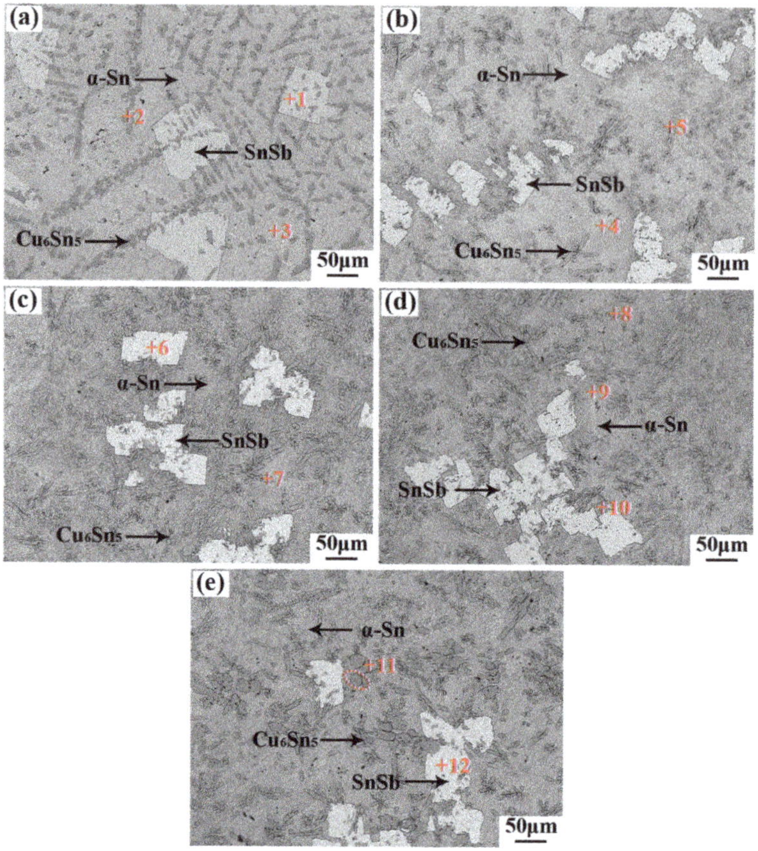

Figure 3. BSE microstructure of Babbitt alloy SnSb11Cu6 with different Co contents: (**a**) 0 wt.% Co; (**b**) 0.5 wt.% Co; (**c**) 1.0 wt.% Co; (**d**) 1.5 wt.% Co; (**e**) 2.0 wt.% Co.

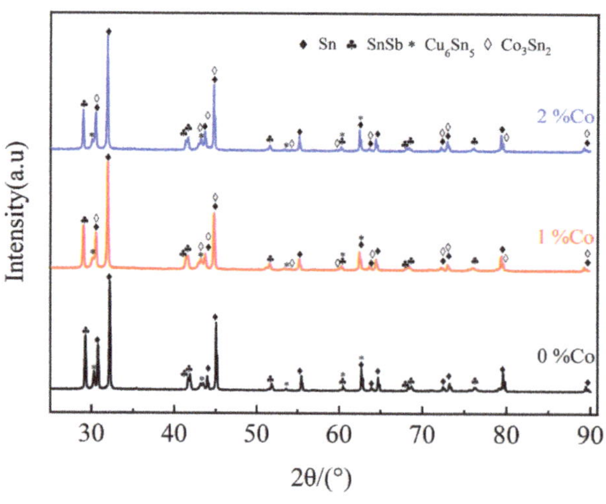

Figure 4. XRD pattern of Babbitt alloy SnSb11Cu6 with different Co contents (0 wt.%, 1.0 wt.%, 2.0 wt.%).

During the early solidification nucleation process, the Cu_6Sn_5 phase first solidified and precipitated. Subsequently, as the crystallization core of the SnSb phase, it exists in the liquid phase melt in the form of a skeleton to prevent the lighter square SnSb phase from floating up. As depicted in Figure 2b–e, the addition of 0.5–2 wt.%. Co passivates the sharp phase angle of SnSb in the microstructure, causing a gradual transformation from square blocks to ellipsoids. Additionally, there is a tendency for these ellipsoids to connect with each other, forming structures resembling flowers. Concurrently, there is a gradual decrease in the content of the SnSb phase. Furthermore, the needle-shaped Cu_6Sn_5 undergoes a transformation into small worm-like particles, which are then evenly distributed within the α-Sn solid solution matrix. When the carbon content exceeds 1.5 wt.%. The size of the fine worm-like Cu_6Sn_5 particles in the structure begins to coarsen, and the SnSb phase also changes from the trend of mutual aggregation and connection into a star to a dispersed distribution in the matrix, as depicted in Figure 2e.

The structure of the electron microscope was subjected to further analysis, as depicted in Figure 3. The combination of the EDS results in Table 3 and the backscattered electron imaging (BEI) results depicted in Figure 5 for the SnSb11Cu6-1.5Co alloy indicates a uniform distribution of the four elements within the alloy, with no discernible agglomeration. Furthermore, the distribution of Co in the microstructure aligns with that of Cu. Additionally, a portion of Sb undergoes a reaction with Sn to create a brittle SnSb phase, while the remaining portion dissolves into the α-Sn based solid solution to enhance the matrix.

When the added amount of the Co element reaches 2.0 wt.%, a gray-black ellipsoidal compound appears in the tissue, as shown in Figure 2 at point 11. Based on its elemental composition and the Cu-Sn phase diagram, as well as the XRD pattern depicted in Figure 4, the compound is identified as Co_3Sn_2. Owing to its high melting point of 1495 °C, Co is expected to be the primary element to precipitate as the crystallization core during the solidification process. At the identical solidification temperature, the non-spontaneous nucleation substrate on which the post-precipitated atoms can attach is relatively increased. This reduces the diffusion distance and nucleation resistance of the post-precipitated atoms, which is conducive to a large number of nucleation in the dendrite region. Subsequently, the Cu_6Sn_5 phase crystallizes from the melt in the form of a network. When the temperature is further reduced, the square-shaped β phase (SnSb) begins to precipitate, and the network-like pre-precipitated phase can inhibit the floating of the lighter β phase (SnSb). Then, the tin-based α solid solution begins to solidify. Upon solidification, the alloy's microstructure becomes finer and more uniform, resulting in a significant refinement of the size of the SnSb phase.

Table 3. Elemental composition of EDS at each point in Figure 3 (at.%).

Point	Sn	Sb	Cu	Co	Possible Phase
1	57.5	42.5	-	-	SnSb
2	44.9	3.7	51.5	-	Cu_6Sn_5
3	86.1	13.9	-	-	(Sn)
4	87.4	12.6	-	-	(Sn)
5	42.3	5.5	43.6	8.5	Cu_6Sn_5
6	56.6	43.4	-	-	SnSb
7	44.0	7.6	35.9	12.4	Cu_6Sn_5
8	39.4	8.0	36.8	15.7	Cu_6Sn_5
9	81.4	18.6	-	-	(Sn)
10	44.0	4.1	48.7	3.1	Cu_6Sn_5
11	35.0	11.3	22.9	30.8	Co_3Sn_2, Cu_6Sn_5
12	57.8	42.2	-	-	SnSb

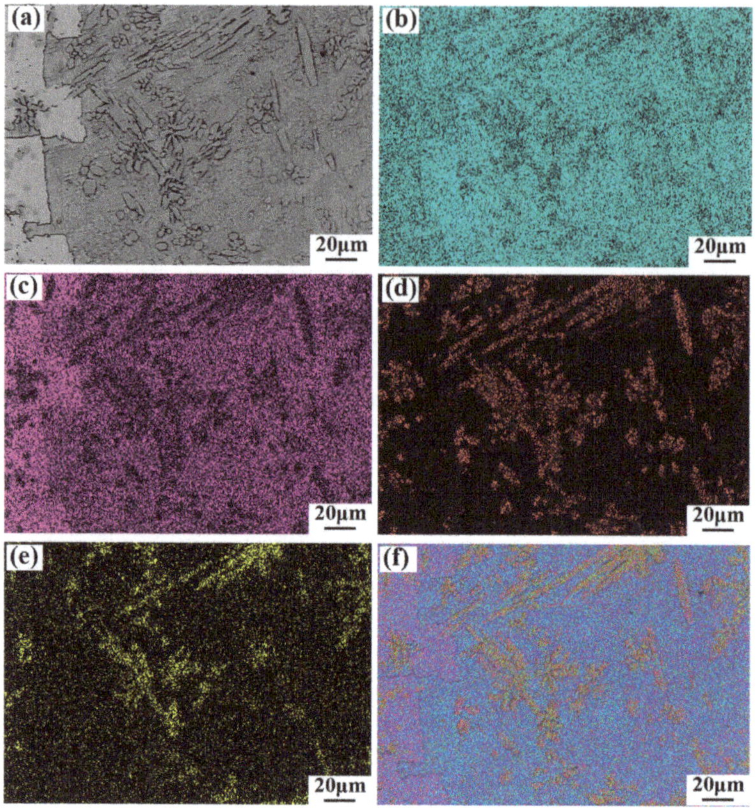

Figure 5. Elemental distribution in microstructure of SnSb11Cu6-1.5Co alloy using BEI and EDS: (**a**) BEI; (**b**) Sn map; (**c**) Sb map (**d**); Cu map; (**e**) Co map; and (**f**) combined elemental composition map.

3.2. *Analysis of Tensile and Compressive Strength*

As a bearing liner, the Babbitt alloy must possess sufficient strength and hardness to endure the substantial pressure exerted by the shaft diameter. The tensile strength and compressive strength of the Babbitt alloy SnSb11Cu6 with 0–2.0 wt.% Co were evaluated using a universal tensile testing machine, and the results are presented in Figures 6 and 7.

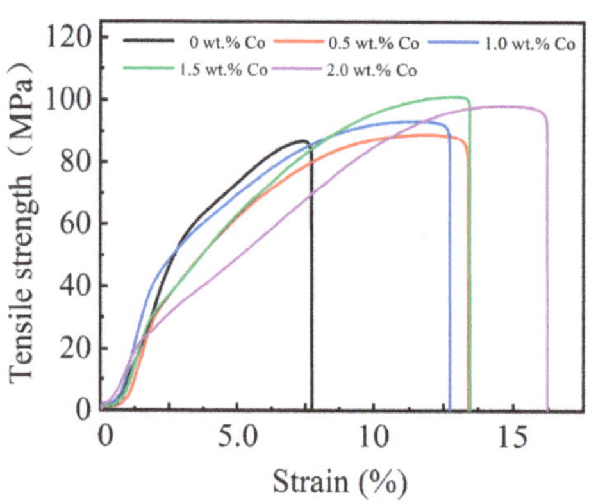

Figure 6. Co content effect on tensile strength of Babbitt alloy SnSb11Cu6 with 0–2.0 wt.% Co.

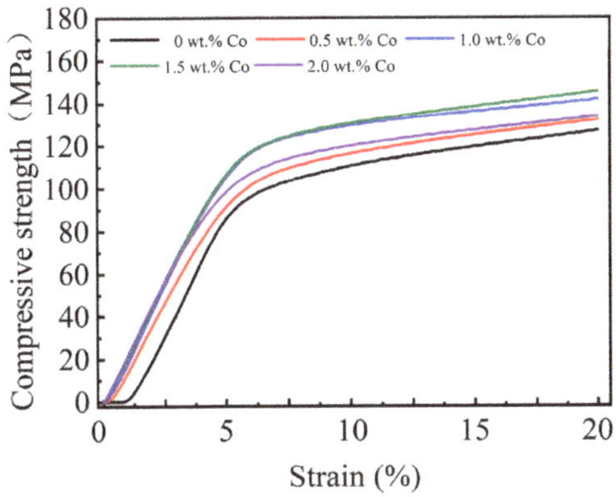

Figure 7. Co content effect on compressive strength of Babbitt alloy SnSb11Cu6 with 0–2.0 wt.% Co.

The addition of trace Co has been observed to result in a significant increase in the tensile strength and compressive strength of the Babbitt alloy. When the Co content is approximately 1.5 wt.%, the tensile and compressive strengths peak at 101.2 MPa and 141.9 MPa, respectively. These values are 16.8% and 11.6% higher than those of the Babbitt alloy SnSb11Cu6 (86.8 MPa and 127.1 MPa). As the Co content increases to 2.0 wt.%, there is a decrease in both the tensile strength and compressive strength. The observed phenomenon can be attributed to the incorporation of an optimal quantity of Co. On the one hand, part of the Sn-based solid solution has a solid solution strengthening effect. Meanwhile, the remaining portion of Co regulates the Cu_6Sn_5 compound from a continuous star to a point-like uniform distribution in the entire matrix. This, in turn, enhances the uniformity of Cu_6Sn_5 distribution within the structure. When subjected to external load, the strengthened solid solution and refined grains will increase the resistance of dislocation movement, making it difficult to advance slip. Therefore, the strength of the alloy is increased. After adding too much Co element, the size of the hard brittle phase Cu_6Sn_5 compound in the structure increases significantly. At the same time, multilateral Co_3Sn_2 compounds also appeared in the structure. This leads to an increase in the overall hardness and brittleness of the structure.

When subjected to an external load, the less coarse grains in the unit volume, the less the plastic deformation generated by the external load can be dispersed to each grain. As a result, it is difficult to uniformly transfer the internal deformation of the metal, resulting in local stress concentration. Then, the strength decreases and breaks first. Therefore, it is essential to manage the quantity of Co in order to modulate the dimensions of Cu_6Sn_5 and the concentration of Co_3Sn_2 in the structures.

3.3. Analysis of Brinell Hardness

The hardness of the Babbitt layer plays a crucial role in its resistance to deformation during operation and has a direct impact on the friction and wear characteristics of the workpiece. A Brinell hardness tester was utilized for assessing the hardness of the Babbitt alloy, and the corresponding results are presented in Figure 8. The Brinell hardness of SnSb11Cu6 alloy was determined to be 26.8 in the absence of Co. The hardness exhibited a gradual increase with the increase in Co content. At a Co content of 2.0 wt.%, the hardness reached 28.3, representing a 5.6% increase compared to that of SnSb11Cu6. The increase in hardness is attributed to the synergistic effect of solid solution strengthening and dispersion strengthening, as indicated by the analysis of the microstructure in Figure 2.

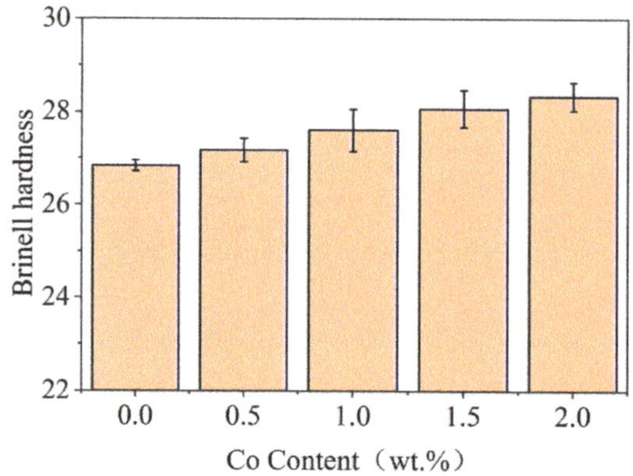

Figure 8. Co content effect on Brinell hardness of Babbitt alloy SnSb11Cu6 with 0–2.0 wt.% Co.

After the addition of the Co element, the SnSb phases in the structure have a tendency to connect with each other to form a flower shape. With the rise in Co content, the hard brittle phase Cu_6Sn_5 compound in the structure changes from a slender strip to an elliptical worm-like shape, the size continues to increase, and the distribution gradually becomes uniform. Furthermore, the emergence of the new Co_3Sn_2 phase within the microstructure is expected to contribute to the enhanced hardness of the alloy. Generally, the proportion of the hard phase in the alloy gradually increases following the addition of the Co element. When subjected to an external load, the hard phase of the protrusion increases relatively, so that the hardness gradually increases with the increase in Co content.

3.4. Analysis of Wear Properties

The friction coefficient between the modified Babbitt alloy and the stainless steel ball was tested by a friction and wear tester. The results of the friction coefficient for dry friction are depicted in Figure 9. It has been observed that under dry friction conditions, the four friction coefficient curves show minimal variation when the Co content ranges from 0 wt.% to 1.5 wt.%. The coefficients range between 0.37 and 0.40. When the Co content reaches 2.0 wt.%, the friction system stabilizes at approximately 0.4 within 0 min to 7 min. The friction coefficient rises to approximately 0.56 and then reaches a stable state as the duration of friction increases.

The wear resistance of tin-based Babbit alloys is closely related to the thickness and distribution of hard phases in the microstructure [22]. During the wear process, the SnSb phase and Cu_6Sn_5 phase, as hard particles, bulge upward to play a supporting role, while the soft matrix α-Sn phase sags downward to play a latent role. When the SnSb phase and Cu_6Sn_5 phase are present in small, uniformly distributed quantities within the eutectic matrix, it is possible to achieve outstanding service performance. Combined with the microstructure in Figure 2, it can be seen that the size of the square SnSb phase in the microstructure decreases after the addition of 0–1.5 wt.% Co element, and the hard and brittle Cu_6Sn_5 phase is fine and evenly distributed in the whole matrix. When the content of the Co element reaches 2.0 wt.%, the size of the hard Cu_6Sn_5 phase in the microstructure increases significantly and is densely distributed. This leads to a decrease in the content of the soft phase in the matrix and reduces the plasticity of the material. Therefore, the embedment and compliance of the material are reduced, which easily causes shedding of the hard phase in the Babbitt alloy and worsens the wear. According to Archard's law, there is a direct proportionality between the wear resistance of a material and its hardness. The greater the hardness of a material, the greater its resistance to deformation, and consequently, its wear resistance is also increased [23]. During the friction process, the embedding performance of the brittle hard particle phase is reduced, which increases the risk of falling off from the matrix and forming abrasive wear. Therefore, it is necessary to control the size and distribution of the hard phase in the tissue.

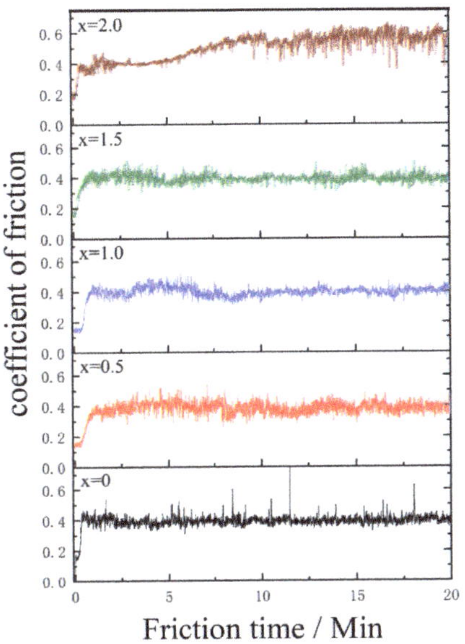

Figure 9. Co content effect on friction coefficient of Babbitt alloy SnSb11Cu6 with 0–2.0 wt.% Co.

3.5. Analysis of Interface Morphology

The Babbitt alloy containing 1.5 wt.% of Co was fabricated into alloy wires with a diameter of 1.6 mm. Subsequently, a CMT welding machine was utilized to perform overlay welding on the steel plate, and the morphology and element distribution of the Babbitt alloy/steel interface are shown in Figures 10 and 11.

It can be seen from Figure 10a,b that the two Babbitt/steel interfaces are continuous and well formed. There are no macroscopic defects such as pores, slag inclusions, and cracks. From the electron microscope morphology of the Babbitt alloy/steel interface in Figure 10c, it can be seen that a relatively flat metallurgical bonding line is formed between the Babbitt alloy and the steel matrix, and there is a serrated compound relying on the interface growth at the Babbitt alloy/steel interface. Further EDS line scanning was performed on the interface, and it was found that the element content from the steel side interface to the additive layer showed a certain concentration gradient. The content of iron decreased gradually, and the content of Sn, Sb and Cu increased gradually, as shown in Figure 11a. Following the addition of 1.5 wt.% of Co, the thickness of the compound layer at the Babbitt alloy/steel interface increased from 0.95 μm to 2.14 μm. The size of the serrated compound is notably larger, as depicted in Figure 10d. When subjected to external load, these serrated compounds can play a pinning role by inserting into the Babbitt alloy matrix relying on the interface layer to improve the holding force at the interface [24]. The energy spectrum analysis of the interface is carried out, and the results are shown in Figure 11b. It was found that from the steel side interface to the additive layer, the iron element gradually decreased, the content of tin, antimony and copper gradually increased, and the Co element had a convex peak at the interface. This indicates that the Co element is enriched at the interface, indicating that the addition of Co element can promote the growth of interfacial compounds.

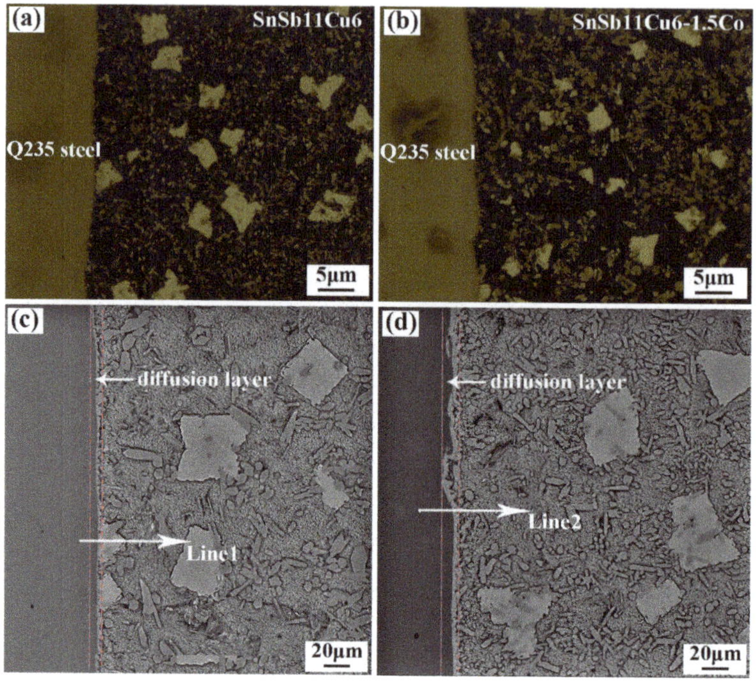

Figure 10. Microstructural interface of Q235 steel/SnSb11Cu6 and Q235 steel/SnSb11Cu6-1.5Co Babbitt alloys: (**a**,**b**) OM; (**c**,**d**) SEM.

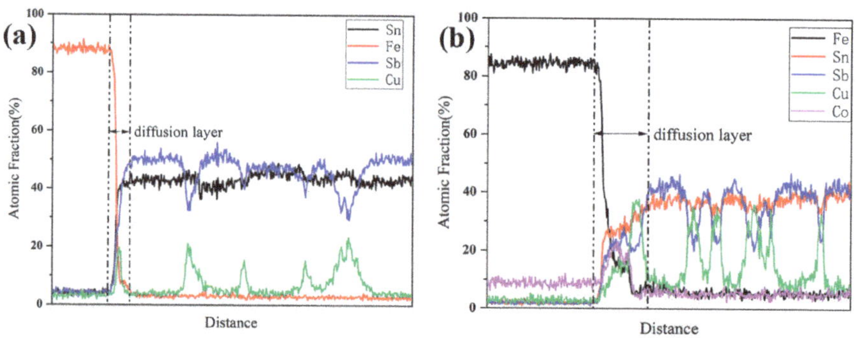

Figure 11. Elemental distribution analysis of Q235 steel/SnSb11Cu6 and Q235 steel/SnSb11Cu6-1.5Co Babbitt alloys: (**a**) SnSb11Cu6; (**b**) SnSb11Cu6-1.5Co.

4. Discussion

In this study, the microstructure, tensile and compressive strength, hardness, and wear properties of the Babbitt alloy SnSb11Cu6 with different contents of cobalt (Co) were comprehensively analyzed. Then, the experimental results were further compared with the findings of other researchers to verify the trends and phenomena in this study.

In terms of the microstructure, with the increase in Co content, the shape of the SnSb phase gradually changes from square blocks to ellipses. And there is a trend of connecting to form a flower-like structure. This observation is consistent with that of Wang [25] and Liu [16] that adding trace amounts of Co to Sn-0.7Cu alloy can alter the morphology of the phase, thus similarly, affecting the mechanical properties of the alloy. In addition, we found that when the Co content exceeds 1.5 wt.%, the size of the Cu_6Sn_5 phase begins to coarsen.

An analysis of the interface morphology revealed that the Babbitt alloy with 1.5 wt.% Co added formed a thicker compound layer at the interface with the steel substrate, which is consistent with the research results of Tang [26]. They pointed out that the addition of Co can promote the growth of interfacial compounds, thereby improving the binding strength of the interface.

The mechanical properties of the Co-modified Babbitt alloy were studied. When the Co content was about 1.5 wt.%, the tensile and compressive strength of the Babbitt alloy reached the peak. This finding is consistent with the research results of Zhang [27–29], who found that the strength of the Babbitt alloy can be significantly improved by adding an appropriate amount of Co. When the Co content increased to 2.0 wt.%, the strength decreased. This may be due to the increase in the size of the hard and brittle Cu_6Sn_5 phase and the formation of a new Co_3Sn_2 phase. Excessive Co will lead to an increase in brittleness of the alloy. In addition, with an increase in Co content, the hardness of the Babbitt alloy gradually increases, which is consistent with the research of Fan [29], who found that the addition of Co can improve the hardness of the alloy through the synergistic effect of solution strengthening and dispersion strengthening. Our results further confirm this. In particular, when the Co content is 2.0 wt.%, the increase in hardness is most significant.

By analyzing the wear properties, we found that the friction coefficient increased significantly when the Co content was 2.0 wt.%. This is because excessive Co will lead to coarsening of the Cu_6Sn_5 phase, which will affect the wear resistance of the alloy. This result is consistent with the research of Zhu and Hao [30,31]. They point out that an increase in the hard phase size will lead to an increase in the friction coefficient. Our study further shows that the size and distribution of the hard phase have an important effect on the wear properties of the Babbitt alloy. This is consistent with Archard's law [23]: the hardness of the material is proportional to its wear resistance.

In summary, our research findings are consistent with those of other researchers, further verifying the effect of Co addition on the microstructure and properties of Babbitt alloys. This study provides a valuable reference for the optimization of Babbitt alloys.

5. Conclusions

(1) The Co element can significantly refine the microstructure of the Babbitt alloy, improve the uniformity of the structure, and exist in the matrix as Co_3Sn_2 compound. As the concentration of Co content increases, the sharp angle passivation of β-SnSb phase transforms into ellipsoidal connection, and the content of SnSb phase decreases gradually. The needle-like Cu_6Sn_5 phase is transformed into small worm-like particles, which are dispersed in the matrix α solid solution. When the Co content exceeds 1.5 wt.%, the Cu_6Sn_5 phase is significantly coarsened. The presence of Co can enhance the formation of interfacial compounds. The interface exhibits a compact structure and strong metallurgical bonding. The serrated compound at the interface depends on the diffusion layer to elongate and expand towards the Babbitt alloy layer.

(2) The tensile strength and compressive strength of Babbitt alloy exhibit a pattern of initial increase followed by decrease as Co content increases. The maximum values of 101.2 MPa and 141.9 MPa were attained at a Co content of 1.5 wt.%, representing an increase of 16.8% and 11.6%, respectively, compared to Babbitt alloy SnSb11Cu6 (86.8 MPa). The Brinell hardness gradually increases with the increase in Co element content. At a Co content of 2.0 wt.%, the alloy achieved its peak value (28.3), representing a 5.6% increase compared to Babbitt alloy SnSb11Cu6.

(3) The trace addition of Co element has minimal impact on the dry friction coefficient of Babbitt alloy. When Co content exceeds 1.5 wt.%, there is a significant deterioration in friction performance.

(4) In general, the addition of 1.5 wt.% of Co results in Babbitt alloy SnSb11Cu6 exhibiting favorable comprehensive properties.

Author Contributions: Conceptualization, Z.C. and T.Z. and M.W.; methodology, M.W.; validation, N.L., M.W. and B.W.; formal analysis, Z.C. and J.Z.; investigation, L.Z.; resources, T.Z.; data curation, M.W.; writing—original draft preparation, M.W.; writing—review and editing, B.W.; visualization, J.Z. and F.J.; supervision, B.W.; project administration, N.L.; funding acquisition, Z.C. All authors have read and agreed to the published version of the manuscript.

Funding: This research was funded by the National Key R&D Program of China (2021YFB3401100) and the 2020 Ningbo "3315 Talent Introduction Plan" Innovative Team (C-Class) and the Natural Science Foundation Project of Ningbo City, China (2022J052).

Institutional Review Board Statement: Not applicable.

Informed Consent Statement: Not applicable.

Data Availability Statement: The original contributions presented in the study are included in the article, further inquiries can be directed to the corresponding authors.

Conflicts of Interest: Authors Zhan Cheng, Meng Wang, Bo Wang, Lei Zhang, Ningbo Li, Jifa Zhou and Fei Jia were employed by the company Ningbo Intelligent Machine Tool Research Institute Co., Ltd., author Ting Zhu was employed by the company Shenke Slide Bearing Co., Ltd. The authors declare that the research was conducted in the absence of any commercial or financial relationships that could be construed as a potential conflict of interest.

References

1. Liu, X.; Zhong, S.; Jing, W.; Zhang, G.; Chang, Y. Research progress on preparation technology and composition optimization modification of tin-based babbitt alloy. *Weld. Join.* **2023**, *2*, 44–52.
2. Naeimian, H.; Mofid, M. TLP bonding of Ti-6Al-4V to Al 2024 using thermal spray Babbitt alloy interlayer. *Trans. Nonferrous Met. Soc. China* **2020**, *30*, 1267–1276. [CrossRef]
3. Wang, J.; Xue, Y.; Zhang, Y.; Wang, S. Creep Mechanical Properties of Babbitt SnSb11Cu6. *Rare Met. Mater. Eng.* **2015**, *44*, 1432–1438.
4. Song, Z.; Peng, Z.; Yan, M.; Li, Z.; Dong, G.; Zhang, L. Microstructures and Mechanical Properties of Tin-Based Babbitt Bearings with Different Fabrication Methods. *Rare Met.* **2021**, *45*, 378–384.
5. Liu, Y.; Yin, P.; Zhang, F.; Wang, J. Research progress of babbitt alloy bearing bonding technology. *Heavy Mach.* **2021**, *6*, 1–7.
6. Que, M.; Bai, X.; Zhou, X.; Zhang, H. Effect of Surface Texturing on the Interfacial Microstructure and Bonding Strength of Babbitt Alloy Cladding by MIG Welding. *Surf. Technol.* **2021**, *50*, 239–245.
7. Fan, J.; Wang, X.; Wu, S.; Wang, X.; Gao, H.; Liu, J. Effect of Nickel Layer on Microstructure Formation and Mechanical Property of Aging Treated Copper-based Bearing. *Mater. Rev.* **2017**, *31*, 68–72.
8. Wang, F.; Li, Q.; Wang, T.; Guo, W.; Zhou, B. Failure Reason Analysis of Babbitt Alloy on the Surface of Bearing Bush. *Mater. Prot.* **2022**, *55*, 227–232.
9. Zhao, X.; Dai, M.; Yu, X.; Wang, Y.; Xu, Z.; Xie, Z. Reasons for the detachment of babbitt alloy for sealing pads used in power plant generators. *Phys. Test. Chem. Anal. Part A Phys. Test.* **2024**, *60*, 53–57+61.
10. Chen, J.; Huang, W.; Li, F.; Liao, D. Application of Arc-Additive Manufacturing Technology in the Field of Sliding Bearings. *Bear* **2021**, *500*, 1–7.
11. Zhang, X.; Zhang, G.; Dai, D.; Cheng, J.; Shao, S. The Overview and Market Potential Analysis of the Babbitt Alloy Composition and Wire Preparation Technology for Sliding Bearing with Additive Manufacturing. *World Nonferrous Met.* **2023**, *6*, 40–44.
12. Zhao, J.; Sun, K.; Liang, G.; Xu, C.; Zhao, J.; Xue, F.; Zhou, J. Effect of Zn additions on the microstructure and mechanical properties of Sn-Babbitt alloys fabricated by arc deposition. *J. Mater. Res. Technol.* **2021**, *15*, 6726–6735. [CrossRef]
13. Nikolay, V.; Roman, S.; Maria, B.; Nikolay, P. Effect of carbon nanotubes on structure and properties of the antifriction coatings produced by plasma cladding. *Fuller. Nanotub. Carbon Nanostruct.* **2020**, *28*, 515–520.
14. Ramadan, M.; Alghamdi, A.; Subhani, T.; Halim, K. Fabrication and Characterization of Sn-Based Babbitt Alloy Nanocomposite Reinforced with Al_2O_3 Nanoparticles/Carbon Steel Bimetallic Material. *Materials* **2020**, *13*, 2759. [CrossRef]
15. Dong, Q.; Yin, Z.; Li, H.; Zhang, X.; Jiang, D.; Zhong, N. Effects of Ag micro-addition on structure and mechanical properties of SnSb11Cu6 Babbitt. *Mater. Sci. Eng. A* **2018**, *722*, 225–230. [CrossRef]
16. Fan, J.; Liu, Z.; Zhai, H.; Wang, X.; Wang, Y.; Li, Y.; Zhou, X.; Wu, S.; Liu, J. Effect of Co content on the microstructure, spreadability, conductivity and corrosion resistance of Sn-0.7Cu alloy. *Microelectron. Reliab.* **2020**, *107*, 1–10. [CrossRef]
17. Zou, C.; Gao, Y.; Yang, B.; Zhai, Q.; Andersson, C.; Liu, J. Melting temperature depression of Sn-0.4Co-0.7Cu lead-free solder nanoparticles. *Solder. Surf. Mt. Technol.* **2009**, *21*, 9–13. [CrossRef]
18. Chuang, T.; Jain, C.; Wu, H. Intermetallic Reactions in Sn-0.4Co-0.7Cu Solder BGA Packages with an ENIG Surface Finish. *J. Electron. Mater.* **2008**, *37*, 1734–1741. [CrossRef]
19. *GB/T 228.1-2010*; Metallic Materials—Tensile Testing—Part 1: Method of Test at Room Temperature. National Standard of the People's Republic of China: Beijing, China, 2010.

20. *GB/T 16748-1997*; Plain Bearings—Compression Testing of Metallic Bearing Materials. National Standard of the People's Republic of China: Beijing, China, 1997.
21. *GB/T 231.1-2009*; Metallic Materials—Brinell Hardness Test—Part 1: Test Method. National Standard of the People's Republic of China: Beijing, China, 2009.
22. He, L.; Zhu, J. Tribological Characteristics of Short Carbon Fiber-reinforced Sn-based Babbitt Alloy. *Chin. J. Nonferrous Met.* **1998**, *2*, 36–40.
23. Liang, C.; Wang, C.; Zhang, K.; Liang, M.; Xie, Y.; Liu, W.; Yang, J.; Zhou, S. Nucleation and strengthening mechanism of laser cladding aluminum alloy by Ni-Cr-B-Si alloy powder based on rare earth control. *J. Mater. Process. Technol.* **2021**, *294*, 117145. [CrossRef]
24. Wang, M.; Zhang, G.; Zhong, S.; Shen, Y.; Long, W.; Dong, H.; Liu, X. Effect of Low-melt Alloy Powder on Brazing Process of Flux-cored Silver Brazing Filler Metal. *Rare Met. Mater. Eng.* **2021**, *50*, 2859–2866.
25. Wang, M.; Zhong, S.; Zhang, G.; Dong, H.; Si, H. The effect of Ni, Cr, Co Elements on the Microstructure and Microhardness of Sn-0.7Cu Solder. Light Ind. *Sci. Technol.* **2021**, *37*, 60–61+74.
26. Tang, L.; Li, R.; Huang, H.; Han, H.; Duan, Z.; Li, L. Study on the Effect of Co Element on the Structure, Melting Point, Wettability, and Reliability of Sn-0.7Cu Leadless Solder Alloy. *Shanxi Metall.* **2024**, *47*, 17–19.
27. Ma, L.; Tai, F.; Shen, H.; Guo, F. Effects of Co Element on Microstructural Evolution and Service Reliability of Sn-3.0Ag-0.5Cu Solder Alloy. *Rare Met. Mater. Eng.* **2010**, *39* (Suppl. S1), 341–345.
28. Dong, C.; Wang, F.; Ding, H.; Du, C. Effect of Addition of Co Element on Properties of Sn-Bi Eutectic Solder. *Hot Work. Technol.* **2015**, *44*, 190–192.
29. Fan, J.; Liu, Z.; Wu, S.; Li, Y.; Wang, Y.; Wang, X.; Liu, J. Effects of Co Content on Microstructure Evolution and Mechanical Properties of Sn-0.7Cu Alloy. *Hot Work. Technol.* **2018**, *47*, 65–69.
30. Zhu, J.; Cui, Z.; Wang, Q. Effect of Laser Power on Microstructure and Properties of Automotive Babbitt Coatings. *Appl. Laser* **2023**, *43*, 9–15.
31. Hao, Y.; Wang, J.; Yang, P.; Wang, Y.; Liang, X.; Gao, J. Microstructures and Properties of Tin-Based Babbitt Metal Prepared by Laser Cladding Deposition. *Chin. J. Lasers* **2020**, *47*, 116–125.

materials

Article

On the Prediction and Optimisation of Processing Parameters in Directed Energy Deposition of SS316L via Finite Element Simulation and Machine Learning

Mehran Ghasempour-Mouziraji [1,2,*], Daniel Afonso [1,2] and Ricardo Alves de Sousa [1,2]

[1] TEMA—Centre for Mechanical Technology and Automation, Department of Mechanical Engineering, University of Aveiro, 3810-193 Aveiro, Portugal; dan@ua.pt (D.A.)
[2] LASI—Intelligent Systems Associate Laboratory, 4800-058 Guimarães, Portugal
* Correspondence: mehran@ua.pt

Abstract: In the current study, the integration of finite element simulation and machine learning is used to find the optimal combination of processing parameters in the directed energy deposition of SS316L. To achieve this, the FE simulation was validated against previously implemented research, and a series of simulations were conducted. Three inputs, namely laser power, scanning speed, and laser beam radius, and two outputs, namely residual stress and displacement, were considered. To run the machine learning model, artificial neural networks and a non-dominated sorting genetic algorithm were applied to determine the optimal combination of processing parameters. In addition, the current study underscores the novelty of combining FE simulation and machine learning methods, which provides enhanced precision and efficiency in controlling residual stress and displacement (geometrical deviation) in the Directed Energy Deposition (DED) process. Then, the results obtained via machine learning were validated with confirmatory tests and reported. The findings offer a practical solution for process parameter optimization, contributing to the progression of additive manufacturing technologies.

Keywords: directed energy deposition; SS316L; finite element simulation; machine learning

1. Introduction

Directed Energy Deposition (DED) is an Additive Manufacturing (AM) method that involves melting materials and depositing them layer by layer to produce, repair, or enhance the mechanical and microstructural properties of parts [1]. In this advanced manufacturing process, the molten material is fed in the form of powder or wire onto the substrate to create a melt pool by using different energy sources such as laser, electron beam or plasma arc. This method is highly beneficial, as it reduces material waste and can work with a wide range of materials, including metals, composites and ceramics. DED is widely used in various applications including the aerospace [2], medical [3], automotive [4], and tooling industries [5], fabricating functionally graded structures [5], complex [6], and customized parts, as well as its suitability for repairing damaged components [7].

Despite these advantages, such as high deposition rate, utilizing a wide range of materials, and design flexibility, DED presents some challenges that restrict the fabrication process and efficiency. The first drawback of this method is the poor surface quality [8] of deposited parts which mostly requires post-processing such as machining and grinding. This extra step enhances the time and cost associated with DED. Another challenge in this process is thermal gradient [9], which creates residual stress that can lead to warping and if

 https://doi.org/10.3390/ma18051039

not properly controlled, affecting the dimensional accuracy and mechanical performance of the fabricated part [10].

In addition, cracking, and other structural defects may occur due to the accumulation of residual stress [11]. These defects can compromise the mechanical integrity of the final product. Although this optimized fabrication method offers various advantages, DED process optimization is a challenging issue that is affected by processing parameters such as laser power, scanning speed, laser beam, and layer thickness. These parameters affect the mechanical and microstructural properties and surface finish. The traditional trial-and-error method for optimization is a time-consuming process and highlights the need for more efficient and systematic approaches. Among various optimization approaches, Machine Learning (ML) has recently been accentuated by industries and academics. Machine learning offers advanced solutions to these challenges by enabling data-driven prediction and optimization of processing parameters [12]. ML algorithms analyze large datasets to find patterns and relationships that are not readily apparent through conventional analytical methods. Regarding DED, ML can be employed to model the complex interactions between processing parameters and resultant material properties, thereby facilitating the prediction and control of the deposition process. Much research has been carried out in the field of optimization of manufacturing processes, particularly the DED process. M. Moradi et al. [13] investigated the Direct Laser Metal Deposition (DLMD) method for the additive manufacturing of Inconel 718 Ni-based superalloy using a full factorial design approach. They focused on how three process parameters, namely laser scanning speed, powder feed rate and scanning strategy, affect the geometrical dimensions, microhardness, and stability of the manufactured walls. They found that lower scanning speeds and higher powder feed rates resulted in increased wall height and width. The microhardness elucidated variability due to microstructural phases. The optimal combination was found to be at low scanning speeds with a unidirectional scanning pattern, accounting for enhanced melting time and smoother surfaces. P. Pant et al. [14] investigated the Laser Metal Deposition (LMD) process for surface coating utilizing SS316L material. Artificial Neural Networks (ANNs) and a combined Particle Swarm Optimization (PSO)-ANN model were used to predict the main outputs such as clad layer height, width, and powder capture efficiency. Before running optimization via (PSO)-ANN, Response Surface Methodology (RSM) was used to gather data. These findings demonstrated that the PSO-ANN model offers superior reliability and accuracy for predicting and optimizing LMD process parameters compared to traditional empirical methods and simple ANN models. M. Pellizzari et al. [15] optimized the process of depositing single-layer H13 tool steel cladding on a copper-beryllium (CuBe) alloy substrate using direct laser metal deposition (DLMD), by considering laser power, powder feed rate, overlap ratio, and substrate temperature. The quality of cladding was evaluated by defects, interfacial adhesion, geometrical features and substrate over-ageing, and it is worth noting that it was affected by CuBe's high reflectivity. Optimal combination of parameters fabricated dense cladding, with substrate preheating preventing cracks but causing softening at higher temperatures: 13% hardness reduction at 150 °C and 40% at 250 °C. R. Mahamood et al. [16] explored the effect of processing parameters on material usage efficiency in LMD of Ti6Al4V using a two-level full factorial design of experiments. Laser power, scanning speed, powder flow rate and gas flow rate are the processing parameters. The optimal combination of processing parameters is as follows: laser power of 3.2 kW, a scanning speed of 0.06 m/s, a powder flow rate of 2 g/min, and a gas flow rate of 3 L/min.

R. Li et al. [17] examined the deformation distribution and residual stress in the fabrication of thin-walled Al-Cu alloy components using a dual-robot collaborative wire-arc additive manufacturing (WAAM) ((FANUC M-20iA, Echternach, Luxembourg) paired with a welding unit (Fronius CMT Advanced 4000R, Wels, Austria)). To validate FE simulation, infrared thermography and structured light sensors were used. Both systems showed symmetric stress distributions, with stress concentrations achieved at the junction between the substrate and the thin wall. In addition, the dual-robot WAAM system demonstrates superior performance over the single-robot system, offering reduced temperature gradients, more uniform stress distributions, and less deformation.

Chi Wu et al. [18] used an ML-based design for biomedical engineering ceramic with additive manufacturing to optimize functionally graded tissue scaffolds with Triply Periodic Minimal Surfaces (TPMSs) for bone regeneration. By combining Bayesian optimization and Lithography-based Ceramic Manufacturing (LCM), the framework achieves improved bone ingrowth and precise geometric qualities, offering a novel tool for practical design optimization in ceramic AM. H. Mu et al. [19] investigated an adaptive online simulation model for Wire Arc Additive Manufacturing (WAAM), addressing challenges in distortion prediction for building a Digital Twin (DT). Utilizing Vector Quantized Variational AutoEncoder coupled with Generative Adversarial Network (VQVAE-GAN) architecture for spatial feature extraction and an RNN for time-scale fusion, the model estimates distortion fields online with high accuracy, significantly outperforming traditional FEM and ANN methods. T. Loreau et al. [20] addressed the complexity of simulating porosity in Wire Laser Metal Deposition (WLMD) due to the intricate interactions between the laser and material and the varying scales of time and space involved, which significantly affect simulation efficiency and performance. To tackle this challenge, they proposed utilizing neural networks to model porosity as a function of processing parameters. Firstly, the relevant parameters were found and analyzed for their effect on porosity. Once the correlations were confirmed, these parameters were used as inputs for constructing and training a Multilayer Perceptron (MLP) neural network with part of the available data. The model's performance was then evaluated using both the training dataset and a separate test dataset to assess its predictive capabilities. In the current study, the prediction and optimization of DED processing parameters are achieved using an ML approach and finite element simulation. The primary objective of this study is to develop robust predictive models for the key output variables of residual stress and geometrical deflection (nodal displacement) and to optimize the process parameters to achieve the desired performance outcomes. Figure 1 shows the flow chart of the current study.

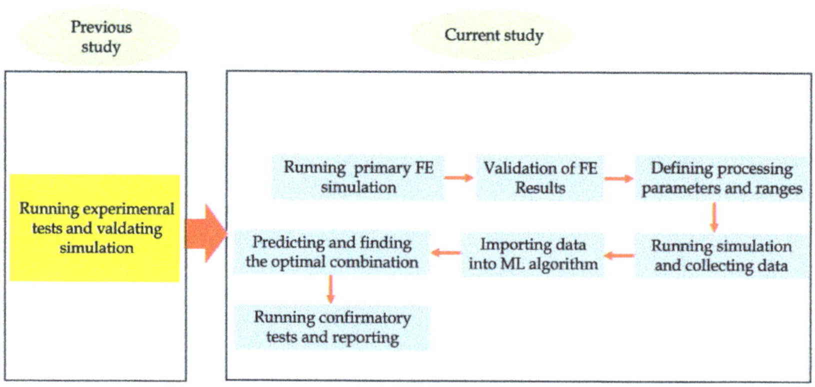

Figure 1. Flow chart of the current study.

2. Finite Element Simulation and Validation

To simulate the DED process, MSC APEX 2023 and Simufact software 2023 were used. Firstly, the parts were designed in MSC Apex and then imported into the Simufact

Welding Module, which applies temperature-dependent material properties that alter the microstructural characteristics with each heating cycle. The cubic element size for the substrate is $2.91 \times 3 \times 2.91$ mm^3 with a total substrate size of $100 \times 100 \times 6$ mm^3 and the element size for layers is $0.6 \times 1.2 \times 0.715$ mm. The current study defines a range for the processing parameters, ensuring that the chosen values are representative operating conditions. By using this range, we aim to explore and identify optimal combinations of processing parameters. During the simulation, thermo-mechanical behavior was modeled throughout the transient deposition process using a quasi-static element formulation. The melt pool and Marangoni effect were modeled using a low-stiffness solid approach and artificial thermal conductivity, respectively. Tables 1 and 2 show the chemical and mechanical properties of SS316 [21–23]. Also, Figure 2 shows the Temperature-dependent material properties of SS316L. The part was built by deposition of 20 layers with different processing parameters, which are given in Table 3. The process parameters included in Table 3 were selected based on a previous study used to validate the simulation

Table 1. Chemical composition of ss316 [24].

Grade	C	Mn	Si	P	S	Cr	Mo	Ni	N
min	-	-	-	-	-	16.0	2.00	10.0	-
max	0.03	2.0	0.75	0.045	0.03	18.0	3.00	14.0	0.10

Table 2. Mechanical properties of SS316 [10].

Parameter	Value
Solidus temperature	1279 °C
Melting temperature	1450 °C
Density	7966 Kg/m^3
Latent heat for melting	25,400 J/Kg
Poison ratio	0.3
Convective heat transfer for first and remaining layers	10, 30 (W/m^{2}K)
Emissivity for first and remaining layers	0.6, 0.6
Elastic module	200 GPa
Ultimate Tensile Strength	580 MPa
Yield Strength	290 MPa
Specific heat	2051.531 J/kg. °C
Thermal conduction	0.225 W/cm. °C

Table 3. Process parameters.

Parameter	Value
Laser power	400–800 W
Scanning speed	10–20 mm/s
Laser beam diameter	0.6–0.8 mm
Powder feed rate	7.5 g/min
Powder size	45–90 μm

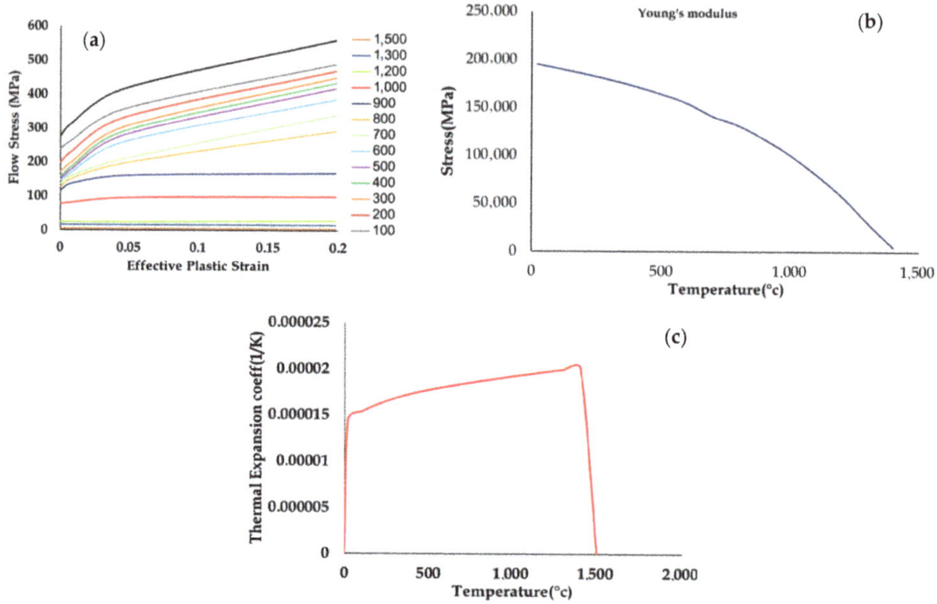

Figure 2. (**a**) Temperature-dependent material properties of SS316L adapted from Simufact Welding (Flow stress) [25], (**b**) Temperature-dependent material properties of SS316L adapted from Simufact Welding (Young's Modulus), (**c**) Temperature-dependent material properties of SS316L adapted from Simufact Welding (thermal expansion) [25].

To validate the simulation, a 50 mm bead was deposited using the same process conditions as previous research [26]. Since in the current study, residual stress and maximum nodal displacement on the sample (deflection) were used, the FE simulation results for those outputs are given in Figure 3.

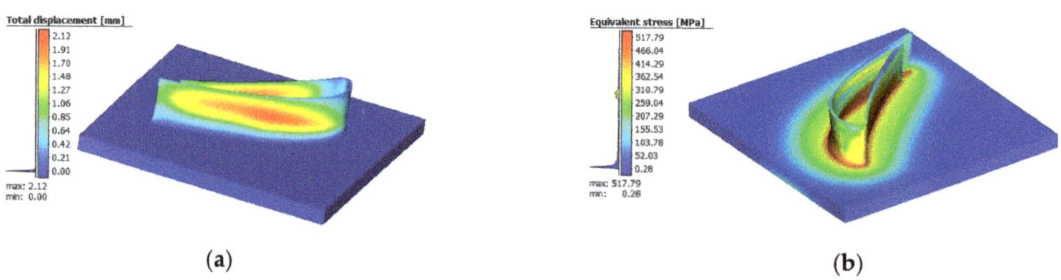

Figure 3. FE simulation of current study (**a**) displacement and (**b**) residual stress.

3. Machine Learning Procedure

Machine Learning (ML) is a subset of Artificial Intelligence (AI) that focuses on developing algorithms and statistical models enabling computers to learn from and make decisions based on data [27]. It has been accentuated as a powerful method in many fields, such as banking [27], healthcare [28], image processing and identification [29,30] and engineering [31,32]. Traditional programming requires developing the rules and instructions to handle the tasks specifically, but ML has no need for any explicit programming, and computers recognize the patterns, make choices, and predict and optimize the data automatically. ML consists of various methods and techniques for training algorithms to learn from data and make predictions or decisions autonomously, as shown in Figure 4.

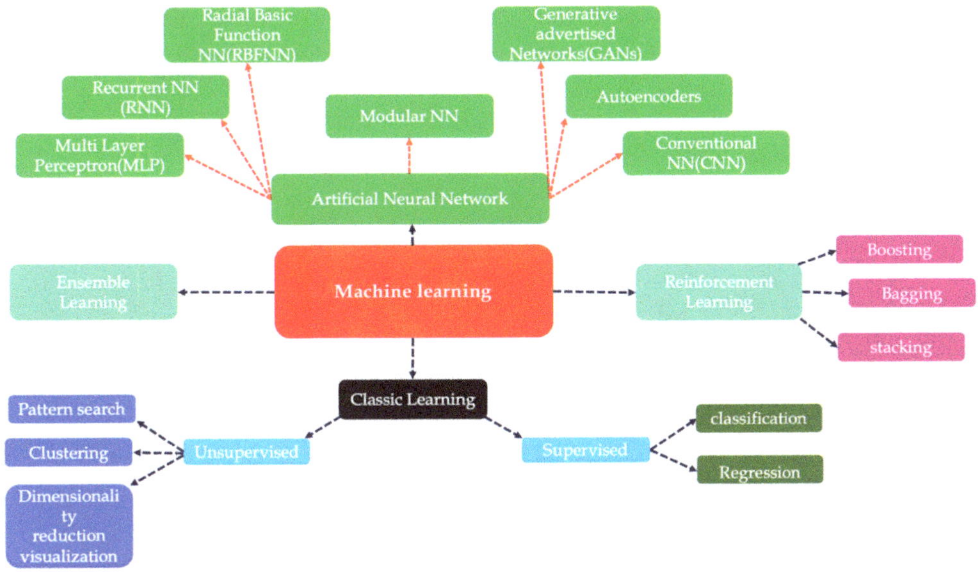

Figure 4. Methods of machine learning.

4. Multi-Objective Optimization of Processing via ML

Multi-objective optimization has enormous practical importance. Almost all optimization challenges that occur in realistic situations may be characterized as having several different objectives. Usually, there is no single optimal solution in multi-objective optimization, also known as vector optimization, that simultaneously fulfills all objective functions. Thus, a set of solutions known as Pareto optimal solutions emerges. Pareto optimal solutions represent the best possible values for the objectives without favoring one objective over another.

In Pareto-based multi-objective optimization, the population is arranged onto the Pareto front, and the individuals are rated according to how superior they are. This process can be defined as [33]:

$$F\ (X) = [f_1(X),\ f_2(X), \ldots,\ f_k(X)]^T \tag{1}$$

represents the vector of objective functions, and $X^* \in \Re^n$ is the vector of design variables. Subject to m inequality constraints [34]:

$$L_i\ (X) \leq 0, \quad i = 1\ \ to\ \ m \tag{2}$$

and p equality constraints [34]:

$$H_j\ (X) = 0, \quad j = 1\ \ to\ \ p \tag{3}$$

When Solution A obtains a lower value (in the case of minimization) for at least one target and does not perform worse than Solution B for the remaining objectives, it outperforms Solution B. On the other hand, a point (Ω is a feasible region in $\Re^n$ satisfying Equations (2) and (3)) dominated all $X \in \Omega$ if and only if [33]:

$$\forall i \in \{1,\ 2, \ldots,\ k\}, \quad \forall X \in \Omega - \{X^*\} \quad f_i\ (X^*) \leq f_i\ (X)\ \wedge\ \exists j \in \{1,\ 2, \ldots,\ k\}: \quad f_j\ (X^*) < f_j\ (X). \tag{4}$$

According to Figure 5, non-dominated solutions are the ones that are much closer to the optimal solution compared to feasible solutions. Moreover, individuals laid on the "Pareto front" have no superiority over each other.

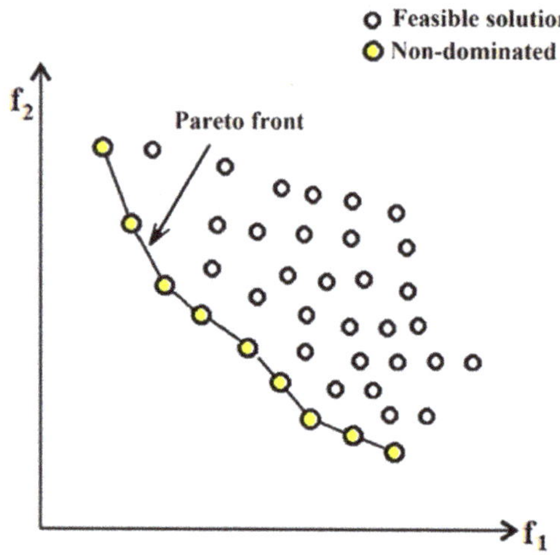

Figure 5. Pareto front in multi-objective optimization algorithms.

5. Artificial Neural Network (ANN)

ANN is commonly used as a model to address engineering challenges. Neural networks are composed of an array of neurons, which serve as simple processing units. Typically, an ANN consists of input, output, and hidden layers, with the number of hidden layers varying based on the process and structure of the network. In this setup, neurons are interconnected to neurons in the next layers via direct connection links with assigned weights. It is worth noting that the weight values are adjusted to learn and find the optimal solution for each problem [35]. The number of neurons in both the input and output layers corresponds to the number of physical variables and targets. However, in the hidden layer, the number of neurons may vary based on the complexity of the problem. Determining the optimal network requires assigning an appropriate number of neurons to each hidden layer. In the literature, a feedforward neural network is a commonly applied model, with the Back Propagation (BP) network being widely used in feedforward architectures. A schematic view of ANN is shown in Figure 6.

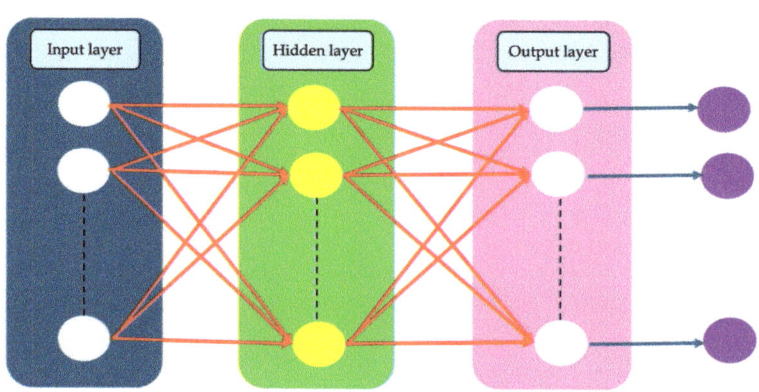

Figure 6. A schematic view of a one-hidden-layer artificial neural network (ANN).

6. Data-Driven Multi-Objective Optimization of Artificial Neural Network

One of the most challenging issues is finding the best ANN with the lowest training and testing errors. In this case, Yarmohammadi et al. [36,37] utilized multi-objective optimization to determine the most optimal ANN with minimal training and testing errors. The primary goal of this section is to construct a multilayer feedforward neural network using the Back Propagation (BP) learning algorithm to predict optimal input parameters while optimizing two conflicting objectives: training error and testing error. The achievement is shown in Tables 4 and 5. In the input layer of ANNs, there are three input parameters, namely laser power, laser beam radius, and scanning speed, of which the range is given in Table 3. In the current study, the input parameters of ANN and their variables are presented in Table 2. The output parameters include residual stress and displacement, which are considered the first target and second target, respectively. The layers between the input and output layers are known as hidden layers. Initially, several attempts were made to find an optimal ANN configuration that minimized all targets simultaneously; however, this approach resulted in high test and training errors. To address this issue, a unique ANN was developed for each output, which helped reduce both training and testing errors.

Table 4. Variables in multi-objective optimization of artificial neural network for the first target.

	Back Propagation Algorithm	Number of Neurones in 1st Layer	Number of Neurones in 2nd Layer	Transfer Function in 1st Hidden Layer	Transfer Function in 2nd Hidden Layer	Train Error (%)	Test Error (%)
A	TrainLM	24	29	Tan-sig	Log-sig	4.326	7.760
B	TrainBR	20	10	Tan-sig	Log-sig	4.412	7.160
C	TrainCGF	20	9	Tan-sig	Log-sig	4.332	7.566

Table 5. Variables in multi-objective optimization of artificial neural network for the second target.

	Back Propagation Algorithm	Number of Neurones in 1st Layer	Number of Neurones in 2nd Layer	Transfer Function in 1st Hidden Layer	Transfer Function in 2nd Hidden Layer	Train Error (%)	Test Error (%)
A	TrainCGF	13	20	Tan-sig	Log-sig	2.518	3.782

In this study, a two-layer network with a maximum number of neurons of 30 per layer is proposed as it has the capability to find the required optimum ANN configuration. Raising the number of layers as well as neurons could result in a negligible improvement in the result, but could also lead to significant increases in the computation cost. There were two optimum configurations for the first target and only one configuration for the second target. In the case of the first target, comparing the output of these two configurations did not show any meaningful difference. Thus, an output of only one of them was reported in this research.

The given information in Table 4 is crucial for training the ANN model and affects its performance in terms of both training and test error. The back propagation algorithm is used to train the ANN, which consists of three approaches TrainLM (Levenberg–Marquardt), TrainBR (Bayesian Regularization) and TrainCGF (Conjugate Gradient with Fletcher–Reeves). TrainLM (Levenberg–Marquardt) is a fast, widely used algorithm for training networks, typically providing high accuracy with fewer iterations. The number of Neurons in 1st Layer refers to the number of neurons in the first hidden layer of the network. Each neuron in this layer processes a different feature of the input data. The number of Neurons in 2nd Layer specifies the neurons in the second hidden layer, where the network starts to learn more complex features from the data. Transfer Function in 1st Hidden Layer defines how the output of each neuron in the hidden layer is calculated. In the current study, the **Tan-sig** (tangent sigmoid) transfer function is utilized in the first layer, which outputs values between -1 and 1, helping the network to learn complex patterns. Transfer Function in 2nd Hidden Layer used in the second hidden layer is **Log-sig** (logistic sigmoid), which outputs values between 0 and 1. This function is particularly useful for binary classification tasks or probabilities. Train Error (%) represents the error (in percentage) during the training phase, showing how well the network fits the training data. The lower the training error is the better in learning patterns. Test Error (%) shows the error (in percentage) when the network is examined based on the data that have not been used in the training section. This aids in assessing how well the model generalizes to new, unseen data.

The performance and train, validation, and test regression graph for the first target is illustrated in Figure 7.

Also, the information for ANN for the second target is shown in Table 5 and related information about the performance of ANN is shown in Table 5. Also, the performance of second target is shown in Figure 8.

The first stage of optimization is generating a random population through iteration toward the best solution. The detailed methodology to determine the optimal ANN is given in [36,37], where the authors give detailed information about the involved steps. The related processing parameters are shown in Table 6.

Table 6. Detailed information about the NSGA.

Parameter	Value
population size	40 chromosomes
mutation rate	0.1
crossover rate	0.85 over 40 generations
Execution time for each network	10 times

To find the final objective of each individual chromosome, 10 independent runs were carried out. This approach helps to eliminate the unstable chromosome from consideration. The evolutionary process of selecting the optimal network variable vectors to identify the Pareto front of two conflicting objectives (training and testing errors) is conducted using the NSGA approach. The optimal artificial neural networks were identified for each output individually, and their corresponding training and testing errors are presented in Tables 4 and 5.

(**a**)

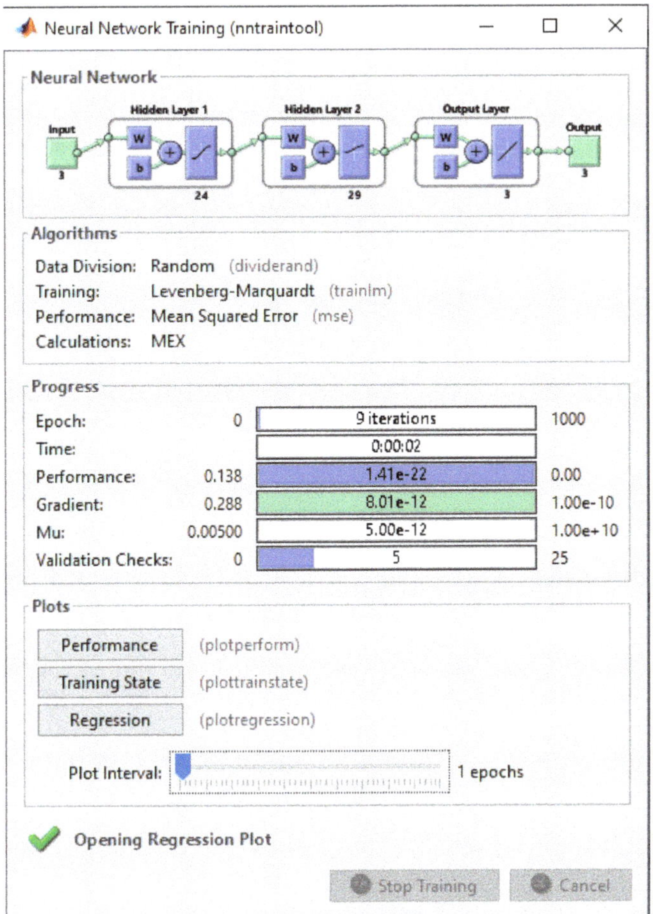

(**b**)

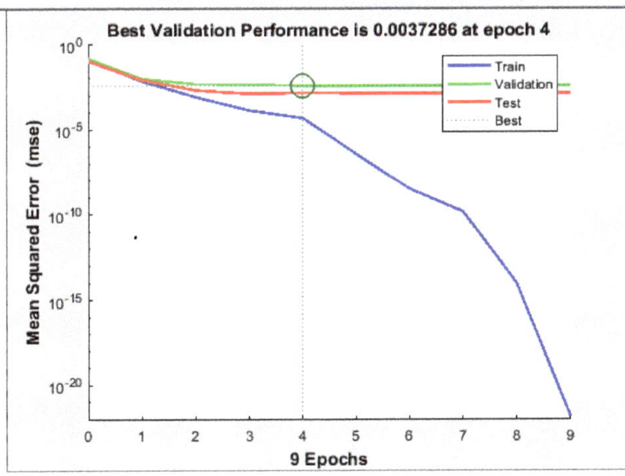

Figure 7. *Cont.*

(c)

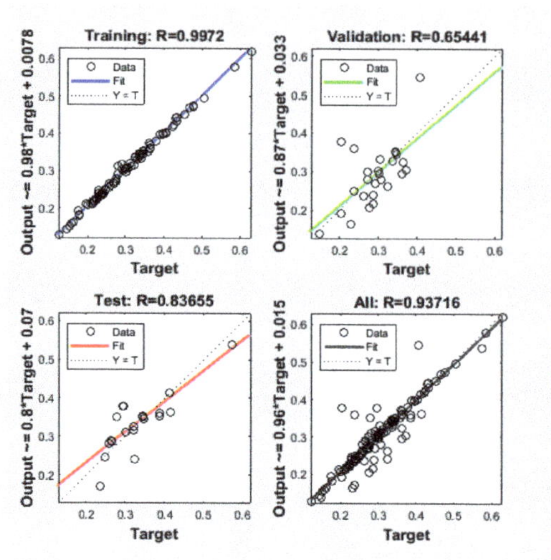

(d)

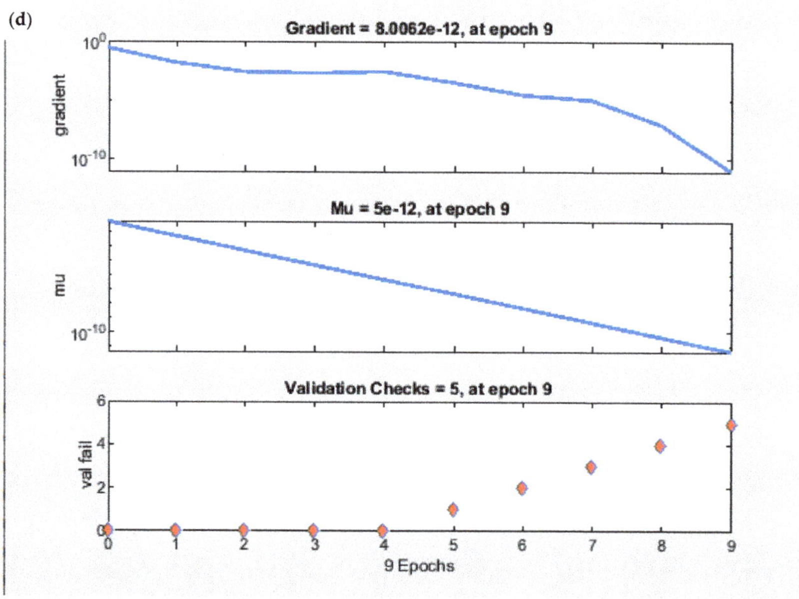

Figure 7. (**a**) Performance of ANN, (**b**), best validation performance, (**c**) Train, validation, and test regression of ANN, (**d**), training state performance plot (for first target).

(a)

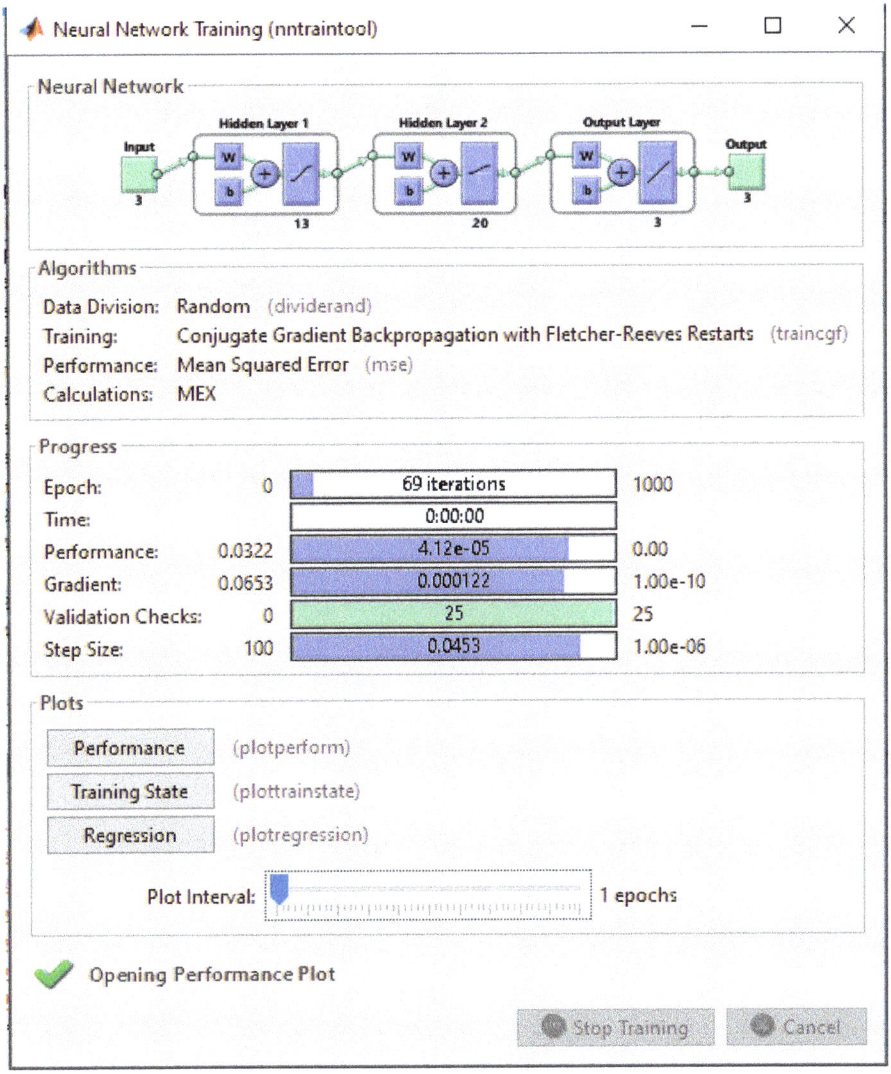

Figure 8. *Cont.*

(b)

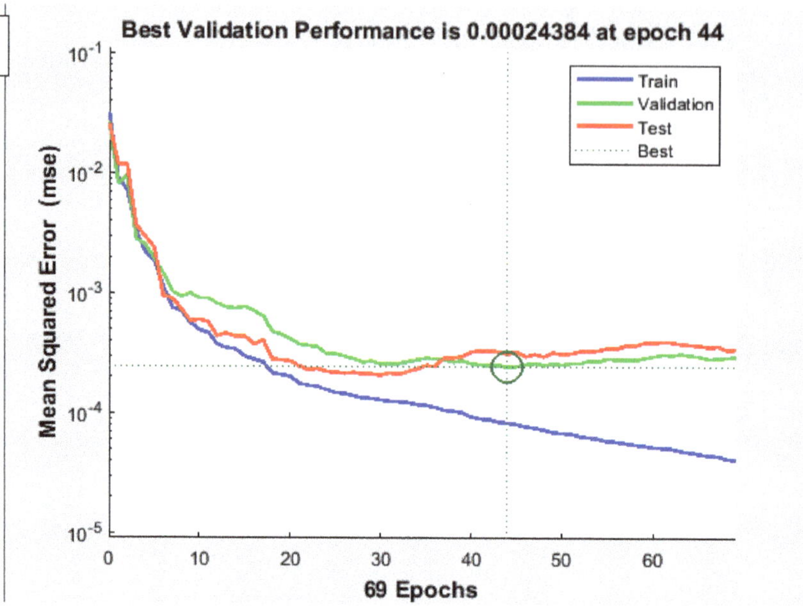

(c)

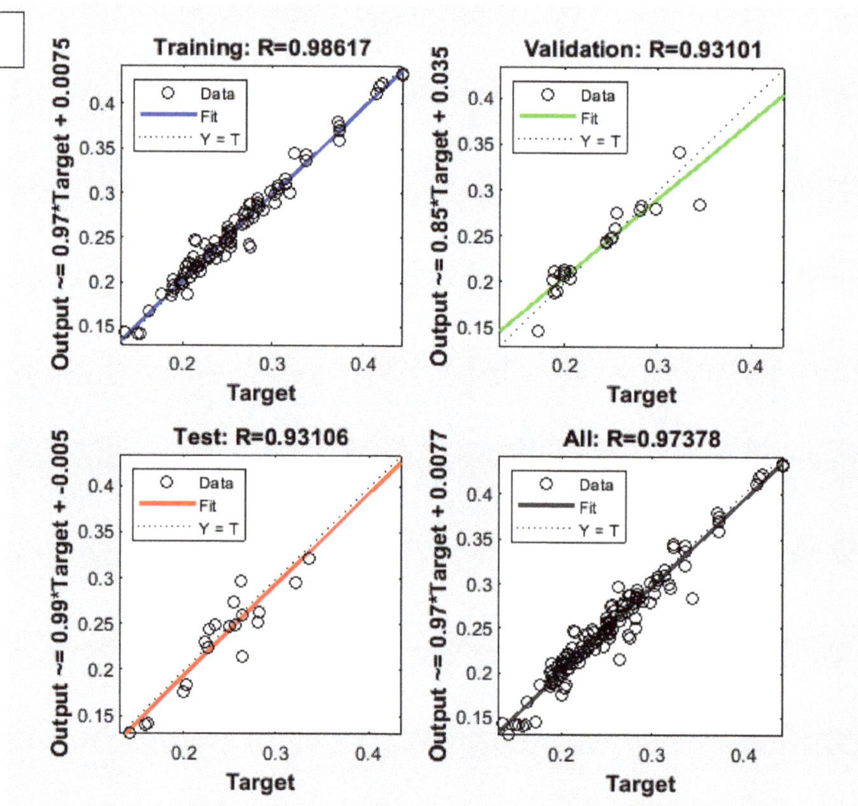

Figure 8. *Cont.*

(d)

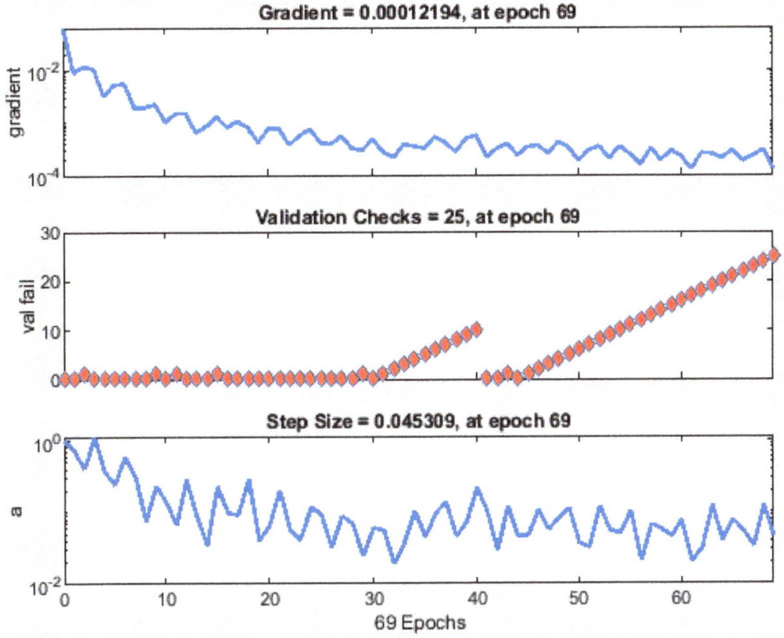

Figure 8. (**a**) Performance of ANN, (**b**), best validation performance, (**c**) train, validation, and test regression of ANN, (**d**), training state performance plot (for second target).

7. Multi-Objective Optimization of Processing Parameters in DED

Having found the optimal ANN for each target, in the next step, the best results (lowest values) are found in the continuous domain bounded. Thus, NSGA was utilized to determine the optimal inputs, resulting in the optimal combination of laser power, scanning speed, and laser beam radius. Figure 9 illustrates the NSGA algorithm utilized in this study. The non-dominated sorting genetic algorithm, NSGA-II, is one of the best evolutionary algorithms for the optimization of multi-objective problems, which has been employed in several studies. Since the sorting method employed in this algorithm introduces all possible scenarios, NSGA guarantees the obtainment of the appropriate combination of parameters leading to the optimum output. The algorithm starts generating a random initial population. Then, using operators such as crossover and mutation, it tries to find the best input parameters until satisfying the termination criteria. The termination criteria could be a certain number of iterations (as used in this study), reaching a certain and acceptable output, or any other criteria a designer determines. Additionally, Table 7 demonstrates the results achieved through the ANN. To select the best option among these individuals, their values referring to the targets were compared and scored 1 to 4 based on their strengths.

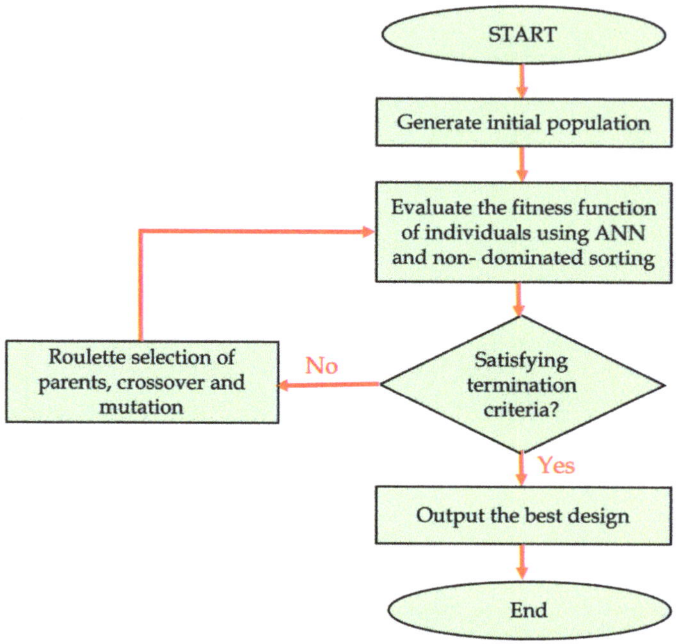

Figure 9. The NSGA-II algorithm.

Table 7. Obtained results through the NSGA and validation.

Laser Power (W)	Scanning Speed (mm/s)	Laser Beam Radius (mm)	Displacement (mm) (NSGA)	Residual Stress (NSGA)	Displacement (mm) (FE Simulation)	Residual Stress (FE Simulation)
684.93	10.46	0.59	1.56	450.57	1.62	489
713.12	10.73	0.57	1.46	457.17	1.51	491

Table 7 presents the optimal values for the current research derived from NSGA. These values provide information about the best combination of processing parameters that lead to an optimal output for the DED process. It is worth noting that the achieved results are in the predefined range of input processing parameters and corresponding outputs according to Table 3. This ensures that the optimization process stays within feasible and practical limits.

Regarding the effect of processing parameters, increasing laser power enhances the melt pool size which increases the higher material deposition volume in the deposition area [38]. This leads to overbuilding, which causes geometrical deviation. In addition, higher laser power causes deeper penetration on the substrate or previous layers causing irregularities on the surface [39]. Furthermore, keyholing may occur, leading to defects such as pores and voids. The effect of laser power on residual stress can be attributed to the high thermal gradient between the melt pool and surrounding materials, which increases residual stress [40]. Higher cooling rates and rapid solidification associated with high laser power further contribute to the formation of residual stress.

By enhancing scanning speed, the delivered energy to the substrate per unit decreases, which creates a smaller thermal gradient followed by lower residual stress due to a small difference between the melt pool and surrounding materials [41,42]. In addition, while less material is deposited to the melt pool, incomplete fusion happens between the layers, causing defects such as porosity or gaps leading to geometrical deviation. Also, a higher scanning speed can reduce heat accumulation to control distortion.

By increasing the laser beam radius, the residual stress decreases, because the larger laser beam spreads the energy on the wider area which leads to lower energy density [43]. With lower energy density, the thermal gradient is reduced, followed by more uniform heating and cooling. On the other hand, increasing the laser beam radius is likely to create a higher deviation in the deposited area. The larger beam size increases the size of the melt pool which decreases the precision and increases geometrical deviation.

8. Conclusions

In the current study, the integration of FE simulation and machine learning, particularly Artificial Neural Networks (ANN) and Non-sorting Genetic Algorithms (NSGAs), was used to find the optimal combination of processing parameters in the directed energy deposition of SS316L to evaluate the residual stress and deflection (displacement). The focus was to assess residual stress and displacement during the process.

1. **Key findings:** Firstly, the validation process was successfully implemented and then a series of tests were carried out to create the data for ANN and NSGA to find the optimal combination of laser power, scanning speed and laser beam radius to reduce the residual stress and displacement. The application of ANN and NSGA enabled the identification of an optimal combination of processing parameters
2. **Main Advantages:** The achievement from the current study illustrated significant enhancement in controlling residual stress and geometrical deviation. This highlights the potential of combining FE simulation with ML algorithms to enhance the efficiency and precision of the DED process, offering a more accurate approach to process optimization.
3. **Implications for Future Research:** The obtained results serve as an applicable reference for further process validation and refinement, particularly when compared to experimental data. This research contributes to a deeper understanding of how different processing parameters affect the overall thermo-mechanical behavior of the DED process, providing insight for future advancements in additive manufacturing.

Author Contributions: M.G.-M.: Conceptualization, Data gathering, methodology, formal analysis, investigation, writing—original draft preparation. D.A.: Supervision, Conceptualization, Funding acquisition, writing—review and editing. R.A.d.S.: Supervision, Conceptualization, writing—review and editing, project administration. All authors have read and agreed to the published version of the manuscript.

Funding: The authors greatly acknowledge the work which was supported by the projects UIDB/00481/2020 and UIDP/00481/2020—Fundação para a Ciência e a Tecnologia; and CENTRO-01-0145-FEDER-022083—Centro Portugal Regional Operational Programme (Centro2020), under the PORTUGAL 2020 Partnership Agreement, through the European Regional Development Fund. Also, thanks to FCT for Ph.D. scholarships No. UI/BD/151258/2021.

Institutional Review Board Statement: Not applicable.

Informed Consent Statement: Not applicable.

Data Availability Statement: The original contributions presented in this study are included in the article. Further inquiries can be directed to the corresponding author.

Conflicts of Interest: The authors declare no conflict of interest.

Nomenclature

Directed Energy Deposition	DED	Non-dominated Sorting Genetic Algorithm	NSGA
Machine Learning	ML	Laser Metal Deposition	LMD
Finite Element	FE	Direct Laser Metal Deposition	DLMD
Artificial Neural Network	ANN	Particle Swarm Optimization	PSO
Multilayer Perceptron	MLP	Wire Laser Metal Deposition	WLMD
Artificial Intelligence	AI	Back Propagation	BP

References

1. Ghasempour-Mouziraji, M.; Lagarinhos, J.; Afonso, D. A review study on metal powder materials and processing parameters in Laser Metal Deposition. *Opt. Laser Technol.* **2024**, *170*, 110226. [CrossRef]
2. Shah, K.; Pinkerton, A.J.; Salman, A.; Li, L. Effects of melt pool variables and process parameters in laser direct metal deposition of aerospace alloys. *Mater. Manuf. Process.* **2010**, *25*, 1372–1380. [CrossRef]
3. Kulkarni, M.; Mazare, A.; Schmuki, P.; Iglič, A. Biomaterial surface modification of titanium and titanium alloys for medical applications. In *Nanomedicine*; One Central Press: Manchester, UK, 2014; pp. 111–615.
4. Miguel, F.; Ramalho, Q. Study of the Laser Metal Deposition (LMD) Technology in the Automotive Industry. Master's Thesis, Instituto Politecnico de Leiria, Leiria, Portugal, 2019.
5. Costa, L.; Vilar, R.; Reti, T.; Deus, A.M. Rapid tooling by laser powder deposition: Process simulation using finite element analysis. *Acta Mater.* **2005**, *53*, 3987–3999. [CrossRef]
6. Liu, C.; Li, Y.; Hao, X. An adaptive machining approach based on in-process inspection of interim machining states for large-scaled and thin-walled complex parts. *Int. J. Adv. Manuf. Technol.* **2017**, *90*, 3119–3128. [CrossRef]
7. Song, L.; Zeng, G.; Xiao, H.; Xiao, X.; Li, S. Repair of 304 stainless steel by laser cladding with 316L stainless steel powders followed by laser surface alloying with WC powders. *J. Manuf. Process.* **2016**, *24*, 116–124. [CrossRef]
8. Rombouts, M.; Maes, G.; Hendrix, W.; Delarbre, E.; Motmans, F. Surface finish after laser metal deposition. *Phys. Procedia* **2013**, *41*, 810–814. [CrossRef]
9. Ghasempour-Mouziraji, M.; Afonso, D.; Hosseinzadeh, S.; Goulas, C.; Najafizadeh, M.; Hosseinzadeh, M.; Ganji, D.; de Sousa, R.A. Modeling the effect of processing parameters on temperature history in Directed Energy Deposition: An analytical and finite element approach. *Rapid Prototyp. J.* **2023**, *30*, 338–349. [CrossRef]
10. Ghasempour, M.; Afonso, D.; Torcato, R. Numerical investigation of deposition strategies on the residual stress and geometrical deviation in Laser Metal Deposition. *IOP Conf. Ser. Mater. Sci. Eng.* **2021**, *1193*, 012095. [CrossRef]
11. Mazzarisi, M.; Angelastro, A.; Latte, M.; Colucci, T.; Palano, F.; Campanelli, S.L. Thermal monitoring of laser metal deposition strategies using infrared thermography. *J. Manuf. Process.* **2023**, *85*, 594–611. [CrossRef]
12. Ghasempour-Mouziraji, M.; Hosseinzadeh, M.; Hajimiri, H.; Najafizadeh, M.; Marzban Shirkharkolaei, E. Machine learning-based optimization of geometrical accuracy in wire cut drilling. *Int. J. Adv. Manuf. Technol.* **2022**, *123*, 4265–4276. [CrossRef]
13. Moradi, M.; Hasani, A.; Pourmand, Z.; Lawrence, J. Direct laser metal deposition additive manufacturing of Inconel 718 superalloy: Statistical modelling and optimization by design of experiments. *Opt. Laser Technol.* **2021**, *144*, 107380. [CrossRef]
14. Pant, P.; Chatterjee, D. Prediction of clad characteristics using ANN and combined PSO-ANN algorithms in laser metal deposition process. *Surf. Interfaces* **2020**, *21*, 100699. [CrossRef]
15. Pellizzari, M.; Zhao, Z.; Bosetti, P.; Perini, M. Optimizing direct laser metal deposition of H13 cladding on CuBe alloy substrate. *Surf. Coat. Technol.* **2022**, *432*, 128084. [CrossRef]
16. Mahamood, R.M.; Akinlabi, E.T. Processing Parameters Optimization for Material Deposition Efficiency in Laser Metal Deposited Titanium Alloy. *Lasers Manuf. Mater. Process.* **2016**, *3*, 9–21. [CrossRef]
17. Li, R.; Ju, G.; Zhao, X.; Zhang, Y.; Li, Y.; Hu, G.; Yan, M.; Wu, Y.; Lin, D. Simulation of residual stress and distortion evolution in dual-robot collaborative wire-arc additive manufactured Al-Cu alloys. *Virtual Phys. Prototyp.* **2024**, *19*, e2409390. [CrossRef]
18. Wu, C.; Wan, B.; Entezari, A.; Fang, J.; Xu, Y.; Li, Q. Machine learning-based design for additive manufacturing in biomedical engineering. *Int. J. Mech. Sci.* **2024**, *266*, 108828. [CrossRef]
19. Mu, H.; He, F.; Yuan, L.; Hatamian, H.; Commins, P.; Pan, Z. Online distortion simulation using generative machine learning models: A step toward digital twin of metallic additive manufacturing. *J. Ind. Inf. Integr.* **2024**, *38*, 100563. [CrossRef]
20. Loreau, T.; Champaney, V.; Hascoet, N.; Lambarri, J.; Madarieta, M.; Garmendia, I.; Chinesta, F. Parametric analysis and machine learning-based parametric modeling of wire laser metal deposition induced porosity. *Int. J. Mater. Form.* **2022**, *15*, 33. [CrossRef]
21. Israr, R.; Buhl, J.; Elze, L.; Bambach, M. Simulation of different path for WAAM with Lagrangian finite element methods. In Proceedings of the 15th LS-DYNA Forum 2018, Bamberg, Germany, 15–17 October 2018.
22. Lampa, C.; Kaplan, A.F.H.; Powell, J.; Magnusson, C. An analytical thermodynamic model of laser welding. *J. Phys. D Appl. Phys.* **1997**, *30*, 1293–1299. [CrossRef]

23. Alimardani, M.; Toyserkani, E.; Huissoon, J.P. A 3D dynamic numerical approach for temperature and thermal stress distributions in multilayer laser solid freeform fabrication process. *Opt. Lasers Eng.* **2007**, *45*, 1115–1130. [CrossRef]
24. Stainless Steel 316: What Is It? How Is It Made? Grades n.d. Available online: https://www.iqsdirectory.com/articles/stainless-steel/stainless-steel-316.html (accessed on 6 November 2022).
25. Kiranmayi, A.V. Characterising High Energy Beam Welding In Structural Steels with Numerical Simulation and Validation. Ph.D. Thesis, University of Bristol, Bristol, UK, 2015.
26. Biegler, M.; Graf, B.; Rethmeier, M. In-situ distortions in LMD additive manufacturing walls can be measured with digital image correlation and predicted using numerical simulations. *Addit. Manuf.* **2018**, *20*, 101–110. [CrossRef]
27. El Naqa, I.; Murphy, M.J. Machine Learning in Radiation Oncology. In *Machine Learning in Radiation Oncology*; Springer: Berlin/Heidelberg, Germany, 2015; pp. 3–11. [CrossRef]
28. Wiens, J.; Shenoy, E.S. Machine Learning for Healthcare: On the Verge of a Major Shift in Healthcare Epidemiology. *Clin. Infect. Dis.* **2018**, *66*, 149–153. [CrossRef] [PubMed]
29. Wäldchen, J.; Mäder, P. Machine learning for image based species identification. *Methods Ecol. Evol.* **2018**, *9*, 2216–2225. [CrossRef]
30. Thiede, S.; Ghafoorpoor, P.; Sullivan, B.P.; Bienia, S.; Demes, M. Potentials and technical implications of tag based and AI enabled optical real-time location systems (RTLS) for manufacturing use cases. *CIRP Ann.* **2022**, *71*, 401–404. [CrossRef]
31. Nasiri, S.; Khosravani, M.R. Machine learning in predicting mechanical behavior of additively manufactured parts. *J. Mater. Res. Technol.* **2021**, *14*, 1137–1153. [CrossRef]
32. Schweidtmann, A.M.; Esche, E.; Fischer, A.; Kloft, M.; Repke, J.U.; Sager, S.; Mitsos, A. Machine Learning in Chemical Engineering: A Perspective. *Chem.-Ing.-Tech.* **2021**, *93*, 2029–2039. [CrossRef]
33. Reddy, M.J.; Kumar, D.N. Multi-Objective Optimization using Evolutionary Algorithms. *Water Resour. Manag.* **2006**, *20*, 861–878. [CrossRef]
34. Boyd, S.; Vandenberghe, L. *Convex Optimization*; Cambridge University Press: Cambridge, UK, 2013.
35. Burke, E.K.; Graham, K. *Search Methodologies: Introductory Tutorials in Optimization and Decision Support Techniques*, 2nd ed.; Springer: Berlin/Heidelberg, Germany, 2014. [CrossRef]
36. Yarmohammadi, S.; Mohammadzadeh, K.; Farhadi, M.; Hajimiri, H.; Modir, A. Multi-objective optimization of thermal and flow characteristics of R-404A evaporation through corrugated tubes. *J. Energy Storage* **2020**, *27*, 101137. [CrossRef]
37. Yarmohammadi, S.; Farhadi, M.; Hajimiri, H.; Salimpour, M.R. Multi-Objective Optimization of R-404A Vapor Condensation in Swirling Flow Using Genetic Algorithms. *Heat. Transf. Eng.* **2017**, *38*, 137–148. [CrossRef]
38. Ferreira, A.A.; Fernando, M.; Vieira, G.; Rosanete, A.; Reis, L. Effects of Processing Parameters on Direct Laser Deposited Materials for Industrial Components Repair. Ph.D. Thesis, University of Porto, Porto, Portugal, 2022.
39. Kusuma, C. The Effect of Laser Power and Scan Speed on Melt Pool Characteristics of Pure Titanium and Ti-6Al-4V Alloy for Selective Laser Melting. Master's Thesis, Wright State University, Dayton, OH, USA, 2016.
40. Lu, X.; Lin, X.; Chiumenti, M.; Cervera, M.; Hu, Y.; Ji, X.; Ma, L.; Yang, H.; Huang, W. Residual stress and distortion of rectangular and S-shaped Ti-6Al-4V parts by Directed Energy Deposition: Modelling and experimental calibration. *Addit. Manuf.* **2019**, *26*, 166–179. [CrossRef]
41. Woo, S.S.; Lee, K.Y.; Baek, G.Y.; Kim, J.W. Residual stress variations in substrate (AISI D2) during directed energy deposition process of high-speed tool steel (AISI M4) powder. *J. Alloy. Metall. Syst.* **2024**, *6*, 100080. [CrossRef]
42. Xiao, Z.; Chen, C.; Zhu, H.; Hu, Z.; Nagarajan, B.; Guo, L.; Zeng, X. Study of residual stress in selective laser melting of Ti6Al4V. *Mater. Des.* **2020**, *193*, 108846. [CrossRef]
43. Sow, M.C.; De Terris, T.; Castelnau, O.; Hamouche, Z.; Coste, F.; Fabbro, R.; Peyre, P. Influence of beam diameter on Laser Powder Bed Fusion (L-PBF) process. *Addit. Manuf.* **2020**, *36*, 101532. [CrossRef]

materials

MDPI

Article

Rapid-Optimized Process Parameters of 1080 Carbon Steel Additively Manufactured via Laser Powder Bed Fusion on High-Throughput Mechanical Property Testing

Jianyu Feng [1,2], Meiling Jiang [2,3], Guoliang Huang [2,3], Xudong Wu [2,3] and Ke Huang [2,3,*]

1 College of Pittsburgh, Sichuan University, Chengdu 610200, China; fengjianyu@stu.scu.edu.cn
2 College of Materials Science & Engineering, Sichuan University, Chengdu 610065, China; jml057589@163.com (M.J.); hgl_sugar@163.com (G.H.); m17311044998@163.com (X.W.)
3 Key Laboratory of Advanced Special Materials & Technology, Ministry of Education, Chengdu 610065, China
* Correspondence: kehuang84@gmail.com

Abstract

To ensure the sustainability of alloy-based strategies, both compositional design and processing routes must be simplified. Metal additive manufacturing (AM), with its exceptionally rapid, non-equilibrium solidification, offers a unique platform to produce tailored microstructures in simple alloys that deliver superior mechanical properties. In this study, we employ laser powder bed fusion (LPBF) to fabricate 1080 plain carbon steel, a binary alloy comprising only iron and carbon. Deviating from conventional process optimization focusing primarily on density, we optimize LPBF parameters for mechanical performance. We systematically varied key parameters (laser power and scan speed) to produce batches of tensile specimens, which were then evaluated on a high-throughput mechanical testing platform (HTP). Using response surface methodology (RSM), we developed predictive models correlating these parameters with yield strength (YS) and elongation. The RSM models identified optimal and suboptimal parameter sets. Specimens printed under the predicted optimal conditions achieved YS of 1543.5 MPa and elongation of 7.58%, closely matching RSM predictions (1595.3 MPa and 8.32%) with deviations of −3.25% and −8.89% for YS and elongation, respectively, thus validating model accuracy. Comprehensive microstructural characterization, including metallographic analysis and fracture surface examination, revealed the microstructural origins of performance differences and the underlying strengthening mechanisms. This methodology enables rapid evaluation and optimization of LPBF parameters for 1080 carbon steel and can be generalized as an efficient framework for robust LPBF process development.

Keywords: laser powder bed fusion; high-throughput mechanical testing; response surface methodology; 1080 carbon steel

1. Introduction

1080 carbon steel has long been recognized as a critically important engineering material due to its exceptional hardness and wear resistance [1,2]. It finds widespread application in machining, tool manufacturing, automotive components, and related industries [3]. For instance, taps and reamers used in machining leverage the high hardness of 1080 steel to ensure cutting precision and tool longevity [4,5]. Similarly, springs in automotive suspension systems rely on its excellent elasticity and fatigue resistance to maintain stable operation under repeated compressive and tensile loading [3].

Despite its advantages, the production of 1080 plain carbon steel suffers from fundamental limitations due to challenges in precise process control and resultant microstructural heterogeneity. These issues impede its ability to satisfy modern industrial requirements for compositional accuracy, performance stability, environmental sustainability, and cost efficiency. Consequently, achieving the optimal combination of high strength, hardness, and wear resistance typically requires complex, multi-step processing strategies, such as advanced heat-treatment sequences and specialized welding protocols, that have been employed in prior research to finely tailor the steel's microstructure. Mishra et al. [6] employed a typical cyclic heat treatment to achieve notably high strength (UTS = 2016 MPa) and reasonable ductility (elongation = 11%); Fujii et al. [7] successfully enhanced the toughness and ductility of AISI 1080 steel using liquid CO_2 during friction stir welding; and Mishra et al. [8] simultaneously improved both the plasticity and strength of AISI 1080 steel via cyclic heat treatment.

In recent years, laser powder bed fusion (LPBF), a core metal additive manufacturing (AM) technology, has demonstrated significant advantages in the efficient, near-net-shape fabrication of complex and precision metal components [9–11]. Based on the principle of discrete deposition and layer-by-layer stacking, LPBF utilizes a high-energy laser beam to selectively melt metal powder according to a predefined 3D model path. Crucially, the inherently high cooling rates associated with LPBF promote rapid solidification of the molten metal within the melt pool and produce tailored microstructures in simple alloys that deliver superior mechanical properties [11–13]. The core of the LPBF process lies in the precise control of key parameters, such as laser power, scan speed, hatch spacing, and layer thickness, to regulate heating/cooling rates, melt pool morphology, dimensions, overall thermal history, etc. This control ultimately dictates the final part's density, microstructure, and resultant mechanical properties [14–17]. Azizi et al. [18] achieved high relative density and optimal surface roughness in Cu–Cr–Zr alloy through systematic LPBF process parameter optimization. Fereiduni et al. [19] fabricated $CuNi_2SiCr$ alloy via LPBF and investigated the influence of process parameters on its thermal conductivity. Huang et al. [20] systematically investigated the regulation of fracture toughness and fatigue crack propagation in LPBF-fabricated tantalum through process parameter modulation. Lin et al. [21] investigated the effects of process parameters on grain boundary characteristics and crack formation mechanisms during LPBF of nickel-based superalloys.

Minor deviations in LPBF parameter combinations can readily introduce defects such as porosity and microcracks or engender substantial microstructural variations (e.g., grain size, morphology, phase composition). These imperfections markedly degrade dimensional accuracy, density, and—critically—the reliability and consistency of mechanical properties including strength, ductility, and toughness. Such sensitivity is especially acute in processing-sensitive alloys like high-carbon and high-strength steels, which possess inherently narrow processing windows requiring exceptionally precise parameter control. Therefore, a comprehensive understanding and systematic optimization of LPBF process parameters are essential preconditions for the stable, repeatable manufacture of 1080 carbon steel through the LPBF process.

This study investigates 1080 carbon steel, a material known for its high strength and relatively simple chemical composition. While LPBF parameter optimization is well-established, conventional approaches predominantly focus on achieving high density and minimizing defects. Traditional LPBF process optimization has predominantly focused on achieving high density and reducing defects, with the assumption that high density itself ensures superior mechanical properties. However, in practice, components with density exceeding 99% often exhibit low strength, ductility, or fatigue resistance. This is because mechanical properties are determined by microstructural characteristics such as grain size and dislocation density, rather than density alone. In contrast, this study takes mechanical properties as the core optimization target and constructs a parameter optimization framework to achieve customized performance for the narrow processing window of 1080 steel, filling the gap in existing research. Furthermore, systematic optimization specifically targeting mechanical performance metrics, particularly for processing-sensitive high-carbon steels like 1080 with inherently narrow processing windows, remains less explored. So, the objective is to achieve an optimal balance between yield strength and elongation through precise control of key LPBF process parameters. A systematic variation of laser power and scan speed was employed to fabricate batches of tensile specimens, which were subsequently evaluated using a high-throughput mechanical testing platform (HTP). The RSM method was utilized to develop predictive models that quantitatively correlate processing parameters with mechanical properties, specifically yield strength (YS) and elongation. RSM was further applied to identify optimal and suboptimal parameter sets that maximize mechanical performance and was verified by actual experiments. Comprehensive microstructural characterization, including metallographic analysis and fracture surface examination, elucidated the microstructural origins of the observed mechanical behavior and revealed the underlying strengthening mechanisms. This work presents an efficient and reliable framework for optimizing LPBF processing of high-performance 1080 carbon steel, enabling targeted mechanical property enhancement and offering generalizability to other alloy systems and processes.

2. Materials and Experimental Methods

2.1. Raw Materials

The 1080 carbon steel powder used in this study was supplied by Hebei Chuangying Metal Materials Co., Ltd. (Xingtai, China) and produced via gas atomization. It exhibits a Hall flow rate of 13.2 s per 50 g, with chemical composition detailed in Table 1; the powder particles display predominantly spherical morphology with uniform size distribution. The high sphericity (≥98.5%) and D50 particle size of 20.81 μm render this powder highly suitable for LPBF processing (Figure 1).

Table 1. Chemical composition of 1080 plain carbon steel.

Elements	C	Si	Mn	S	P	Fe
Wt%	0.75	0.19	0.79	0.01	0.01	Bal.

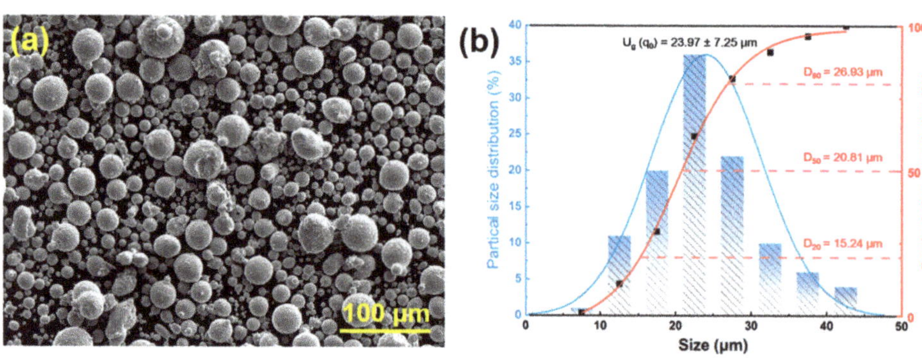

Figure 1. Morphology and size distribution of powder: (**a**) SEM, (**b**) size distribution.

2.2. Experimental Methods

LPBF processing was performed using an HBD-80 system (Hanbang Union 3D Technology Co., Ltd., Shanghai, China) equipped with a YLR-500 fiber laser (beam diameter: 33 μm; wavelength: 1060 nm). Prior to fabrication, 1080 steel powder was dried for 8 h in an argon-purged desiccator to eliminate residual moisture and prevent pore formation. During printing, continuous argon flow maintained oxygen levels below 5 ppm to suppress interlayer oxidation (Figure 2 illustrates the LPBF scheme). Based on previous research, process parameter development employed the following ranges: laser power (P) = 100–375 W, scan speed (V) = 400–600 mm/s, hatch spacing (H) = 120 μm, and layer thickness (t) = 30 μm. The volumetric energy density (VED), calculated as VED = P/(V·H·t), ranged from 46.3 to 260.4 J/mm^3, providing a preliminary indicator of energy input. Table 2 details the 16 parameter combinations and corresponding VED values used for specimen fabrication. Conventional mechanical testing approaches would incur prohibitive time and resource costs for this parameter matrix. Consequently, we employed our self-developed high-throughput testing platform [22] to efficiently acquire tensile property data, with three tests conducted per set and the results averaged.

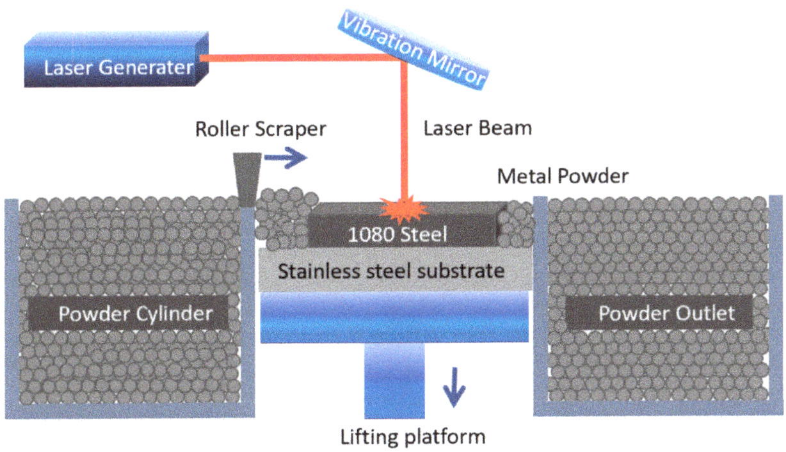

Figure 2. Schematic diagram of LPBF technology.

To identify the optimal process parameters for maximizing mechanical performance in high-density 1080 carbon steel, an analytical model based on RSM was established. Laser power (P), scan speed (v), and hatch spacing (h) were selected as the key experimental factors. Utilizing high-throughput mechanical testing data, the Design-Expert 13 software was employed to construct a statistically robust experimental matrix. Yield strength (YS) and elongation (EL) were designated as the response variables to guide optimization. A quadratic polynomial regression model [23] (Equation (1)) was implemented to construct a mathematical relationship approximating the parameter property mapping between experimental factors and response targets.

$$y = \alpha_0 + \sum_{i=1}^{k} \alpha_i x_i + \sum_{i=1}^{k} \alpha_{ii} x_i^2 + \sum_{i<j}^{k} \alpha_{ij} x_i x_j + \omega \quad (1)$$

where y represents the response variable, x represents the predictor variable, k represents the number of experimental factors, α_0 is the constant term, α_i is the coefficient of the first-order term, α_{ii} is the coefficient of the second-order term, α_{ij} is the coefficient of the interaction term, and w represents the random error.

Table 2. Process parameters of specimens.

No.	P (W)	V (mm/s)	h (μm)	VED (J/mm^3)
1	100	600	120	46.3
2	125	600	120	57.9
3	150	600	120	69.4
4	175	600	120	81.0
5	200	600	120	92.6
6	225	600	120	104.2
7	250	600	120	115.7
8	275	600	120	127.3
9	300	600	120	138.9
10	325	600	120	150.5
11	350	600	120	162.0
12	375	600	120	173.6
13	350	400	120	243.1
14	375	400	120	260.4
15	300	400	120	208.3
16	325	400	120	225.7

2.3. Material Characterization Techniques

Sample density was measured via Archimedes' principle using a MAY-D80 automated density balance (MAYZUM Scientific Instruments (Shenzhen) Co., Ltd., Shenzhen, China). Metallographic examination of the macrostructure was performed with a Zeiss Axio Imager 40MAT optical microscope (Carl Zeiss (China) Co., Ltd., Suzhou, China). For electron backscatter diffraction (EBSD) analysis, samples underwent sequential preparation: electrochemical polishing in 10% $HClO_4$ ethanol solution, followed by broad-beam ion milling using a Gatan PECS II 685 system. EBSD orientation maps were then acquired on a Zeiss Gemini 300 scanning electron microscope Carl Zeiss AG, Oberkochen, Germany) equipped with an Oxford Instruments Symmetry EBSD detector. All EBSD data were processed using AZtecCrystal software (v2.1).

3. Results

3.1. Mechanical Properties from High-Throughput Tensile Testing

Strength and ductility, as critical metrics for assessing mechanical performance, are often significantly enhanced in LPBF-fabricated components due to rapid solidification and thermal cycling effects. Leveraging our self-developed high-throughput tensile platform with automated parallel testing capability, we efficiently evaluated these properties. The corresponding stress–strain curves are presented in Figure 3, while key mechanical property data are summarized in Table 3.

3.2. Process-Property Analytical Model

Leveraging the tensile performance dataset, we employed response surface methodology (RSM) via Design-Expert 13 software to establish predictive relationships between process parameters and mechanical properties. Figure 4 (adapted from [24]) illustrates the RSM optimization workflow. Regression analysis revealed quadratic polynomial models provided the best fit for the experimental data. The two-variable polynomial equations relating yield strength (YS) and elongation (EL) to laser power (P) and scan speed (V) are expressed as follows:

$$YS = +3271.77605 - 5.88014 \times P - 3.72528 \times V + 0.012212 \times P \times V \quad (2)$$

$$EL = 5.05 + 2.85 \times P + 0.0515 \times V - 0.0255 \times P \times V - 0.9211 \times P^2 \tag{3}$$

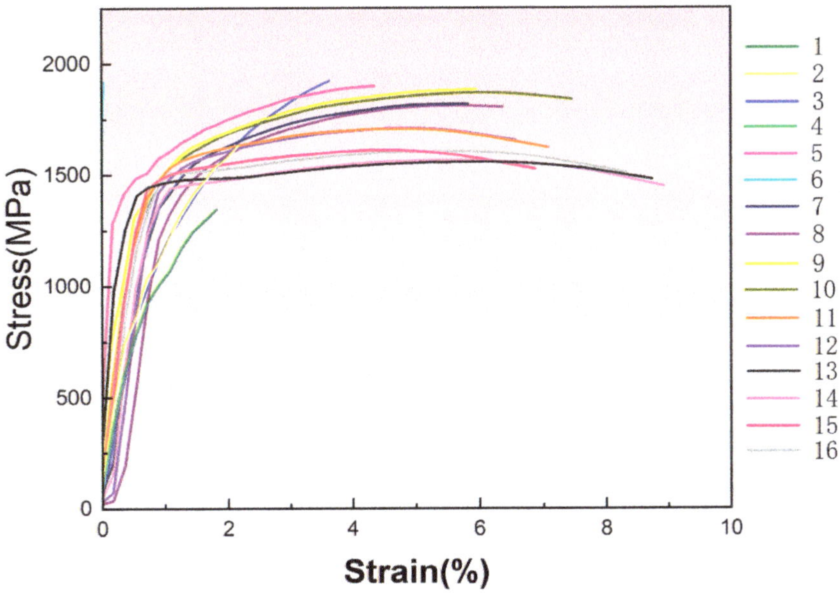

Figure 3. Stress–strain curves of 1080 specimens through high-throughput tensile testing.

Table 3. High-throughput tensile test results.

No.	P (W)	V (mm/s)	h (μm)	Yield Strength (MPa)	Elongation (%)
1	100	600	120	1195.3 ± 5.8	1.4 ± 0.1
2	125	600	120	1203.0 ± 6.6	2.1 ± 0.2
3	150	600	120	1237.2 ± 13.1	2.8 ± 0.5
4	175	600	120	1262.4 ± 17.1	4.0 ± 0.4
5	200	600	120	1359.5 ± 16.7	3.9 ± 0.6
6	225	600	120	1333.6 ± 13.0	4.7 ± 0.1
7	250	600	120	1415.1 ± 20.2	5.4 ± 0.4
8	275	600	120	1467.9 ± 18.7	5.8 ± 0.5
9	300	600	120	1487.2 ± 15.6	6.3 ± 0.2
10	325	600	120	1528.4 ± 12.6	6.7 ± 0.5
11	350	600	120	1530.6 ± 10.9	6.6 ± 0.4
12	375	600	120	1543.5 ± 8.0	7.0 ± 0.1
13	350	400	120	1404.7 ± 19.2	6.5 ± 1.4
14	375	400	120	1420.1 ± 12.9	7.1 ± 0.5
15	300	400	120	1477.8 ± 7.1	6.2 ± 0.8
16	325	400	120	1480.5 ± 15.8	6.5 ± 0.8

Figure 5a presents the 3D response surface for yield strength (YS), revealing a nonlinear relationship between YS and the laser power/scan speed parameters. The response surface exhibits a distinct maximum (red region) corresponding to peak YS values and a minimum (blue region) associated with lower strength. Figure 5b displays the corresponding contour plot, demonstrating the combined effect of laser power and scan speed on YS. Analysis indicates the optimal parameter window for maximizing yield strength occurs within P ≈ 250–350 W and V ≈ 500–550 mm/s, where contour lines converge toward the response maximum.

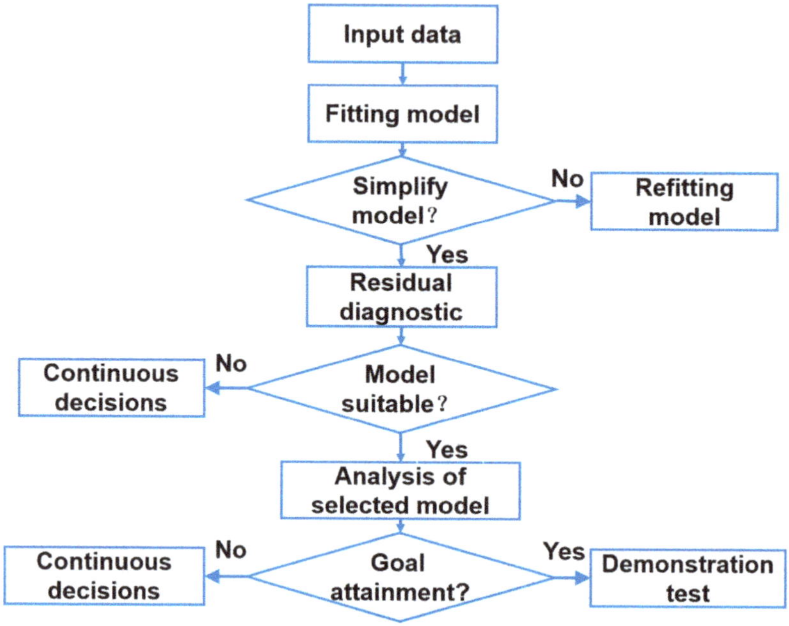

Figure 4. RSM parameter optimization process [17].

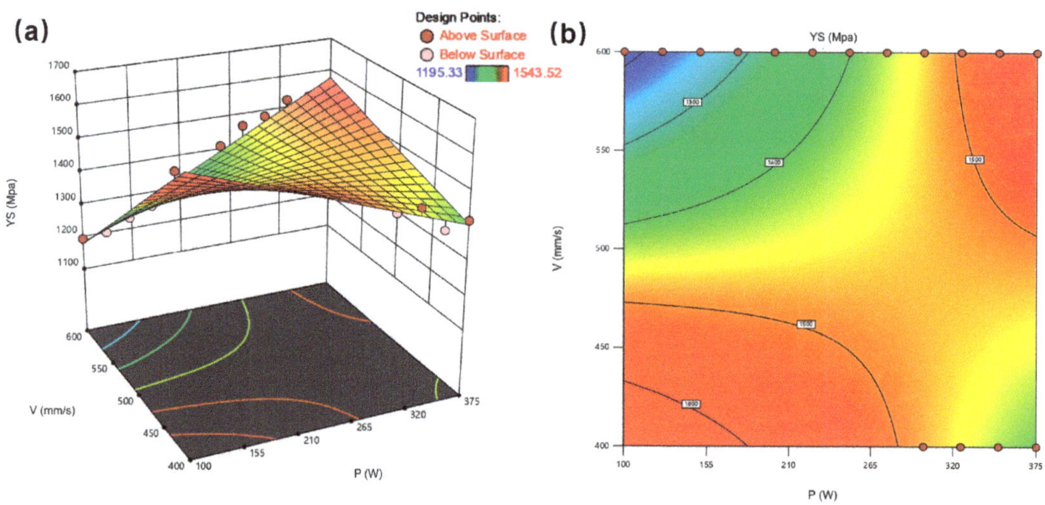

Figure 5. Influence of laser power and laser velocity on yield stress of LPBF 108: (**a**) 3D surface; (**b**) contour maps illustrate the effect of power and velocity on the YS.

Figure 6a presents the 3D response surface for elongation (EL), demonstrating a positive correlation with increasing laser power and decreasing scan speed. The monotonic ascending trend indicates enhanced material ductility at higher energy input regimes. Figure 6b displays the corresponding contour plot, revealing a region of maximum elongation (red contours) concentrated at high laser power and low scan speed combinations. This parameter space corresponds to volumetric energy densities exceeding 180 J/mm^3, where improved inter-track fusion reduces void formation and enhances plastic deformation capacity.

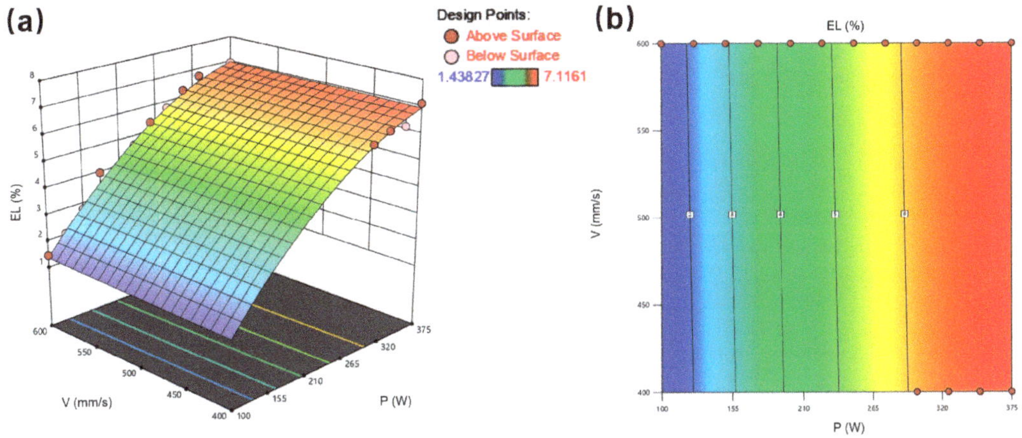

Figure 6. Influence of laser power and laser velocity on elongation of LPBF 1080: (**a**) 3D surface; (**b**) contour maps illustrate the effect of power and velocity on the EL.

Tables 4 and 5 present the analysis of variance (ANOVA) results for the yield strength and elongation regression models. We employed ANOVA to evaluate model significance and confidence levels [25], where the F-statistic and *p*-value determined the statistical significance of model terms [26,27]. The F-statistic quantifies the ratio of model variance to residual variance, with higher values indicating greater model significance. The *p*-value represents the probability of obtaining an F-statistic at least as extreme under the null hypothesis, with $p < 0.05$ denoting statistical significance and $p < 0.01$ indicating high significance [28]. Larger *p*-values corresponded to increased probability of model error. Furthermore, ANOVA revealed comparable influence magnitudes of laser power (P) and scan speed (V) on yield strength and elongation, with all linear and quadratic terms exhibiting $p < 0.01$.

Table 4. Variance analysis of regression equation terms of YS.

Source	Sum of Squares	df	Mean Square	F-Value	*p*-Value Prob > F	Significant
Model	203,124.9099	3	67,708.3033	96.6238	<0.0001	Significant
P	616.6186	1	616.6186	0.8780	0.3667	
V	7703.0096	1	7703.0097	10.9926	0.0061	
PV	18,012.3838	1	18,012.3838	25.7047	0.0002	
Residual	8408.9021	12	700.74183			
Cor Total	211,533.8119	15				

Table 5. Variance analysis of regression equation terms of EL.

Source	Sum of Squares	df	Mean Square	F-Value	*p*-Value Prob > F	Significant
Model	49.31	4	12.33	262.50	<0.0001	Significant
P	4.21	1	4.21	89.70	<0.0001	
V	0.0026	1	0.0026	0.0556	0.8179	
P × V	0.0003	1	0.0003	0.0072	0.9340	
P^2	1.24	1	1.24	26.43	0.0003	
V^2	0.0000	0				
Residual	0.5166	11	0.0470			
Cor Total	49.83	15				

Table 6 presents comprehensive model reliability metrics. The coefficient of determination (R^2) and adjusted R^2 (adj-R^2) approach unity (0.9602/0.9503 for YS; 0.9896/0.9859 for EL), confirming exceptional agreement between experimental and predicted values [28]. A coefficient of variation (C.V.) of 1.90% for YS and 4.17% for EL indicates minimal data dispersion, while adequate precision ratios (YS: 30.0679; EL: 46.6148) substantially exceed the recommended threshold of 4. These statistically robust validation metrics demonstrate the model's efficacy for predicting mechanical properties in LPBF-processed components.

Table 6. Reliability analysis of regression variance of YS and EL.

	Std. Dev.	Mean	C.V.%	R^2	Adjusted R^2	Predicted R^2	Adeq Precision
YS	26.47	1396.68	1.90	0.9602	0.9503	0.9312	30.0679
EL	0.2167	5.19	4.17	0.9896	0.9859	0.9769	46.6148

To evaluate model fidelity and predictive accuracy, we conducted comparative analysis between experimental measurements and model-predicted values. Figure 7 presents the experimental vs. predicted plots for yield strength (YS) and elongation (EL) in LPBF-processed 1080 carbon steel. Data points demonstrate tight clustering about the parity line with minimal deviation. This alignment validates the model's robustness and high predictive precision within the investigated parameter space.

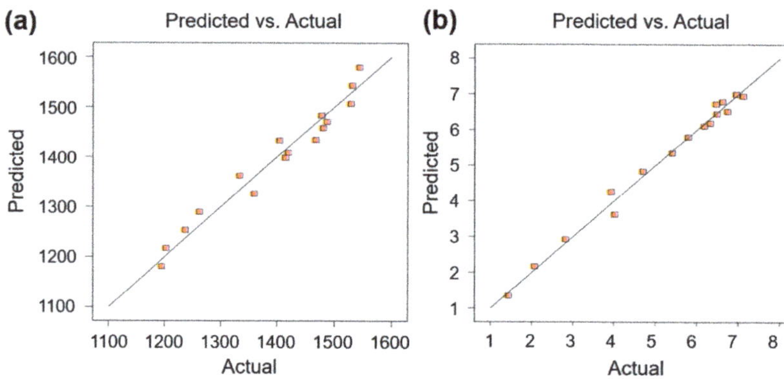

Figure 7. Comparison of predicted and actual values of (**a**) yield stress and (**b**) EL of LPBF 1080 steel.

3.3. Process Optimization and Verification

To validate the optimization methodology, experimental verification was performed using two parameter sets recommended by the RSM software (Design-Expert 13): optimal parameters (P = 375 W, v = 600 mm/s, h = 120 μm) and sub-optimal parameters (P = 350 W, v = 400 mm/s, h = 120 μm), as presented in Table 7 and Figure 8. The deviation rate is defined as the ratio of the actual average value to the predicted value (Table 7), with a prediction deemed reliable when this rate is less than 10% [29]. For yield strength (YS), the deviation rates of the optimal and sub-optimal groups are 3.25% and −4.07%, respectively—both well within the 10% threshold. Analysis of the coefficient of variation (C.V.) further confirms the high repeatability of mechanical test results: the optimal group exhibits a YS C.V. of only 0.52%, while the sub-optimal group shows 2.79%. This close alignment between predicted and experimental YS values underscores the predictive model's accuracy and reliability. Regarding elongation (EL), the deviation rates for both parameter sets are also below 10% (optimal: −8.89%; sub-optimal: −6.90%), validating the model's effectiveness even for this property, which is highly sensitive to minor surface scratches, internal microcracks, or localized residual stress concentrations. These factors can act as initiation sites for deformation localization under tensile loading; however, the model demonstrates robust predictive capability despite such sensitivities.

Table 7. P, V, H of predicted YS-optimal specimen and EL-optimal specimen.

		YS (MPa)	**EL (%)**
Optimal	Predicted	1595.33	8.32
	Actual-1	1553.56	6.76
	Actual-2	1543.13	7.25
	Actual-3	1533.88	8.74
	Actual Mean	1543.52	7.58
	Actual STD	8.04	0.84
	Actual C. V	0.52%	11.1%
	Deviation rate	−3.25%	−8.89%
		YS (MPa)	EL (%)
Sub-optimal	Predicted	1464.34	7.25
	Actual-1	1363.72	6.74
	Actual-2	1392.89	6.58
	Actual-3	1457.5	6.93
	Actual Mean	1404.70	6.75
	Actual STD	39.19	0.14
	Actual C. V	2.79%	2.12%
	Deviation rate	−4.07%	−6.90%

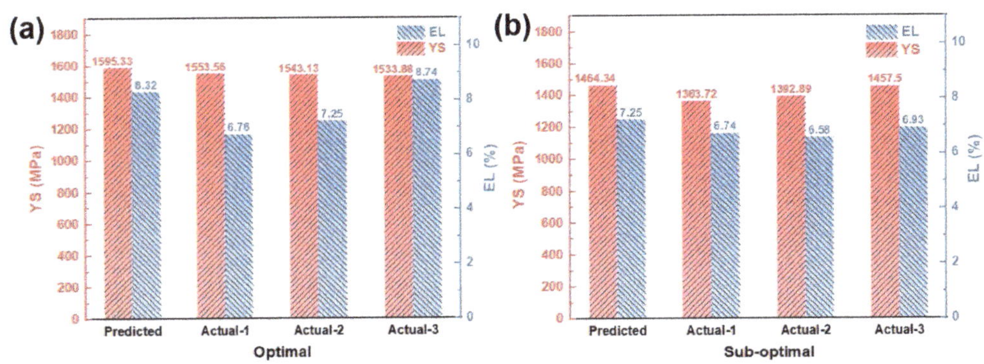

Figure 8. Comparison of predicted and actual values of yield stress and EL of LPBF 1080 steel: (**a**) optimal; (**b**) sub-optimal.

Crucially, metallographic analysis (Figure 9) demonstrates near-full density (>99.2%) with negligible defect formation in both parameter sets. This microstructural evidence, combined with the model's predictive accuracy for yield strength, validates the efficacy of our RSM-based approach for optimizing LPBF processing of 1080 carbon steel.

Figure 10 presents tensile fracture morphologies of specimens fabricated under (a) optimal and (b) sub-optimal parameters. Both exhibit ductility-dominated mixed fracture modes yet display distinct plastic deformation heterogeneity: the sub-optimal specimen (Figure 10b) shows widely dispersed dimples (including elongated variants) with tear ridges, cleavage steps, and river patterns, indicating localized stress concentration and heterogeneous plastic constraints. Conversely, the optimal specimen (Figure 10a) features uniformly distributed equiaxed dimples with narrow size distribution, demonstrating superior energy dissipation during plastic deformation. This morphological contrast directly correlates with differentiated fracture mechanics: the homogeneous dimple structure in Figure 10a creates higher energy barriers against crack propagation, enhancing fracture toughness and elongation. Crucially, neither specimen reveals pores or inclusions, confirming exceptional density that effectively suppresses stress concentration initiation, thereby providing microstructural assurance for enhanced tensile performance.

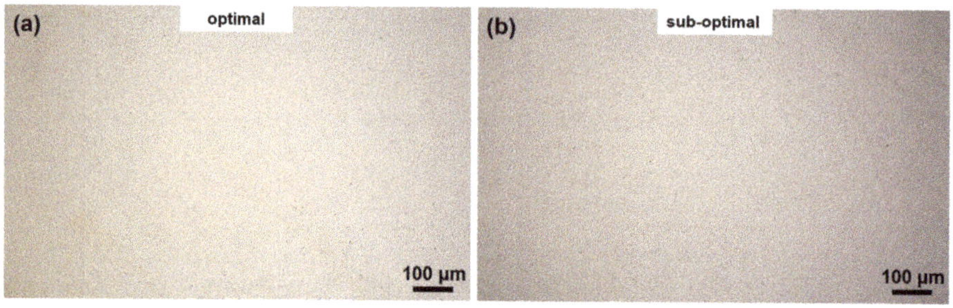

Figure 9. Metallographic micrographs of 1080 steel: (**a**) optimal, (**b**) sub-optimal.

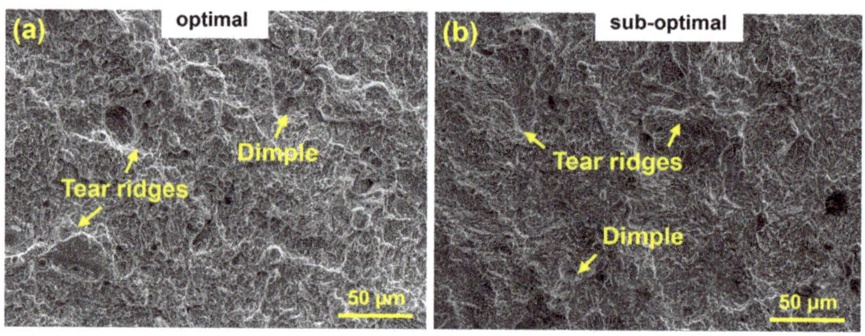

Figure 10. Scanning electron microscope (SEM) images of tensile fracture surfaces: (**a**) optimal, (**b**) sub-optimal.

4. Discussion

4.1. Optimal vs. Sub-Optimal Parameters: Process–Microstructure–Property Linkages

To elucidate the underlying mechanisms responsible for the performance divergence between optimal and sub-optimal process sets, detailed characterization and analysis were conducted on both specimens.

4.1.1. Grain Size and Morphology

Figure 11 presents the size distribution figures of LPBF-processed 1080 carbon steel fabricated under (a) optimal (P = 375 W, h = 120 μm, v = 600 mm/s, VED = 173.60 J/mm^3) and (b) sub-optimal (P = 350 W, h = 120 μm, v = 400 mm/s, VED = 243.06 J/mm^3) parameters. Colors represent distinct crystallographic orientations: red = <001>, green = <101>, and blue = <111>. Quantitative analysis reveals refined grain structures with average sizes of 1.49 μm (Figure 11(a1)) and 1.54 μm (Figure 11(b1)). This grain coarsening at higher VED results from reduced nucleation density, where grain growth dominates over nucleation [28], consistent with observations in Invar alloys [30] and steels [31]. According to the Hall–Petch relationship, the finer-grained optimal specimen exhibits enhanced strength due to heightened grain boundary strengthening.

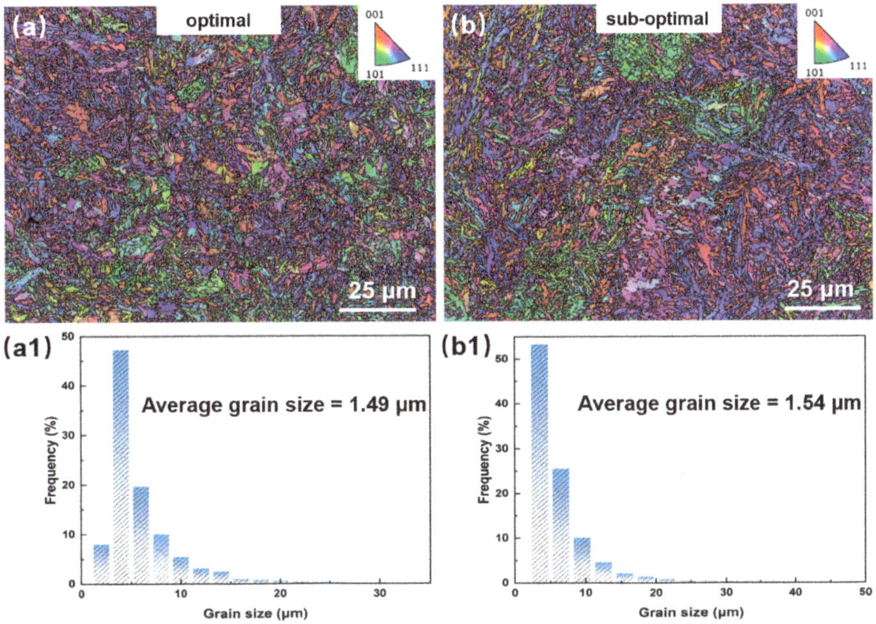

Figure 11. Grain distribution of LPBF- fabricated 1080 specimens during different processes: (**a**,**a1**) optimal specimen; (**b**,**b1**) sub-optimal specimen.

4.1.2. Crystallographic Texture

Figure 12 presents inverse pole figures (IPF) of optimal and sub-optimal specimens, revealing distinct crystallographic texture characteristics. Figure 12a exhibits a strong <111>//Z fiber texture with no pronounced preferential orientations along other axes. In contrast, Figure 12b demonstrates a complex texture comprising the following: strong <111>//Z and <121>//Y components and weaker <111>//X, <101>//Y, and <101>//Z orientations. Quantitative texture analysis confirms significant differences in texture index (1.24 vs. 0.78), demonstrating process-dependent texture evolution in LPBF-processed 1080 steel.

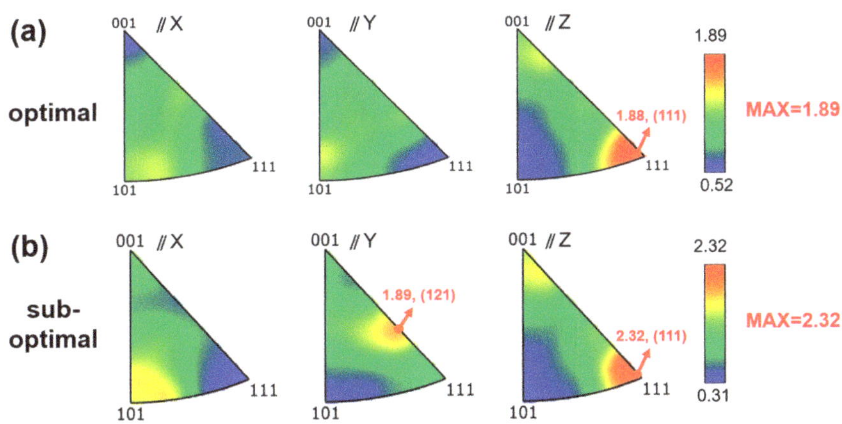

Figure 12. Inverse pole figures (IPFs) of LPBF-fabricated 1080 specimens during different processes: (**a**) optimal specimen; (**b**) sub-optimal specimen.

4.1.3. Geometric Dislocation Density

During LPBF processing, rapid cooling induces significant thermal stress accumulation, promoting plastic deformation within the solidified materials [32]. Kernel average misorientation (KAM) quantifies plastic strain magnitude and indicates localized concentrations of thermal stress or defect density [33]. Figure 13 presents KAM maps and grain boundary character distributions. Regions of elevated plastic deformation appear green in Figure 13a,b, corresponding to higher KAM values. Quantitative analysis reveals a greater grain boundary density (9.75×10^{14} m^{-2}) in the optimal parameter specimen (Figure 13a) vs. 6.85×10^{14} m^{-2} in the sub-optimal specimen (Figure 13b), consistent with the finer grain size observed in IPF analysis. Figure 13(a1,b1) displays grain boundary misorientation distributions, with low-angle boundaries (2–15°) marked red and high-angle boundaries (>15°) blue. Increasing VED from 173.6 J/mm^3 (optimal) to 243.06 J/mm^3 (sub-optimal) correlates with the decreased low-angle boundary fraction of 29.0% to 27.1% and increased high-angle boundary fraction of 71.0% to 72.9%. This transition suggests enhanced recovery processes at higher energy input.

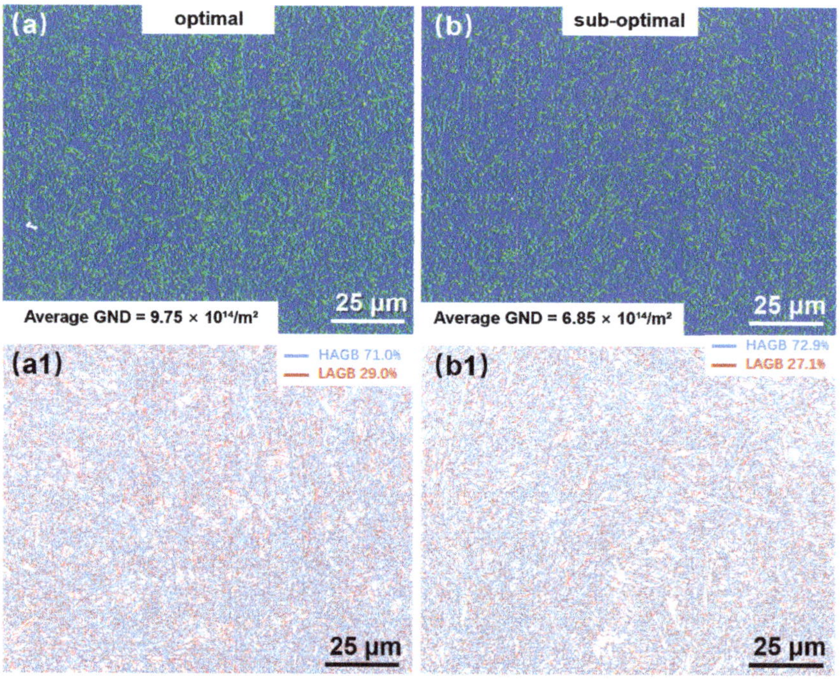

Figure 13. KAM figures (**a**,**b**) and distribution of grain boundary angle (**a1**,**b1**) of LPBF- fabricated 1080 specimens during different processes: (**a**) optimal specimen; (**b**) sub-optimal specimen.

4.1.4. Process–Structure–Property Linkages of the Optimal and Sub-Optimal Specimens

Variations in key process parameters, such as laser power and scan speed, influence the volumetric energy density (VED), which in turn directly governs the microstructural and mechanical properties evolution of 1080 carbon steel fabricated via LPBF. The optimal parameter set characterized with lower VED of 173.6 J/mm^3 promotes refined grains (1.49 μm average size) and elevated geometrically necessary dislocation (GND) density (9.75×10^{14} m^{-2}), with kernel average misorientation (KAM) analysis confirming concentrated plastic deformation zones. Conversely, the sub-optimal set with higher VED of 243.06 J/mm^3 induces grain coarsening (1.54 μm) and reduced GND density (6.85×10^{14} m^{-2}). These microstructural distinctions directly modulate mechanical properties through grain boundary and dislocation strengthening mechanisms. The optimal specimen achieves superior yield strength (1543.52 MPa) and elongation (7.58%) vs. the sub-optimal specimen (1404.70 MPa, 6.75%), thereby conducting a rough process–structure–property (P-S-P) linkage paradigm.

4.2. Advantages of This High-Throughput Testing and Analytical Modeling Approach

4.2.1. High Efficiency

This process optimization framework achieved completion within 50 h, primarily through three synergistic advancements: Firstly, the high-throughput tensile platform enabled automated batch testing with 10-fold efficiency gains over manual methods. Secondly, it eliminated time-intensive polishing and detailed microstructural characterization for all density-qualified specimens. Thirdly, RSM streamlined data analysis, model construction, and parameter optimization into a unified computational workflow.

4.2.2. Fine-Tuning Process Parameters to Achieve Customized Performance

Conventional process optimization predominantly focuses on densification, defect control, and microstructural evolution, with mechanical properties typically measured only at selected parameter sets. Therefore, a traditional approach fundamentally limits systematic establishment of process-property correlations. In contrast, our methodology leveraging 16 distinct process-property datapoints developed an analytical model capable of recommending optimal process parameters for arbitrary customized performance targets within defined ranges. This data-driven framework holds significant implications for demanding applications such as aerospace and medical components where customized performance is critical.

5. Conclusions

This study fabricated 1080 carbon steel using the LPBF technique and proposes a rapid process optimization framework based on high-throughput mechanical testing and response surface methodology (RSM).

Optimal process parameters were determined through response surface methodology (RSM), and the relationship between mechanical properties and microstructure was systematically investigated. The main conclusions are as follows:

(1) This performance-oriented process optimization framework, built upon a high-throughput testing platform and advanced data analysis techniques, demonstrates exceptional efficiency. Within only 50 h, a comprehensive process–property database and predictive model were established, enabling the identification of processing parameters that yield the customized values of yield strength (YS) and elongation (EL).

(2) Under the optimized process conditions, the LPBF-fabricated 1080 carbon steel exhibited a yield strength of 1543.52 MPa and an elongation of 7.58%. These values are in close agreement with the RSM-predicted results (1595.33 MPa and 8.32%), corresponding to relative errors of −3.25% and −8.89%, respectively, thereby validating the accuracy and reliability of the predictive model.

(3) The superior mechanical performance at the optimal condition is primarily attributed to a relatively low volumetric energy density (VED), which results in refined grain structures (average grain size: 1.49 μm), an elevated geometrically necessary dislocation (GND) density (9.75×10^{14} m^{-2}), and grain boundary-mediated dislocation strengthening. These underlying mechanisms were revealed through detailed microstructural characterization.

Author Contributions: Conceptualization, J.F. and K.H.; methodology, J.F. and M.J.; software, J.F. and M.J.; validation, J.F., M.J. and X.W.; formal analysis, J.F. and X.W.; investigation, J.F., G.H. and X.W.; resources, J.F., X.W.; data curation, J.F. and M.J.; writing—original draft preparation, J.F. and M.J.; writing—review and editing, J.F. and M.J.; visualization, J.F. and M.J.; supervision, K.H. and G.H.; project administration, K.H. and G.H.; funding acquisition, K.H. and G.H. All authors have read and agreed to the published version of the manuscript.

Funding: This research received no external funding.

Institutional Review Board Statement: Not applicable.

Informed Consent Statement: Not applicable.

Data Availability Statement: The original contributions presented in this study are included in the article. Further inquiries can be directed to the corresponding authors.

Conflicts of Interest: The authors declare no conflicts of interest.

References

1. Dixit, M.K.; Kumar, P.P.; Banerjee, S. Impact of using high strength low alloy steel on reducing the embodied energy, carbon, and water impacts of building structures: A case study. *Dev. Built Environ.* **2025**, *22*, 100671. [CrossRef]
2. Chen, M.-T.; Gong, Z.; Zhang, T.; Zuo, W.; Zhao, Y.; Zhao, O.; Zhang, G.; Wang, Z. Mechanical behavior of austenitic stainless steels produced by wire arc additive manufacturing. *Thin-Walled Struct.* **2024**, *196*, 111455. [CrossRef]
3. Sirjani, F.K.; Shabani, A.; Raeissi, K.; Nahvi, S.M.; Molavian, A.M. Effect of heat treatment on microstructure and high-temperature oxidation behavior of flame-sprayed Ni-5 wt.% Al coating on low carbon steel. *Results Surf. Interfaces* **2025**, *19*, 100515. [CrossRef]
4. Han, J.M.; Zhang, R.; Ajayi, O.O.; Barber, G.C.; Zou, Q.; Guessous, L.; Schall, D.; Alnabulsi, S. Scuffing behavior of gray iron and 1080 steel in reciprocating and rotational sliding. *Wear* **2010**, *271*, 1854–1861. [CrossRef]
5. Kathleen, C.; Michael, E.; Cox, C.D.; Connor, S.; Jason, C. Optical image and Vickers hardness dataset for repair of 1080 steel using additive friction stir deposition of Aermet 100. *Data Brief* **2022**, *41*, 107862. [CrossRef]
6. Mishra, A.; Mondal, C.; Maity, J. Effect of Combined Cyclic Heat Treatment on AISI 1080 Steel: Part II-Mechanical Property Evaluation. *Steel Res. Int.* **2017**, *88*, 1600215. [CrossRef]
7. Fujii, H.; Chung, Y.D.; Sun, Y.F. Friction stir welding of AISI 1080 steel using liquid CO_2 for enhanced toughness and ductility. *Sci. Technol. Weld. Join.* **2013**, *18*, 500–506. [CrossRef]
8. Mishra, S.; Mishra, A.; Show, B.K.; Maity, J. Simultaneous enhancement of ductility and strength in AISI 1080 steel through a typical cyclic heat treatment. *Mater. Sci. Eng. A* **2017**, *688*, 262–271. [CrossRef]
9. Sun, S.; Brandt, M.; Easton, M. 2-Powder bed fusion processes: An overview. In Laser Additive Manufacturing. Elsevier: Amsterdam, The Netherlands, 2017; pp. 55–77. [CrossRef]
10. Kirchheim, A.; Katrodiya, Y.; Zumofen, L.; Ehrig, F.; Wick, C. Dynamic conformal cooling improves injection molding. *Int. J. Adv. Manuf. Technol.* **2021**, *114*, 107–116. [CrossRef]
11. Liu, J.; Zhao, Z.; Rao, D.; Yang, Y.; Zheng, L.; Wang, C.; Deng, P.; Niu, L.; Huang, Z. Synergistic enhancement of mechanical and tribological properties in WC-reinforced 316L stainless steel matrix composite fabricated by laser powder bed fusion. *J. Mater. Res. Technol.* **2025**, *37*, 617–628. [CrossRef]
12. Sing, S.L.; Wiria, F.E.; Yeong, W.Y. Selective laser melting of lattice structures: A statistical approach to manufacturability and mechanical behavior. *Robot. Comput. Integr. Manuf.* **2018**, *49*, 170–180. [CrossRef]
13. Lee, H.; Lim, C.H.J.; Low, M.J.; Tham, N.; Murukeshan, V.M.; Kim, Y.-J. Lasers in additive manufacturing: A review. *Int. J. Precis. Eng. Manuf.-Green Technol.* **2017**, *4*, 307–322. [CrossRef]
14. Shi, S.; Li, Z.; Yang, C.; Zeng, Z.; Cheng, X.; Gong, X. Multiscale simulation and experimental optimization of process parameters and post-treatment for laser powder bed fusion of high-strength Ni-based superalloy. *J. Mater. Res. Technol.* **2025**, *37*, 700–720. [CrossRef]
15. Dai, K.; He, X.; Zhang, W.; Kong, D.; Guo, R.; Hu, M.; He, K.; Dong, C. Tailoring the microstructure and mechanical properties for Hastelloy X alloy by laser powder bed fusion via scanning strategy. *Mater. Des.* **2023**, *235*, 112386. [CrossRef]
16. Enrique, P.D.; Minasyan, T.; Toyserkani, E. Laser powder bed fusion of difficult-to-print γ' Ni-based superalloys: A review of processing approaches, properties, and remaining challenges. *Addit. Manuf.* **2025**, *106*, 104811. [CrossRef]
17. Deep, A.; Beidokhti, M.M.; Piili, H. Preliminary manufacturability evaluation of complex geometrical parts based on layer thickness in the metal powder bed fusion process. In *Progress in Additive Manufacturing*; Springer: Cham, Switzerland, 2025; pp. 1–21.

18. Azizi, N.; Jabari, E.; Hasanabadi, M.; Toyserkani, E. High-power laser powder bed fusion of Cu–Cr–Zr alloy: A comprehensive study on statistical process optimization, microstructure, and mechanical properties. *J. Manuf. Process.* **2025**, *149*, 931–954. [CrossRef]
19. Fereiduni, E.; Ghasemi, A.; Sargent, N.; Balbaa, M.; Wang, L.; Elbestawi, M.; Sing, S.L.; Xiong, W. Synergy of laser powder bed fusion (LPBF) and heat treatment for CuNi2SiCr alloy enhancement. *Mater. Des.* **2025**, *255*, 114189. [CrossRef]
20. Huang, R.; Shi, W.; Xu, K.; Wang, R.; Xie, H.; Wei, K. The influence of process parameters and building orientation on fracture toughness and fatigue crack growth behavior of laser powder bed fused tantalum. *Int. J. Fatigue* **2025**, *199*, 109086. [CrossRef]
21. Lin, W.; Zhang, Z. Cracking Reduction and Texture Control in Laser Powder Bed Fusion Additive Manufactured Non-weldable Nickel Alloy. *J. Mater. Eng. Perform.* **2025**, 1–14. [CrossRef]
22. Ke, H.; Chris, K.; Adam, U.; Brooke, S. High throughput testing platform and application in wire arc additive manufactured superalloy: Anisotropy investigation and component qualification. *Materialia* **2021**, *16*, 101086. [CrossRef]
23. Liu, J.; Shi, Y. Multi-objective Optimization in Selective Laser Melting of AlSi10Mg Alloy Based on Response Surface Methodology. *J. Mater. Eng. Perform.* **2024**, *34*, 3036–3046. [CrossRef]
24. Zhou, Y.; Fang, Y.; Wang, J.; Xu, L.; Wang, D. Process Parameter Optimization on Selective Laser Melting of CuSn10 Powders Based on Response Surface Methodology. *J. Phys. Conf. Ser.* **2022**, *2262*, 012002. [CrossRef]
25. Zhou, T.; Shi, G.; Wu, Q.; Wang, Z.; Che, J.; Wu, H. Optimization of Cutting Parameters for Cubic Boron Nitride Tool Wear in Hard Turning AISI M2. *J. Mater. Eng. Perform.* **2023**, *33*, 11298–11308. [CrossRef]
26. Guo, C.; He, S.; Yue, H.; Li, Q.; Hao, G. Prediction modelling and process optimization for forming multi-layer cladding structures with laser directed energy deposition. *Opt. Laser Technol.* **2021**, *134*, 106607. [CrossRef]
27. Zhou, T.; Wu, Q.; Wang, Z.; Zhao, G.; Li, S.; Guo, B.; Wu, H. Analysis of machined surface topography of AISI M2 in hard turning based on Box-Behnken Design. *Proc. Inst. Mech. Eng.* **2024**, *238*, 58–71. [CrossRef]
28. Feng, Z.; Tan, H.; Fang, Y.; Lin, X.; Huang, W. Selective laser melting of TiB2/AlSi10Mg composite: Processability, microstructure and fracture behavior. *J. Mater. Process. Tech.* **2022**, *299*, 117386. [CrossRef]
29. Alimohammadi, A.; Khorram, A.; Araee, A. Statistical modeling and optimization of clad geometry in laser cladding of Amdry 365 on Hastelloy X superalloy with response surface methodology. *Vacuum* **2024**, *230*, 113665. [CrossRef]
30. Huang, G.; He, G.; Peng, X.; He, Y.; Chen, H.; Huang, K.; Liu, Y. Effect of processing parameters on the microstructure, mechanical properties and thermal expansion behavior of Invar 36 alloy manufactured by laser powder bed fusion. *Mater. Sci. Eng. A* **2024**, *897*, 146329. [CrossRef]
31. Abele, E.; Stoffregen, H.A.; Kniepkamp, M.; Lang, S.; Hampe, M. Selective laser melting for manufacturing of thin-walled porous elements. *J. Mater. Process. Tech.* **2015**, *215*, 114–122. [CrossRef]
32. Qinyao, S.; Dafan, D.; Lin, H.; Anping, D.; Cheng, Z.; Baode, S. Laser powder bed fusion of a non-equiatomic FeNiCoAl-based high-entropy alloy: Printability, microstructure, and mechanical properties. *J. Alloys Compd.* **2023**, *938*, 168562.
33. Marchese, G.; Parizia, S.; Rashidi, M.; Saboori, A.; Manfredi, D.; Ugues, D.; Lombardi, M.; Hryha, E.; Biamino, S. The role of texturing and microstructure evolution on the tensile behavior of heat-treated Inconel 625 produced via laser powder bed fusion. *Mater. Sci. Eng. A* **2020**, *769*, 138500. [CrossRef]

materials

MDPI

Article

Evaluation of a Method for Determining Material Strength Based on Hardness Measurements: A Case Study of the Ti6Al4V Alloy

Karolina Karolewska [1,*], Mateusz Wirwicki [2] and Bogdan Ligaj [3]

[1] Faculty of Mechanical Engineering, al. PBS Bydgoszcz University of Science and Technology, Kaliskiego 7, 85-796 Bydgoszcz, Poland

[2] Faculty of Mechatronics, ul. UKW Kazimierz Wielki University, Kopernika 1, 85-074 Bydgoszcz, Poland; wirwicki@ukw.edu.pl

[3] BSW University of Bydgoszcz, ul. Unii Lubelskiej 4C, 85-059 Bydgoszcz, Poland; bogdan.ligaj.1974@gmail.com

* Correspondence: karolina.karolewska@pbs.edu.pl

Abstract

The aim of this study was to evaluate the feasibility of estimating the tensile strength of Ti6Al4V alloy, based on HV measurements. The investigation included samples that were manufactured using both additive technology and conventional methods, under various conditions: as-built, heat-treated, and untested mechanically. Static tensile tests and HV measurements were performed to assess the influence of the manufacturing method, heat treatment, and mechanical loading on material performance. The highest tensile strength was recorded for as-built samples, while the greatest ductility was observed in conventionally drawn bar samples. Hardness values generally correlated with tensile strength trends; however, in heat-treated specimens, the relationship between hardness and tensile strength was found to be nonlinear. Specimens that were not subjected to tensile testing exhibited higher HV values than their mechanically tested counterparts, indicating a potential effect of prior deformation on the local material condition. The results confirm that hardness testing can be a useful indirect method for estimating the tensile strength of Ti6Al4V, particularly in materials with controlled and uniform microstructures. For additively manufactured and heat-treated materials, however, the current empirical models may require adjustment or enhancement using advanced predictive approaches. The proposed indirect method offers an alternative to destructive testing, especially in the industrial quality control context for metal AM.

Keywords: Ti6Al4V alloy; tensile strength; hardness; additive manufacturing; DMLS; heat treatment; mechanical properties; microstructure; non-destructive testing; metal AM

1. Introduction

Determining the mechanical properties of engineering materials, particularly their tensile strength, is one of the key tasks in mechanical engineering. This parameter determines whether a given material can be used in structural components, especially in industries with high safety requirements, such as the aerospace, biomedical, and energy sectors. Traditional testing methods such as the tensile test, while providing precise data, are also destructive techniques that require properly prepared samples and specialized equipment. As a result, their application in routine quality control may be limited, due to cost and time consumption [1,2].

https://doi.org/10.3390/ma18163726

In this context, indirect methods for assessing mechanical properties are gaining increasing interest, with hardness analysis being the most commonly used technique. Hardness (HV), which is defined as a measure of a material's resistance to local plastic deformation, can be determined relatively quickly and inexpensively, and its measurement is often non-destructive or minimally invasive. Importantly, numerous studies indicate a correlation between hardness and tensile strength, which creates the potential to use hardness as a predictor of the strength of metallic materials [3,4].

One of the materials for which this issue is of particular importance is the titanium alloy Ti6Al4V, which belongs to the group of $\alpha+\beta$ dual-phase alloys. It is characterized by a high strength-to-weight ratio, excellent corrosion resistance, and outstanding biocompatibility, making it a dominant material in aerospace, implantology, and military applications. At the same time, the mechanical properties of Ti6Al4V are highly dependent on its microstructure, which can be significantly modified by the manufacturing technology used and any subsequent heat treatment processes [5].

In recent years, the additive manufacturing (AM) of metals has been developing rapidly, particularly the direct metal laser sintering (DMLS) method, which belongs to the class of selective laser sintering technologies. This technique enables the tool-free production of geometrically complex components with high dimensional accuracy and efficient material usage. However, structures produced in this way exhibit a microstructure that differs from that of conventionally manufactured materials, including a characteristic grain alignment along the layer-by-layer build direction, the presence of pores, microdefects, and residual stresses. These features have a direct impact on mechanical properties, including hardness and strength [6,7].

Compared to materials produced by traditional methods, such as rolling or drawing, DMLS materials often exhibit greater anisotropy in their properties and a stronger dependence on the build direction. For example, the study by Leuders et al. (2013) demonstrated significant differences in tensile strength and elongation depending on the sample orientation relative to the build axis [8]. For this reason, appropriate post-process heat treatment is crucial in order to homogenize the microstructure, reduce the residual stresses, and eliminate porosity.

Heat treatment, when applied to both DMLS and conventionally produced samples, enables controlled phase transformation and modification of the morphology of the α and β phases. With appropriately selected annealing parameters, a more homogeneous microstructure can be achieved, which translates into stabilized mechanical properties, including improved ductility and fatigue resistance. Another important aspect is the influence of heat treatment on hardness values, which directly affects the accuracy of strength estimation using indirect methods [9].

In this study, four groups of Ti6Al4V alloy samples were used, differing in their manufacturing method and heat treatment condition: DMLS samples in the 'as-built' state, DMLS samples after annealing, drawn bar samples without heat treatment, and bar samples after annealing. This variation in sample types makes it possible to evaluate the influence of the microstructure and manufacturing technology on the relationship between hardness and tensile strength.

Modeling the relationship between hardness and tensile strength can take various forms—from simple linear models to more complex power–law or logarithmic functions. The most commonly cited approach is the model proposed by Cahoon et al. (1971), in which tensile strength is expressed as a function of Vickers hardness and the elastic modulus. While this model is widely used, it is based on an assumption of microstructural homogeneity, which may not be applicable to additively manufactured materials [4].

The literature emphasizes that applying universal correlations between hardness and tensile strength carries the risk of prediction errors, particularly for materials with complex microstructures or significant technological differences [10,11]. Therefore, many research teams recommend developing dedicated, empirical relationships tailored to a specific material, processing condition, and hardness range.

The use of hardness as an indicator of material quality holds significant practical importance—especially in the context of industrial additive manufacturing. The ability to non-invasively estimate tensile strength enables faster production decisions, reduces waste, and lowers the costs associated with destructive testing. Therefore, the development of accurate hardness–strength models tailored to the specific characteristics of DMLS and Ti6Al4V can greatly enhance the efficiency of industrial quality control [12].

At the same time, it should be remembered that hardness measurement is not without its limitations. The hardness value may depend on measurement conditions (load, duration, and surface condition), as well as on local structural inhomogeneities. In the case of Ti6Al4V, especially after additive manufacturing processes, significant hardness fluctuations may occur, even within a single sample [13].

Moreover, the homogeneity of the microstructure plays a key role in the accuracy of estimations. Microdefects, chemical segregations, microstructural gradients, and grain orientation can disrupt the relationship between hardness and strength. Therefore, studies in this area should be conducted on samples representing the full range of material conditions and processing technologies.

The Ti6Al4V alloy is also characterized by the presence of two phases (α and β), the volume fraction, morphology, and distribution of which influence both the hardness and strength. Phase transformations during heat treatment, as well as differences between the microstructure of DMLS and conventionally produced materials, can lead to significant variations in the hardness–strength correlation, which should be taken into account when developing predictive models [5].

A review of the literature shows that although there are numerous studies analyzing the relationship between hardness and strength for various metal alloys, data for the Ti6Al4V alloy—especially in the context of DMLS—remain limited. Studies such as those conducted by Kasperovich and Hausmann (2015) and Qiu et al. (2015) are promising but highlight the need for further work on standardizing approaches and validating models under different technological conditions [14,15].

The aim of this article is, therefore, to evaluate the effectiveness of an indirect method for estimating the tensile strength of the Ti6Al4V alloy, based on hardness test results. The study analyzes samples that were produced using the DMLS method, as well as drawn bar stock, both before and after heat treatment. The obtained results will make it possible to verify the suitability of the selected correlation models and to assess the impact of manufacturing technology and microstructure on prediction accuracy.

2. Research Object

The research object consisted of samples made of Ti6Al4V ELI alloy intended for biomedical applications, with a lower oxygen content. Table 1 shows the chemical composition of the titanium powder used to produce the test samples.

Table 1. Ti6Al4V ELI powder chemistry according to Renishaw's catalog [16], given as a weight percentage (wt.%).

Al	V	Fe	O	C	N	H	Y	Other	Ti
5.5–6.5	3.5–4.5	≤0.25	≤0.13	≤0.08	≤0.05	≤0.012	≤0.005	≤0.4	rest

The test samples were printed on an EOS M290 device. The manufacturing process was characterized by the following parameters: laser power of 400 W, time of the laser beam on the material, 80 μs, hatch distance of 60 μm, layer thickness of 60 μm, and scanning speed of 5000 mm/s. The scanning rotation angle between successive superimposed layers was 0.

Samples with a circular cross-section were adopted for the study of static properties. The geometric and physical forms of the research objects are shown in Figure 1.

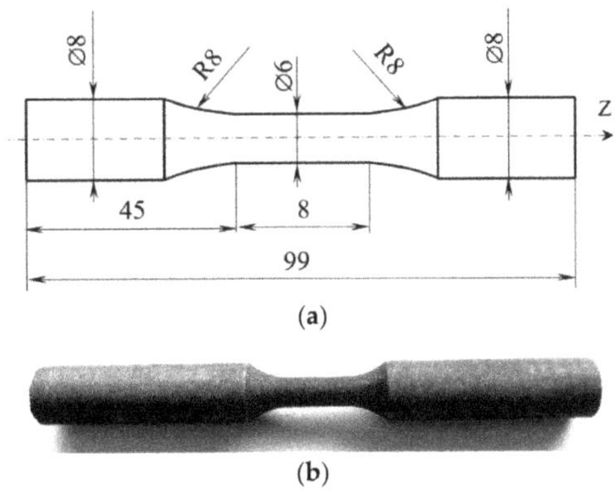

Figure 1. Research material: (**a**) sample diagram with dimensions; (**b**) physical form of the sample. Data from [17].

Two types of samples made by the DMLS method were accepted for testing. The first series was in the form of as-built samples, without additional heat or mechanical treatment. The second series of samples was subjected to a heat treatment process consisting of the supersaturation of the test material at a temperature of 980 °C for 1 h in a Seco Warwick Case Master Evolution single-chamber vacuum furnace. The third group consisted of samples made of annealed drawn bars. The samples were divided into 6 groups of 3. Table 2 shows the designations of the samples for the respective groups.

Table 2. Description and marking of samples taken for testing.

No.	Group Name	Description
1	AS Specimen	A printed sample without heat treatment, subjected to a static tensile test
2	HT Specimen	A printed sample with heat treatment, subjected to a static tensile test
3	DB Specimen	A sample from a drawn bar without heat treatment, subjected to a static tensile test
4	ASW Specimen	A printed sample without heat treatment, not subjected to a static tensile test
5	HTW Specimen	A printed sample with heat treatment, not subjected to a static tensile test
6	DBW Specimen	A sample made of a drawn bar without heat treatment, not subjected to the static tensile test

The AS, HT, and DB samples were tested under static tensile loads. The ASW, HTW, and DBW samples were not subjected to this analysis, but were instead used to create the samples used in the hardness tests and microscopic examinations. They acted as reference (comparative) samples in determining how thermal processes and the process of testing static properties may affect the hardness of the material.

3. Research Methods

3.1. Tensile Properties

The static tensile test made it possible to determine the mechanical properties of the material, i.e., tensile strength S_u, yield point $S_{y0.2}$, Young's modulus E, and elongation A. The tests, conducted under static load conditions, were carried out on an INSTRON 8502 testing machine (Figure 2). The test was carried out, based on a displacement of the machine piston by 0.05 mm/s. The parameters recorded during the tests were force and displacement. The displacement during the tests was determined using the classic INSTRON 2620-601 extensometer with a measuring base of 8 mm and a measuring range of 1 mm, which was used to determine the deformation value. During the static tensile test, five samples from each group (AS, HT, and DB) were used.

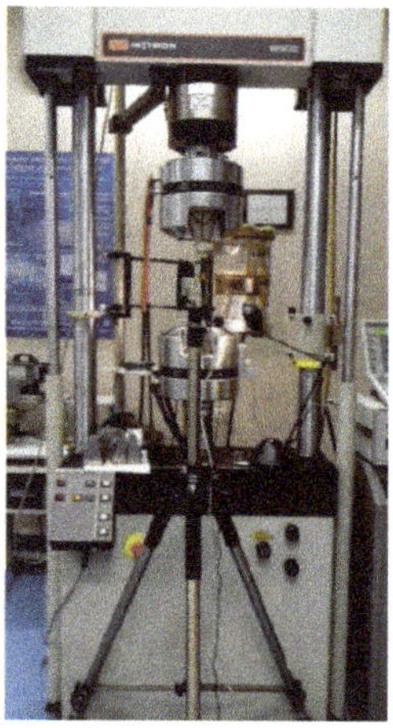

Figure 2. Tensile strength machine – INSTRON 8502, Bydgoszcz, POLAND.

3.2. Material Hardness Tests

The material hardness tests were carried out on a Huatec HV_{10} hardness tester (Figure 3). The hardness test was carried out using the Vickers HV5 scale with a force equal to 49.03 N. The AS, ASW, HT, HTW, DB, and DBW samples were tested. Hardness measurements were made on the cross-section of the sample on the *x*-*y* plane. Figure 4 shows the hardness measurement points for the samples.

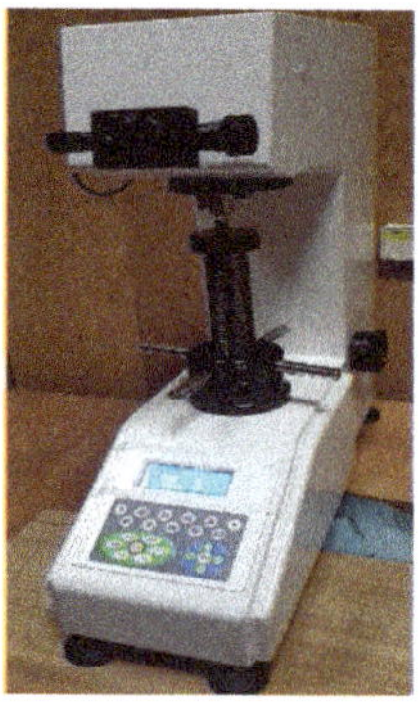

Figure 3. HUATEC HV10 hardness tester, Bydgoszcz, POLAND.

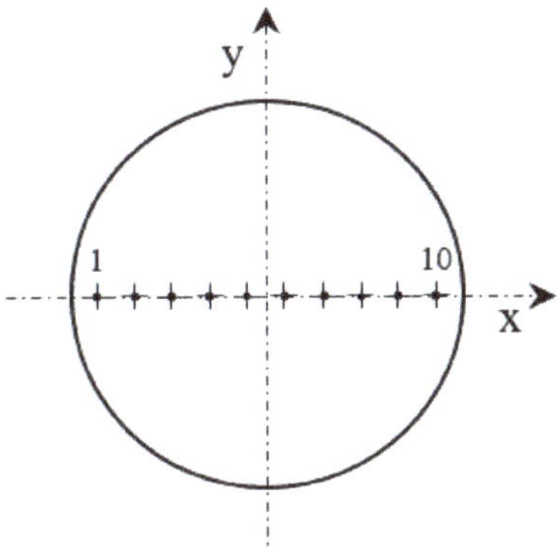

Figure 4. Schematic representation of the hardness measurement method on the *x*-*y* plane.

3.3. *Tests of the Microstructure of Materials*

After the tests of mechanical properties and hardness measurements were conducted, the samples were taken for metallographic tests. Microscopic photos were taken along the *z*-axis in two areas: P1—in the grip part of the sample, and P2—in the location of the fracture of the samples subjected to the static tensile test, which is shown schematically in Figure 5.

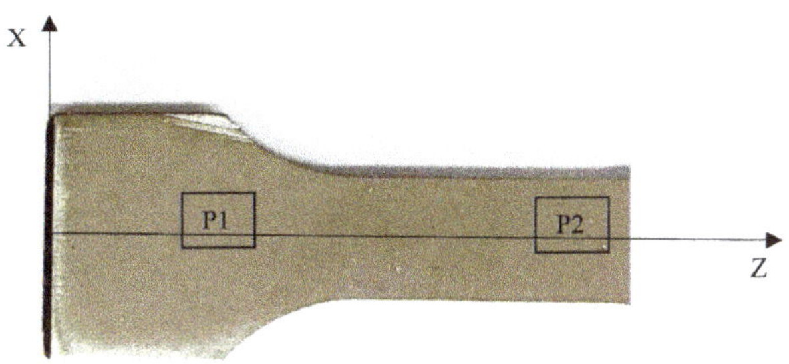

Figure 5. Scheme of the sample used for metallographic tests, with the places for taking microstructural pictures marked.

4. Research Results

4.1. *Tensile Strength Results*

Table 3 shows the test results, which were the mean value with any deviations of the following parameters: tensile strength, proof stress, Young's modulus, and elongation. The results were compared to those obtained for samples made of a drawn bar with the same geometry, as shown in Figure 3.

Table 3. Results of the tests of static tensile strength.

No.	Specimen No	S_u MPa	$S_{y0.2}$ MPa	E MPa	A %
1	2	3	4	5	6
1	AS specimen	1414.3 ± 106.1	1278 ± 95.8	106,300 ± 7972.5	10.4 ± 0.8
2	HT specimen	1050.5 ± 78.8	992 ± 74.4	117,840 ± 8838	2.9 ± 0.2
3	DB specimen	969.7 ± 72.8	927 ± 69.5	112,000 ± 8400	27.4 ± 2.1

In this study, a static tensile test was conducted on three types of Ti6Al4V alloy samples. The test results are presented as a stress–strain graph (Figure 6). Analysis of the graph clearly distinguishes the mechanical characteristics of each material.

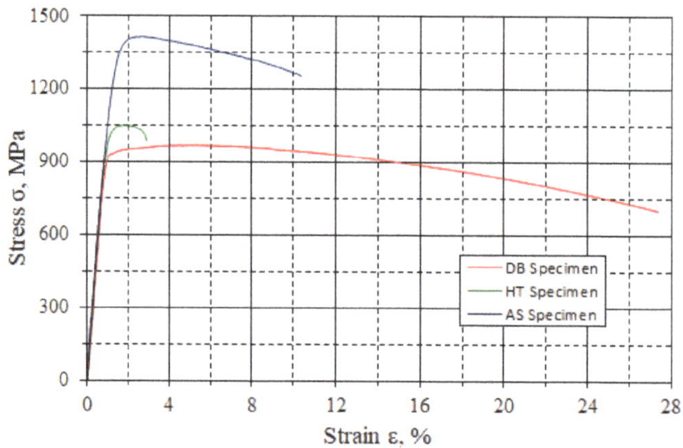

Figure 6. Sample plot of static tension results.

The AS sample exhibited the highest tensile strength (ultimate tensile strength (UTS) ≈ 1450 MPa), accompanied by relatively low elongation (ε ≈ 8–9%). The stress–strain curve is characterized by a steep rise in the maximum stress, followed by a sudden fracture, which is typical for materials with high strength and limited ductility. The high strength of the AS sample results from its specific columnar microstructure, which is formed during rapid solidification in the DMLS process, as well as the presence of residual stresses and unannealed dislocations [7,8].

The HT sample, obtained by heat-treating the AS sample, showed lower tensile strength (UTS ≈ 1050 MPa) and even lower elongation (ε ≈ 4–5%). This reduction is due to microstructural changes occurring during annealing, including stress relaxation, phase transformations, and a decrease in dislocation density [9]. Despite improved structural homogeneity, the material lost part of its initial strength without a noticeable gain in ductility, which may be the result of suboptimal heat treatment parameters (e.g., an excessively high temperature or insufficient annealing time).

The DB sample (drawn bar, without heat treatment) exhibited the lowest tensile strength (UTS ≈ 950 MPa) but the highest elongation, reaching as much as 26–27%. The DB tensile curve features a long plastic plateau, indicating excellent ductility and a stable deformation mechanism. The results confirm the expected behavior of conventional metallic materials when subjected to cold working processes such as drawing, which promote structural uniformity and increase the number of obstacles to dislocation movement [11].

4.2. Analysis of Tensile Strength Results

When analyzing the results of the static properties tests, it can be seen that the samples printed in the as-built condition (AS specimen) are characterized by the highest strength. However, they do not have as much ductility as the samples made of drawn bars (DB specimen), which is also visible in the breakthroughs of the samples after the static tensile test. This material behavior is influenced by the method of producing the material. The printed samples are sintered in each layer, which induces a type of weld to form. This significantly affects not only the high brittleness but also the high strength of elements made with additive technology.

The amount of work that must be applied to damage the sample during the static tensile test was also analyzed. The calculated work is the area under the sample stretching plot, which is shown schematically in Figure 7.

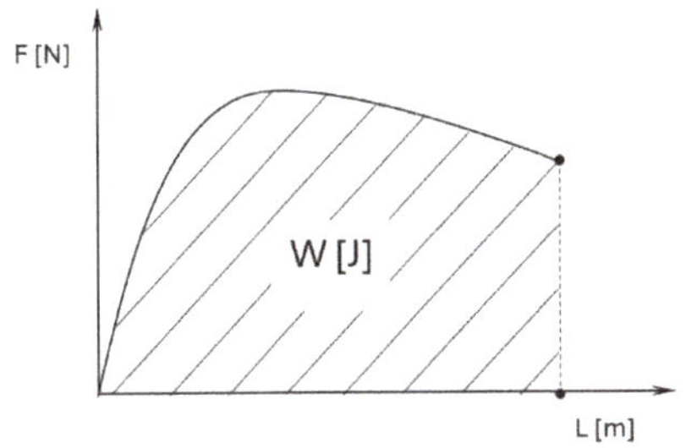

Figure 7. Scheme for determining the energy needed to destroy samples during strength tests.

The heat-treated samples did not improve their properties, which finding contrasts with most of the reports in the literature, where heat treatment is generally shown to enhance ductility and homogenize the microstructure of additively manufactured Ti6Al4V components [9,18–20]. Compared to the as-built samples, they lost both strength and plasticity. The material may exhibit such behavior due to insufficient heat treatment time and a certain temperature, which is higher than the β-transus transformation temperature. The highest ductility and the uniform $\alpha + \beta$ structure for the Ti6Al4V material are obtained concurrently by applying heat treatments above the α-transus and below the β-transus temperatures. The applied heat treatment could also insufficiently eliminate the defects arising in the structure, resulting from the method of producing the sample, i.e., unmelted powder particles or gas being used in the working chamber that was trapped in the molten layer.

The analysis showed that the greatest effort was needed to destroy a sample made of a drawn bar—11.19 J (Figure 8). Much less energy was needed to destroy a sample produced in the laser sintering process—6.20 J—which is 55.4% of the energy needed to destroy a sample made of a drawn bar. The heat-treated printed sample is characterized by the smallest area under the tensile diagram, which translates into the smallest amount of work put into damaging it—1.17 J.

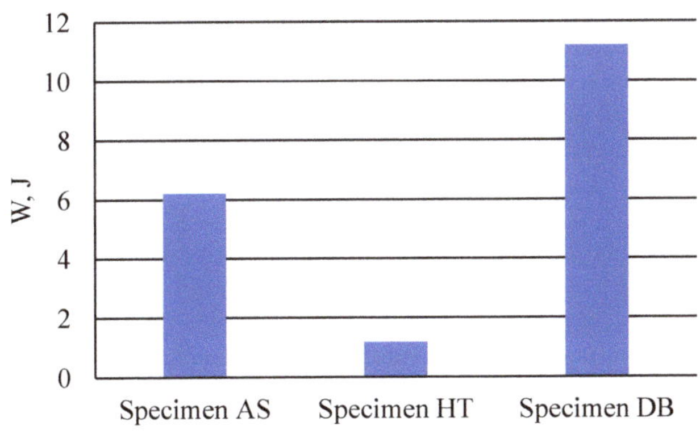

Figure 8. Graph showing the amount of energy needed to damage the samples.

4.3. Material Hardness Results

The hardness results for the measurements carried out on the x-y plane are presented in Table 4 by means of statistical parameters, i.e., the mean, standard deviation, median, and relative standard deviation.

Table 4. Analysis of hardness measurement results on the x–y plane.

Specimen	Number of Measurements	Average [hv]	Standard Deviation [hv]	Median [hv]	Relative Standard Deviation [%]
AS Specimen	10	347.1	12.5	343.5	3.6
HT Specimen	10	392.9	15,8	394.0	4.0
DB Specimen	10	294.7	6.4	294.5	2.2
ASW Specimen	10	412.7	22.8	410.3	5.5
HTW Specimen	10	356.9	20.6	358.0	5.8
DBW Specimen	10	336.7	4.5	337.0	1.3

To verify the potential of using hardness as an indirect indicator for assessing the tensile strength of the Ti6Al4V alloy, Vickers hardness (HV) measurements were carried out for six groups of samples: AS, HT, and DB (samples subjected to tensile testing), as well as ASW, HTW, and DBW (samples not subjected to tensile testing). Each hardness value represents the average of 10 independent measurements, through which it was possible to determine not only the mean value but also the standard deviation, median, and relative standard deviation (%RSD).

The highest average hardness was recorded for the ASW sample (412.7 HV), which was produced using the DMLS method in its as-built state, without annealing and without prior mechanical loading. The hardness of the HTW sample (printed and annealed but not subjected to tensile testing) was 356.9 HV, while the DBW sample (drawn bar, also untested) measured 336.7 HV. These values are slightly higher than those of their corresponding tensile-tested counterparts: AS—347.1 HV, HT—392.9 HV, and DB—294.7 HV. This suggests that performing the tensile test may, in some cases, lead to localized modifications in the material's structure—such as surface work hardening or microscopic deformation—resulting in slight changes (in most cases, reductions) in hardness.

It is worth noting, however, the unusual result observed for the HT sample, which, despite being annealed, shows greater hardness (392.9 HV) than its corresponding untested counterpart, HTW (356.9 HV). This may indicate structural inhomogeneity, local phase transformations, or the influence of micro-deformations that are introduced during the tensile testing process. Possible variations in the specific locations where the hardness measurements were taken on the sample should also be considered.

From the standpoint of result uniformity, the most stable values were obtained for the DBW (1.3% RSD) and DB (2.2% RSD) samples, confirming the high quality and consistency of the microstructure in the material derived from the drawn bar. In contrast, the samples produced using the DMLS method, particularly the HTW and ASW, showed significantly greater result dispersion (%RSD of 5.8% and 5.5%, respectively), which may stem from the inherent anisotropy of additive materials and local differences in density and phase distribution.

A comparison of the hardness results with the tensile strength values enables an analysis of the potential correlations between these properties. The AS sample, characterized by a tensile strength of approximately 1450 MPa, had an average hardness of 347.1 HV, while the DB sample, with the lowest tensile strength (~950 MPa), reached a hardness of 294.7 HV. As expected, higher hardness translated into higher strength, which confirms the validity of using empirical relationships such as the Cahoon model or other classic HV–UTS correlations.

However, the HT sample disrupts this simple correlation. Despite having greater hardness than the AS sample (392.9 HV vs. 347.1 HV), it achieved significantly lower tensile strength (~1050 MPa). This indicates that surface hardness does not fully reflect the strength properties of the heat-treated material. This may result from the fact that heat treatment modifies not only the local hardness but also the internal microstructure, the distribution of α and β phases, dislocation density, and the residual stress distribution—all of which are critical parameters for the material's behavior under tensile loading.

The analyses of both the mean values and standard deviations lead to the conclusion that hardness can be a useful but insufficient indicator for indirectly estimating the strength of Ti6Al4V materials, particularly in their heat-treated state. For the as-built samples (AS and DB), the HV–UTS relationship shows strong consistency, justifying the use of predictive models in the context of additive manufacturing without annealing or when using conventional materials. However, for heat-treated materials, it is necessary to consider additional structural parameters, which may require extending the correlation model or employing more advanced predictive techniques (e.g., machine learning).

Additionally, the fact that the samples not subjected to tensile testing (ASW, HTW, and DBW) exhibited higher hardness compared to their tensile-tested counterparts suggests the potential influence of mechanical deformation on subsequent hardness measurements—even though the HV test itself is considered minimally invasive. This further reinforces the need to perform hardness measurements in a controlled manner, taking into account the material's loading history.

Importantly, the samples not subjected to tensile testing (ASW, HTW, and DBW) exhibited higher hardness values compared to their tensile-tested counterparts. At first glance, this appears contrary to the classical phenomenon of strain hardening; however, additional mechanical and structural effects must be considered. During the tensile test, local relaxation of residual stresses and redistribution of the dislocation structure may occur, leading to reduced post-test hardness. The literature suggests that in materials with a martensitic microstructure (such as the α' phase that is present in the AS samples), these effects can result in strain softening or may even initiate micro-recrystallization within the necking region [15,21] (Brodsky and Thompson, 1984; Qiu et al., 2015). Additionally, local temperature increases during testing (adiabatic heating) may induce micro-annealing [11], and, in additively manufactured structures, partial stress relaxation due to plastic deformation is also possible [14]. As a result, hardness measurements taken after tensile testing may yield lower values, despite the simultaneous occurrence of classical strain hardening. Similar observations for Ti6Al4V were also reported by Bermingham et al. (2011) and Yadollahi and Shamsaei (2017) [9,13].

In summary, the hardness analysis confirms this test's potential as a supplementary indicator for assessing the mechanical properties of Ti6Al4V. However, its effectiveness depends on technological consistency, microstructural homogeneity, and knowledge of the processing conditions. Only under such conditions is it possible to build reliable and stable predictive models of tensile strength based on hardness measurements.

4.4. Determination of Material Strength Based on Hardness Measurement Results

The presented results of the hardness measurements of Ti6Al4V alloy samples made from a drawn bar and the additive method show differences. The DB samples are characterized by the lowest hardness value, which is also reflected in the strength results obtained via the static tensile test. The difference between the specimen under tension (DB specimen) and the as-built sample (DBW specimen) is 42 HV. This difference may indicate changes in the structure of the material resulting from clamping the testing machine's jaws on the sample during the test, as well as the static tensile test itself. Similar differences are observed in the AS and ASW samples, where the difference is 65.6 HV, and in the HT and HTW samples, where the difference is 36 HV.

The analysis of the results of the hardness tests on the x–y plane showed that the ASW and HT samples have the hardest structure, and the difference between them is 19.8 HV. The heat-treated printed samples (HTW) and printed samples without heat treatment (AS) show a similar hardness—the difference is 9.8 HV. The largest value for the dispersion of hardness in the results was observed for the ASW sample, at 22.8 HV.

When analyzing only the hardness results for those specimens taken from the samples not subjected to the static tensile test, where there are no changes in the material caused by tensile forces, it is noted that the ASW sample has the highest hardness, and the DBW has the lowest. The results of the tensile strength tests are similar, with the highest value for the AS sample and the lowest for DB. This highlights the relationship between the hardness and strength of the material.

The performed hardness measurements made it possible to determine the material strength indirectly. To determine the tensile strength in an analytical manner using the results of the hardness measurements, the formula described in Ref. [5] was used, which is presented below:

$$UTS = \frac{H}{C}(1-n)\left(\frac{12.5n}{(1-n)}\right) \quad (1)$$

where:

H—Vickers hardness value or HV,

C—material factor, and

n—strengthening factor.

The next stage used Equation (1) to carry out analytical calculations and to determine the local tensile strengths for the obtained values of hardness measurements on the x-y plane of the sample cross-section. To determine the tensile strength on the basis of the hardness tests, specimens taken from those samples not subjected to the strength test were used and compared with the material strength when determined through a static tensile test.

Figures 9–11 show the material strength results (UTS), calculated on the basis of hardness tests for the ASW, HTW, and DBW samples, respectively. The presented graphs show the strength values in terms of points compliant with the hardness measurements (UTS), their average value, and the average strength value obtained based on the static tensile test, together with the confidence intervals. To calculate the UTS, the cyclic hardening factor for the Ti6Al4V alloy with a value of n = 0.9 was used, as reported in Ref. [22]. Two extreme values of the C coefficient have been proposed, which, after converting the UTS, are within the assumed confidence interval.

Figure 9 presents the charts for the ASW sample for two values of the material coefficient, C = 1.9 and C = 2.1. Figure 10 shows the graphs for the HTW sample for two values of the material factor, C = 2.3 and C = 2.5. Figure 11 presents the graphs for the DBW sample for two values of the material factor, C = 2.3 and C = 2.6. Tables 5–7 show the calculation results of the UTS mean value, standard deviation, median, and relative standard deviation for the ASW, HTW, and DBW samples, respectively.

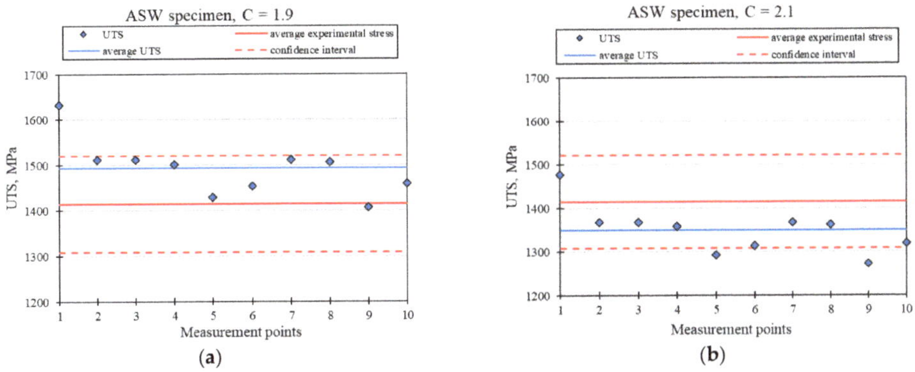

Figure 9. Graph of the calculated UTS for the ASW sample with the adopted value of the material factor at: (**a**) C = 1.9, (**b**) C = 2.1.

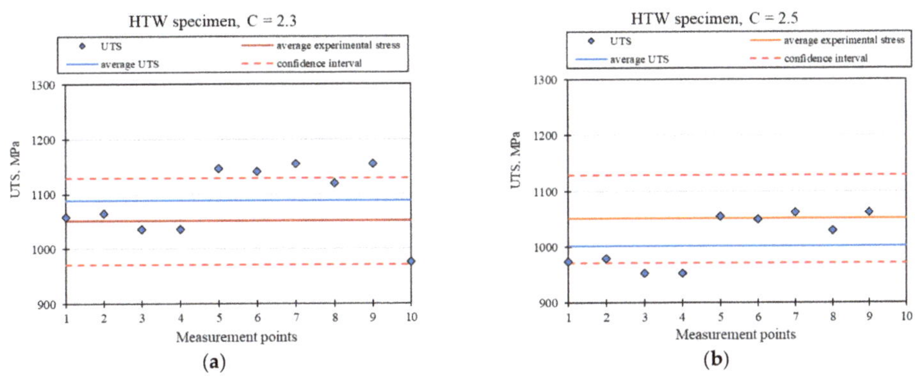

Figure 10. Graph of the calculated UTS for the HTW sample with the adopted value of the material factor at: (**a**) C = 2.3, (**b**) C = 2.5.

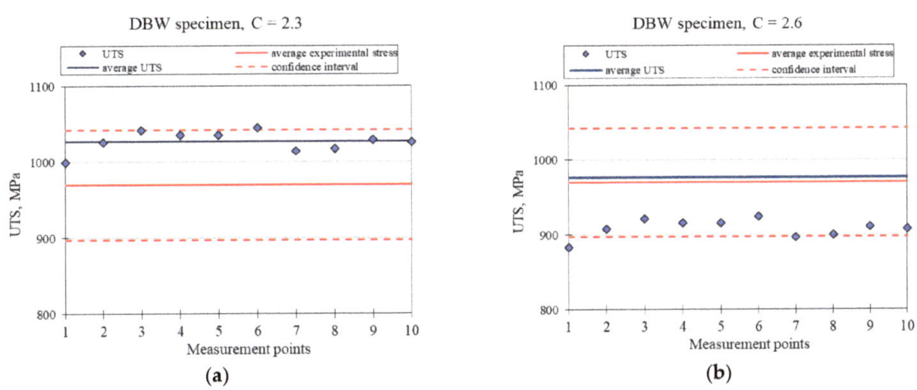

Figure 11. Graph of the calculated UTS for the DBW sample with the adopted value of the material factor at: (**a**) C = 2.3, (**b**) C = 2.6.

Table 5. Analysis of the results of the analytical calculations of tensile strength for the ASW sample.

Material Factor	Number of Measurements	UTS Average [MPa]	Standard Deviation [MPa]	Median [MPa]	Relative Standard Deviation [%]
1.9	10	1492.2	62.3	1503.6	4.2
2.1	10	1350.1	56.4	1360.4	4.2

Table 6. Analysis of the calculation results of the tensile strength analysis for the HTW sample.

Material Factor	Number of Measurements	UTS Average [MPa]	Standard Deviation [MPa]	Median [MPa]	Relative Standard Deviation [%]
2.3	10	1088.6	62.8	1091.9	5.7
2.5	10	1001.5	57.8	1004.6	5.7

Table 7. Analysis of the results of analytical calculations of tensile strength for the DBW sample.

Material Factor	Number of Measurements	UTS Average [MPa]	Standard Deviation [MPa]	Median [MPa]	Relative Standard Deviation [%]
2.3	10	1026.9	13.8	1027.9	1.3
2.6	10	908.5	12.2	909.3	1.3

A comparison of the results of the experimental tests with the results obtained analytically shows that it is possible to use the hardness tests to determine the strength of the material. The presented test results show that the average UTS values determined through the hardness tests were within the accepted confidence intervals, as determined on the basis of the results of the static tensile test. The different values of the C coefficient are due to the effect of the relevant treatment that was applied and the resulting variable material structure. In each of the presented cases, two extreme values of C were given, between which the strength values are close to the experimental average value.

4.5. Analysis of the Material Microstructure Test

Figure 12 shows photos of the individual microstructures of samples made of the Ti6Al4V alloy. Figure 12a,b show the microstructure of a printed sample in the as-built condition, subjected to strength tests (AS specimen). Figure 12c,d show the microstructure of a printed sample, subjected to heat treatment and after the tensile test (HT specimen). Figure 12e,f show the microstructure of a sample made of a pulled bar after testing its mechanical properties (DB specimen). Figure 12g shows the structure of the printed sample in the as-built condition, without undergoing the tensile test (ASW specimen). Figure 12h shows the microstructure before the strength tests of a heat-treated printed sample (HTW specimen). Figure 12a,c,e,g,h show the microstructure in the gripping part (P1). Finally, Figure 12b,d,f show the microstructure in the measuring part (P2).

The AS specimen is characterized by a clearly visible martensitic α' phase structure with a strongly acicular (needle-like) morphology. The grains are elongated in the z-axis direction (the layer-building direction), indicating directional solidification during processing. The α' martensite forms as a result of the very rapid cooling that is inherent to the additive manufacturing process and the absence of any subsequent heat treatment. This microstructure is typical for as-built 3D-printed materials and is responsible not only for their very high hardness and tensile strength but also for their brittleness and limited ductility. The observed fractures in the tensile-tested sample are consistent with this interpretation—the material exhibits no significant plastic deformation before failure.

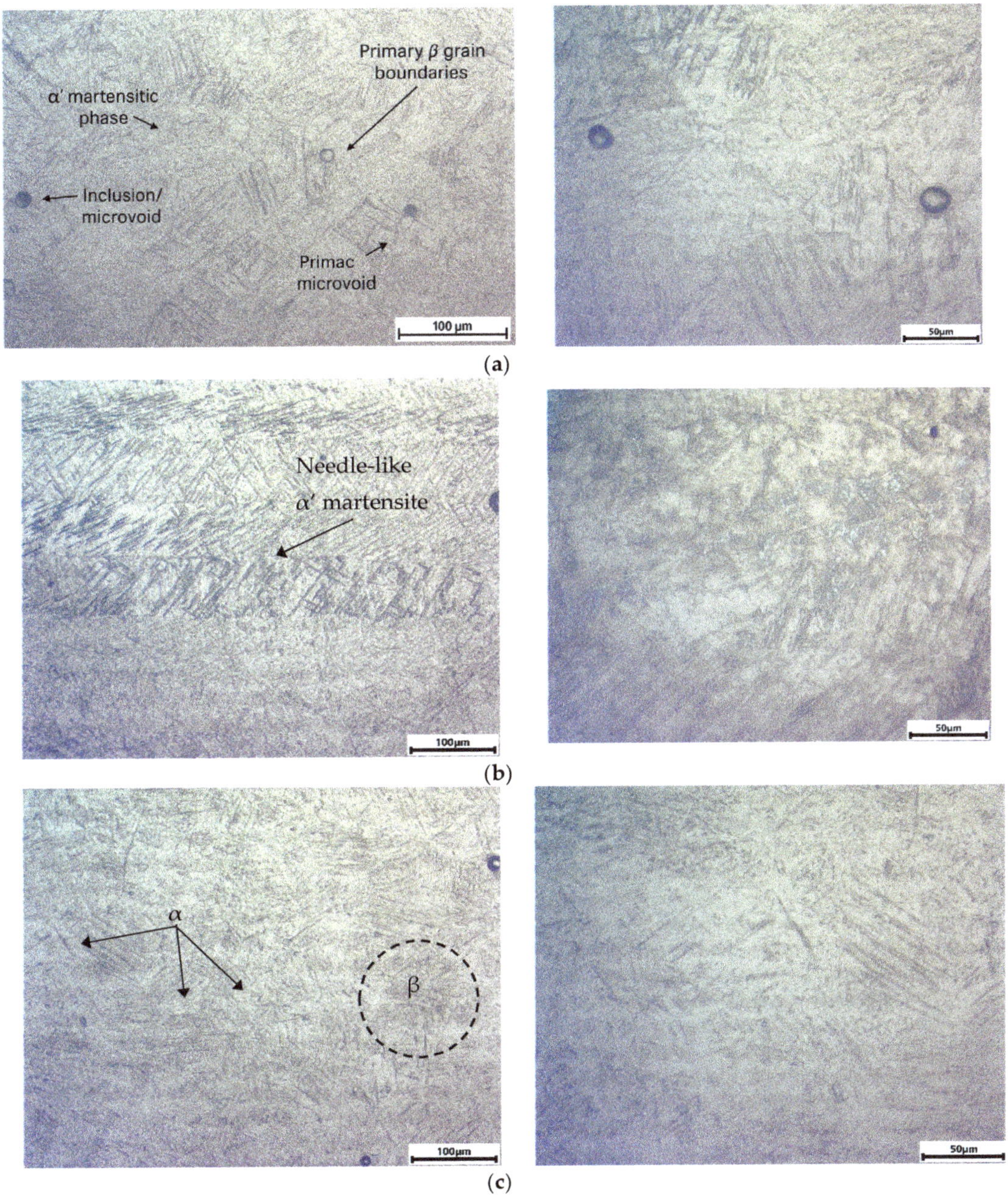

Figure 12. *Cont.*

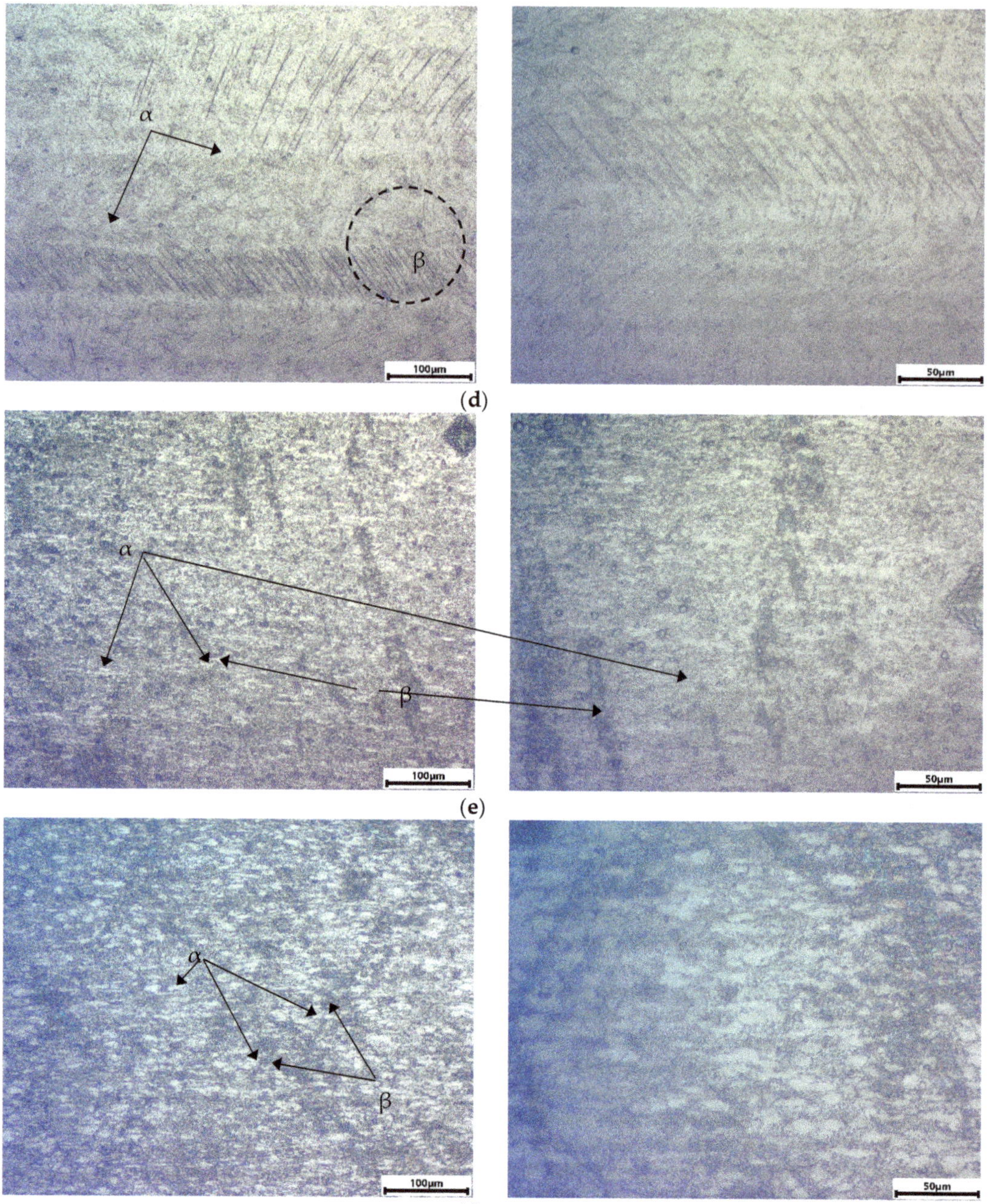

Figure 12. *Cont.*

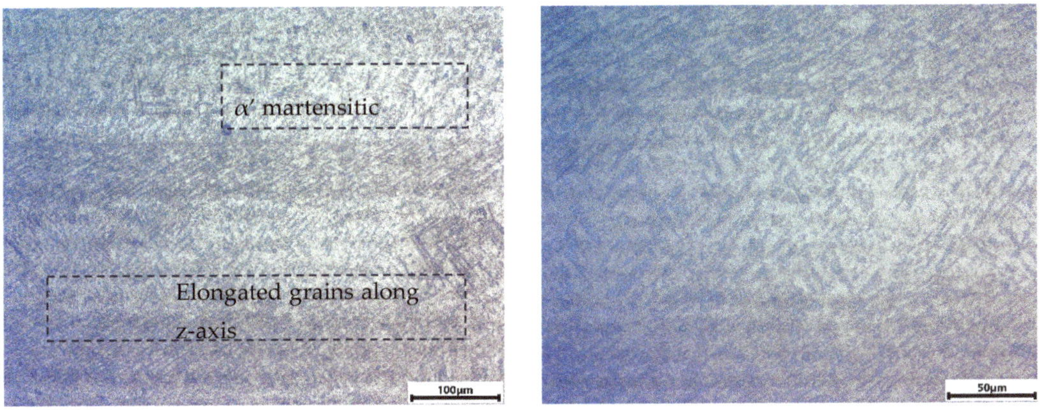

(**g**)

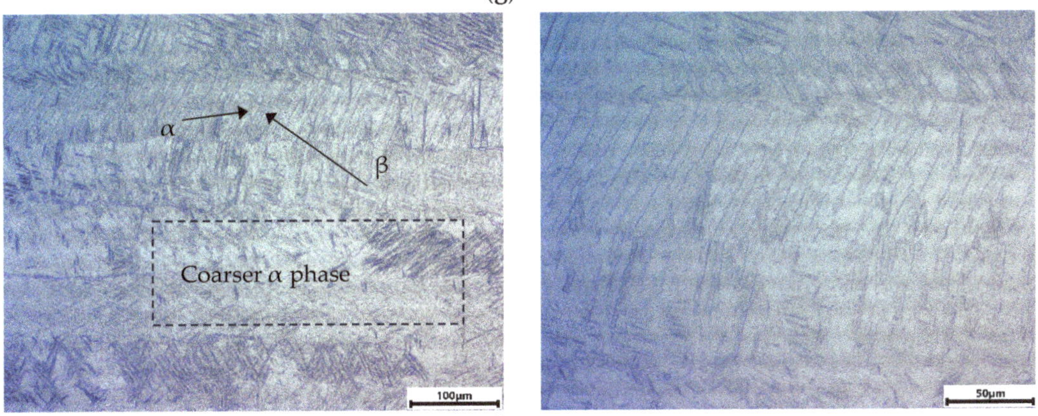

(**h**)

Figure 12. Microstructure pictures of the Ti6Al4V alloy samples on the left and specific area on the right: (**a**) AS specimen—P1, (**b**) AS specimen—P2, (**c**) HT specimen—P1, (**d**) HT specimen—P2, (**e**) DB specimen—P1, (**f**) DB specimen—P2, (**g**) ASW specimen—P1, and (**h**) HTW specimen—P1.

In the case of the HT specimen, a more balanced mixture of α (matrix) and β (darker regions) phases is visible, with grains that are less elongated and more equiaxed. The reduced presence of acicular martensitic α' indicates its decomposition as a result of thermal exposure. Although the microstructure transformed toward a more ductile form, the mechanical properties did not improve. This may be due to an excessive heat treatment temperature (above the β-transus) or an insufficient holding time, which led to incomplete stabilization of the α + β phases and degradation of the microstructure.

The DB specimen exhibits a homogeneous, fine-grained α + β structure that is typical of plastically worked Ti6Al4V products. The α phase appears in plate-like or lamellar form, uniformly distributed within the β matrix. The absence of a grain orientation suggests isotropic mechanical behavior. This microstructure provides well-balanced mechanical properties, combining good strength and ductility. Its uniformity forms a reliable basis for predictive modeling and serves as a reference point for additively manufactured materials.

In the ASW specimen, which is microstructurally identical to the AS specimen but has not been subjected to mechanical deformation, an intense presence of martensitic α' is also observed, with the grain structure aligned along the layer-building direction (z-axis). The absence of deformation marks or residual stress indicators suggests an 'undisturbed' material condition post-printing. This sample enables the evaluation of the isolated effect of tensile testing. When compared to the AS specimen, the microstructural differences are minimal, indicating that neither the HV test nor the tensile test significantly alters the deep internal structure. However, both may influence the residual stress state and local deformation characteristics (e.g., dislocation density).

The HTW specimen, which was printed and heat-treated similarly to the HT specimen but was not subjected to tensile loading, exhibits a visibly $\alpha + \beta$ phase structure composed of larger, less well-defined grains compared to the HT specimen. The absence of mechanical deformation (tensile test) likely resulted in a lack of dynamic grain refinement. The presence of a coarser microstructure may explain both the lower repeatability of hardness measurements and a potential decrease in mechanical performance. This specimen suggests that heat treatment alone may not be sufficient to optimize the material properties—controlled cooling or additional mechanical processing steps might be necessary.

Table 8 presents a comparative summary of the specimens based on their resulting microstructures. The as-built additively manufactured specimens (AS and ASW) exhibit a characteristic martensitic α' structure, which is responsible for high hardness and strength, but also contributes to significant brittleness. Heat treatment (HT and HTW) leads to decomposition of the martensite and the emergence of a more classical $\alpha + \beta$ phase mixture; however, the effectiveness of this transformation depends strongly on the heat treatment parameters. In the present study, the applied thermal processing did not significantly improve the mechanical properties. The specimens made from drawn bar material (DB and DBW) demonstrate the most predictable and balanced microstructure, serving as a reference state for comparison.

Table 8. Microstructural and mechanical characteristics of titanium alloy specimens.

No.	Specimen	Technology	Heat Treatment	Tensile Test State	Microstructure Characteristics	Dominant Phase Features	Impact on Mechanical Properties
1	AS Specimen	DMLS (3D printing)	None	After tensile testing	Martensitic α', needle-like, columnar, Z-orientation	High hardness, brittleness	Very high strength, low ductility
2	HT Specimen	DMLS	Yes	After tensile testing	$\alpha + \beta$, spheroidal, decomposed martensite	Coarser grains, partial transformation	Reduced strength and ductility compared to AS
3	DB Specimen	Drawn bar	None	After tensile testing	$\alpha + \beta$, fine-grained, lamellar, homogeneous	Dominant α phase, regular	Good balance of strength and ductility
4	ASW Specimen	DMLS	None	Without tensile testing	Martensitic α', same as AS, Z-orientation	Similar to AS, no deformation	Baseline for evaluating the effect of tensile testing
5	HTW Specimen	DMLS	Yes	Without tensile testing	$\alpha + \beta$, larger grains than in HT, less ordered	Weaker cohesion, less homogeneous structure	Potentially lower hardness, variable strength

The lack of a noticeable influence of tensile testing on the macroscopic microstructure of the samples indicates that the dominant factors modifying the internal structure are the manufacturing method (additive vs. conventional) and the applied heat treatment, rather than mechanical deformation during testing.

Both the AS specimen (after tensile testing) and the ASW specimen (untested) represent the characteristic microstructure that is formed by the DMLS (direct metal laser sintering) process. A common feature of these samples is the presence of the martensitic α' phase, forming a dense, acicular (needle-like) structure with a columnar arrangement aligned along the *z*-axis—the direction of layer-wise additive manufacturing.

The α' martensite forms as a result of extremely rapid cooling during the metal printing process. It is a metastable phase that is hard and brittle. The directional grain alignment leads to anisotropy in its mechanical properties. The lack of observable microstructural differences between the AS and ASW specimens suggests that tensile loading does not significantly alter the overall grain morphology, although it may locally affect the stress distribution and dislocation density.

The application of heat treatment in the HT and HTW specimens caused significant changes in the material's structure. The transformation of the α' martensite into a classical biphasic $\alpha + \beta$ microstructure is evident. The α grains exhibit a more rounded (spheroidal) morphology, while the β phase appears between them in the intergranular regions.

Although this structure is typical of annealed titanium alloys, the analyzed samples do not display full microstructural homogeneity, and the grains remain relatively coarse. The HT sample, which underwent tensile testing, shows slightly more refined grains compared to the HTW, suggesting that tensile deformation may have promoted partial microstructural reorganization or activated dynamic recrystallization processes.

Despite the phase transformation, the mechanical properties of the HT and HTW specimens deteriorated compared to the as-built condition, which could be attributed to non-optimal annealing parameters, for example, an excessively high temperature leading to overgrowth of the β phase.

The DB specimen (after tensile testing) represents a conventional microstructure that is typical of plastically worked materials. It features a fine lamellar $\alpha + \beta$ structure, uniformly distributed across the sample cross-section. The absence of preferential grain orientation indicates that the material is isotropic, which results in consistent and predictable mechanical behavior.

Comparison of the microstructures in the P1 (grip section) and P2 (gauge section, fracture zone) regions for selected Ti6Al4V specimens enables an assessment of the influence of tensile loading on local structural changes in the material.

In the AS specimen (additively manufactured, without heat treatment), both P1 and P2 reveal a distinct columnar martensitic α' microstructure that is characteristic of DMLS-fabricated materials in the as-built state. However, in region P2—the fracture zone—this microstructure shows local disturbances: the acicular α' phase is partially deformed and disrupted, likely due to tensile stresses and local thermal effects during fracture. Subtle changes in needle orientation and density suggest an increase in dislocation density and the initiation of microcracks. These features indicate that although the overall phase composition remains unchanged, the tensile test introduces localized structural reorganization in the fracture zone.

In the HT specimen (additively manufactured and heat-treated), region P1 displays a balanced $\alpha + \beta$ microstructure with relatively uniform, spheroidal grains that are typical of annealed titanium alloys. In contrast, region P2 exhibits greater irregularity, with disrupted phase boundaries and a more scattered grain morphology. These changes may indicate grain boundary motion or intragranular deformation induced by tensile loading. In both the AS and HT specimens, it is evident that while tensile testing does not cause phase transformation, it does affect the arrangement and integrity of the microstructure—particularly in DMLS materials, where anisotropy and residual stresses are already present, due to the layer-wise manufacturing process.

For the DB specimen (drawn bar, no heat treatment, and subjected to tensile testing), the differences between P1 and P2 are significantly less pronounced than in the printed samples. Region P1 shows a classic fine lamellar $\alpha + \beta$ structure that is uniformly distributed and with no preferential grain orientation, confirming the isotropy and homogeneity of conventionally processed Ti6Al4V. In region P2 (the fracture zone), the same general structure is preserved, although the local elongation of α lamellae and slight blurring of phase boundaries can be observed—likely a result of tensile-induced strain and localized stress concentration.

Compared to the printed specimens, the microstructural changes in region P2 of the DB sample are more subtle, confirming the greater structural resistance of the conventional material to tensile deformation. The drawn bar material exhibits high microstructural stability and resistance to localized damage, which translates into more favorable mechanical properties and reduced scatter in terms of both the hardness and tensile strength results.

Microstructural analysis of the Ti6Al4V alloy specimens revealed significant differences resulting from the manufacturing method, the heat treatment, and the presence of mechanical deformation. Samples produced via additive manufacturing (DMLS) in the as-built condition were characterized by a needle-like, columnar martensitic α' structure, which promotes high hardness and strength but limits ductility and microstructural uniformity. The application of heat treatment led to a transformation into a biphasic $\alpha + \beta$ structure; however, no significant improvement in mechanical properties was observed, indicating the need for optimization of the annealing parameters.

In contrast, the microstructures of specimens prepared from the drawn bar material exhibited the highest degree of homogeneity and stability, with a fine-grained $\alpha + \beta$ structure that is typical of conventionally processed titanium alloys, offering a favorable balance between strength and ductility. The effect of tensile deformation on the microstructure was marginal, primarily involving localized strain-induced modifications.

These findings confirm the critical role of microstructures in shaping the mechanical performance of Ti6Al4V and highlight the importance of carefully selecting the manufacturing method and heat treatment parameters in the context of additive technologies for titanium alloys.

5. Discussion

The results of our research confirm the effectiveness of the indirect method for estimating ultimate tensile strength (UTS) based on the Vickers hardness (HV) value in Ti6Al4V alloy. However, the degree of agreement with the model largely depends on the manufacturing technology and the structural state of the material. Samples printed using the DMLS method in the as-built condition (AS) exhibited the highest strength and high levels of hardness, corroborating the findings of Gong et al. [7], Vrancken et al. [23], and Tammas-Williams et al. [24], who indicated the presence of a martensitic α' structure and grain orientation as the main factors contributing to increased hardness.

For those samples subjected to heat treatment (HT), surprisingly lower UTS and elongation values were obtained compared to the as-built state. This result diverges from the literature reports by Bermingham et al. [9] and Ahmed and Rack [25], wherein annealing below the β-transus temperature led to improved ductility while maintaining moderate strength. One potential cause is the use of an excessively high temperature, resulting in grain coarsening and weakening of the phase boundaries.

Samples made from drawn bar (DB) material exhibited the lowest hardness and strength but the highest ductility. This observation aligns with the results of Semiatin et al. [26], Dieter [11], and Facchini et al. [27], who emphasized that conventional plastic deformation yields a lamellar α+β microstructure and isotropic mechanical properties. At the same time, this confirms the effectiveness of HV-UTS models for materials with a homogeneous structure.

It was observed that samples not subjected to mechanical testing (ASW, HTW, and DBW) had higher average HV values than their tensile-tested counterparts. A similar trend was noted by Leuders et al. [8] and Maskery et al. [28], suggesting that static loading may cause internal stress relaxation or local dislocation movement, thereby affecting the HV.

The analytically determined values (using the Cahoon model) for ASW, HTW, and DBW, after accounting for factors C and n, were consistent with the empirical results. A similar approach was employed by Tammas-Williams et al. [24] and Prashanth et al. [12], who achieved high agreement in HV-UTS models for additively manufactured Ti6Al4V structures.

Discrepancies between hardness and strength in the case of the HT samples indicate that the microstructure (particularly a heterogeneous β phase) has a greater influence than surface HV measurements. This is supported by studies from Frazier [29] and Ghidini et al. [18], which suggest the need to account for texture, chemical segregation, and residual stresses when developing predictive models.

The literature, including the work of Yadollahi and Shamsaei [13], indicates that the implementation of machine learning models can accommodate data irregularities and complex structural variables, potentially increasing prediction accuracy for heat-treated materials. This concept is gaining traction, as seen in publications by Garmestani et al. [30] and Tjong and Ma [31], wherein artificial intelligence supports the modeling of metal properties.

Microscopic analysis confirmed the presence of typical martensitic α' structures in the AS/ASW and $\alpha+\beta$ in the DB/DBW. Similar observations were made by Vilaro et al. [32] and Zhang et al. [33], emphasizing that the microstructure resulting from DMLS is characterized by strong anisotropy, which influences the fracture mechanics and HV result dispersion.

Ultimately, this study confirms the validity of using HV measurements to assess the quality of Ti6Al4V alloy, although there are still limitations stemming from microstructural complexity. It is necessary to introduce corrections to HV-UTS models for heat-treated cases or to apply hybrid methods that combine structural and mechanical data, as proposed by Liu et al. [34] and Yan et al. [35].

6. Conclusions

This study evaluated the feasibility of indirectly determining the tensile strength of Ti6Al4V alloy based on Vickers hardness measurements. The investigation was carried out on specimens produced using both the additive manufacturing method (DMLS) and drawn bar stock, in both the as-built condition and after heat treatment. Some specimens were subjected to tensile testing, while others were used for hardness measurements and microstructural analysis.

The results confirmed that a clear correlation exists between hardness and tensile strength; however, this relationship is strongly dependent on the manufacturing method and the structural state of the specimens. The highest tensile strength was achieved by the as-built DMLS specimens (AS specimens), which featured a characteristic columnar martensitic α' structure. However, these specimens exhibited significantly reduced ductility, as evidenced by their low elongation values and brittle fracture mode. The heat-treated specimens (HT specimens), despite their expected improvement in properties, showed a decrease in both strength and ductility. This is attributed to the application of an excessively high annealing temperature (above the β-transus) and the potentially ineffective elimination of structural defects.

In contrast, conventionally manufactured specimens (DB specimens) exhibited the lowest tensile strength but the highest ductility and the most homogeneous microstructure. The resulting fine-grained $\alpha + \beta$ structure, with no preferential grain orientation, provided stable and isotropic mechanical properties, making this material a reliable reference for further hardness–strength correlation studies.

The hardness measurements revealed that those specimens not subjected to tensile testing (ASW, HTW, and DBW) demonstrated higher average HV values than their tensile-tested counterparts. This may suggest the influence of mechanical testing on the local condition of the material—for example, surface work hardening or relaxation of the dislocation network. The average hardness values obtained for the ASW, HTW, and DBW samples were used to analytically estimate the tensile strength, using established models and selected material coefficients. These calculated values fell within the confidence intervals determined from mechanical testing, confirming the usefulness of the indirect estimation method.

Based on their analysis, it was found that there is a relationship between the strength and hardness of the material and the selection of appropriate material factors. It was found that the strength of a structural element that is produced with additive technology can be determined without needing to use destructive tests, which is economically justified. The analytically obtained results for material strength were comparable with the results obtained for the static tensile test (they fell within the determined confidence interval). For samples printed without heat treatment and not subjected to the static tensile test (ASW) using the proposed analytical method, two material factors were determined: $C = 1.9$ and $C = 2.1$. For printed samples that were not subjected to the static tensile test (HTW), the material factors were $C = 2.3$ and $C = 2.5$. DBW samples were characterized by the coefficients $C = 2.3$ and $C = 2.6$. All analyzed strength results were within the confidence limits.

It should be emphasized, however, that the application of universal hardness–strength correlation models (HV–UTS) may lead to significant predictive errors, especially with heat-treated additively manufactured materials, where the microstructure tends to be heterogeneous and the deformation mechanisms are complex. The variability in hardness values and the limited improvement in mechanical properties after annealing highlight the need to carefully select the heat treatment parameters—particularly temperature and dwell time—ideally keeping them below the β-transus.

Microstructural observations confirmed that the dominant factors affecting both the microstructure and mechanical performance are the manufacturing method and thermal processing. Tensile testing induced only localized changes (e.g., in the fracture zone) without substantially altering the overall phase composition. Therefore, when predicting the tensile strength from hardness data, greater emphasis should be placed on the sample's structural features and processing history than on its mechanical load history.

In summary, the results demonstrate that hardness can serve as an effective indirect indicator for estimating the tensile strength of Ti6Al4V, especially under controlled conditions and in samples with uniform microstructures (e.g., in the as-built state or after plastic deformation). For heat-treated or structurally complex materials, the accuracy of such estimations may be limited and may require the calibration of local material coefficients or the application of more advanced models (e.g., machine learning-based approaches). Despite these limitations, the proposed method represents a viable alternative to conventional destructive testing, supporting efficient quality control in industrial applications—particularly in the additive manufacturing of titanium alloys.

Author Contributions: Conceptualization, B.L. and K.K.; methodology, K.K.; formal analysis, K.K.; writing—original draft preparation, K.K. and M.W.; writing—review and editing, K.K. and M.W.; visualization, K.K. All authors have read and agreed to the published version of the manuscript.

Funding: This work was co-financed by the Polish Minister of Science under the program "Regional Initiative of Excellence".

Institutional Review Board Statement: Not applicable.

Data Availability Statement: The original contributions presented in this study are included in the article material. Further inquiries can be directed to the corresponding author.

Conflicts of Interest: The authors declare no conflict of interest.

References

1. *ISO 6892-1:2019*; Metallic Materials—Tensile Testing—Part 1: Method of Test at Room Temperature. International Organization for Standardization: Geneva, Switzerland, 2019.
2. Callister, W.D.; Rethwisch, D.G. *Materials Science and Engineering: An Introduction*, 10th ed.; Wiley: Hoboken, NJ, USA, 2020.
3. Tabor, D. *The Hardness of Metals*; Oxford University Press: Oxford, UK, 1951.
4. Cahoon, J.R.; Broughton, W.H.; Kutzak, A.R. The determination of yield strength from hardness measurements. *Metall. Trans.* **1971**, *2*, 1979–1983. [CrossRef]
5. Lütjering, G.; Williams, J.C. *Titanium*, 2nd ed.; Springer: Berlin/Heidelberg, Germany, 2007. [CrossRef]
6. Kruth, J.P.; Froyen, L.; Van Vaerenbergh, J.; Mercelis, P.; Rombouts, M.; Lauwers, B. Selective laser melting of iron-based powder. *J. Mater. Process. Technol.* **2005**, *149*, 616–622. [CrossRef]
7. Gong, H.; Rafi, K.; Gu, H.; Janaki Ram, G.D.; Starr, T.; Stucker, B. Influence of defects on mechanical properties of Ti–6Al–4V components produced by selective laser melting and electron beam melting. *Mater. Des.* **2014**, *86*, 545–554. [CrossRef]
8. Leuders, S.; Thöne, M.; Riemer, A.; Niendorf, T.; Tröster, T.; Richard, H.A.; Maier, H.J. On the mechanical behaviour of titanium alloy TiAl6V4 manufactured by selective laser melting: Fatigue resistance and crack growth performance. *Int. J. Fatigue* **2013**, *48*, 300–307. [CrossRef]
9. Bermingham, M.J.; Dargusch, M.S.; StJohn, D.H.; Littlefair, G. Effect of heat treatment on the microstructure and mechanical properties of Ti–6Al–V produced by selective laser melting. *J. Mater. Process. Technol.* **2011**, *211*, 1141–1148. [CrossRef]
10. Bartkowiak, T.; Kaczmarek, Ł. *Metody badań właściwości mechanicznych materiałów konstrukcyjnych*; Wydawnictwo Politechniki Poznańskiej: Poznań, Poland, 2022.
11. Dieter, G.E. *Mechanical Metallurgy*, 3rd ed.; McGraw-Hill: Columbus, OH, USA, 1986.
12. Prashanth, K.G.; Scudino, S.; Klauss, H.J.; Surreddi, K.B.; Löber, L.; Wang, Z.; Chaubey, A.K.; Kühn, U.; Eckert, J. Microstructure and mechanical properties of Al–12Si produced by selective laser melting: Effect of heat treatment. *Mater. Sci. Eng. A* **2016**, *590*, 153–160. [CrossRef]
13. Yadollahi, A.; Shamsaei, N. Additive manufacturing of fatigue resistant materials: Challenges and opportunities. *Int. J. Fatigue* **2017**, *98*, 14–31. [CrossRef]
14. Kasperovich, A.; Hausmann, J. Improvement of fatigue resistance and ductility of TiAl6V4 processed by selective laser melting. *J. Mater. Process. Technol.* **2015**, *220*, 202–214. [CrossRef]
15. Qiu, C.; Adkins, N.J.E.; Attallah, M.M. Microstructure and tensile properties of selectively laser-melted and of HIPed laser-melted Ti–6Al–4V. *Mater. Sci. Eng. A* **2015**, *578*, 230–239. [CrossRef]
16. Data Sheet: Ti6Al4V ELI-0406 Powder for Additive Manufacturing. Available online: https://resources.renishaw.com/en/details/data-sheet-Ti6Al4V-eli-0406-powder-for-additive-manufacturing{-}{-}94700 (accessed on 30 June 2025).
17. Karolewska, K.; Ligaj, B.; Wirwicki, M.; Szala, G. Strength analysis of Ti6Al4V titanium alloy produced by the use of additive manufacturing method under static load conditions. *J. Mater. Res. Technol.* **2020**, *9*, 1365–1379. [CrossRef]
18. Ghidini, T.; Zolesi, V.; Kofod, G. *Additive Manufacturing for Space Applications*; Springer: Berlin/Heidelberg, Germany, 2021.
19. Simonelli, M.; Tse, Y.; Tuck, T. Effect of the build orientation on the mechanical properties and fracture modes of SLM Ti–6Al–4V. *Mater. Sci. Eng. A* **2014**, *616*, 1–11. [CrossRef]
20. Vilaro, T.; Colin, C.; Bartout, J.D. As-fabricated and heat-treated microstructures of the Ti–6Al–4V alloy processed by Selective Laser Melting. *Metall. Mater. Trans. A* **2011**, *42*, 3190–3199. [CrossRef]
21. Brodsky, N.A.; Thompson, A.W. Strain softening and dynamic recovery in titanium alloys. *Metall. Trans. A* **1984**, *15*, 87–94.
22. Chlebus, E.; Kuźnicka, B.; Kurzynowski, T.; Dybała, B. Mechanical properties of Ti–6Al–4V alloy manufactured by direct metal laser sintering. *Mater. Charact.* **2011**, *62*, 488–495. [CrossRef]
23. Vrancken, B.; Thijs, L.; Kruth, J.P.; Van Humbeeck, J. Heat treatment of Ti6Al4V produced by Selective Laser Melting: Microstructure and mechanical properties. *Acta Mater.* **2014**, *68*, 150–158. [CrossRef]
24. Tammas-Williams, S.; Zhao, H.; Léonard, F.; Derguti, F.; Todd, I.; Prangnell, P.B. XCT analysis of the influence of melt strategies on defect population in Ti–6Al–4V components manufactured by Selective Electron Beam Melting. *Mater. Charact.* **2015**, *102*, 47–61. [CrossRef]
25. Ahmed, T.; Rack, H.J. Phase transformations during cooling in $\alpha+\beta$ titanium alloys. *J. Mater. Sci.* **1998**, *243*, 206–211. [CrossRef]
26. Semiatin, S.L.; Seetharaman, V.; Weiss, I. Hot workability of titanium and titanium alloys: An overview. *JOM* **1997**, *49*, 33–39. [CrossRef]
27. Facchini, L.; Magalini, E.; Robotti, P.; Molinari, A.; Höges, S. Ductility of a Ti–6Al–4V alloy produced by selective laser melting of prealloyed powders. *Mater. Sci. Eng. A* **2010**, *528*, 711–718. [CrossRef]
28. Maskery, I.; Aboulkhair, N.T.; Aremu, A.O.; Tuck, C.J.; Ashcroft, I.A.; Everitt, N.M.; Hague, R.J.M. Quantification and characterisation of porosity in selectively laser melted AlSi10Mg using X-ray computed tomography. *Mater. Sci. Eng. A* **2016**, *628*, 230–239. [CrossRef]
29. Frazier, W.E. Metal Additive Manufacturing: A Review. *J. Mater. Eng. Perform.* **2014**, *23*, 1917–1928. [CrossRef]

30. Garmestani, H.; Radhakrishnan, B.; Pagan, D. Multiscale materials modeling for additive manufacturing of metals: Challenges and opportunities. *Comput. Mater. Sci.* **2019**, *161*, 111–123. [CrossRef]
31. Tjong, S.C.; Ma, Z.Y. Microstructural and mechanical characteristics of in situ metal matrix composites. *Mater. Sci. Eng. R Rep.* **2000**, *29*, 49–113. [CrossRef]
32. Vilaro, T.; Colin, C.; Bartout, J.D.; Nazé, L.; Sennour, M. Microstructural and mechanical approaches of the selective laser melting process applied to a nickel-base superalloy. *Mater. Sci. Eng. A* **2012**, *534*, 446–451. [CrossRef]
33. Zhang, Y.; Wu, L.; Guo, X.; Kane, S.; Deng, Y.; Jung, Y.-G.; Lee, J.-H.; Zhang, J. Additive manufacturing of metallic materials: A review. *J. Mater. Eng. Perform.* **2017**, *27*, 1–13. [CrossRef]
34. Inayathullah, S.; Buddala, R. Review of machine learning applications in additive manufacturing. *Results Eng.* **2025**, *25*, 103676. [CrossRef]
35. Yan, W.; Lin, S.; Kafka, L.; Lian, Y.; Yu, C.; Liu, Z.; Yan, J.; Wolff, S.; Wu, H.; Ndip-Agbor, E.; et al. Data-driven multi-scale multi-physics models to derive process—structure—property relationships for additive manufacturing. *Comput. Mech.* **2018**, *61*, 521–541. [CrossRef]

materials

MDPI

Communication

Analysis of the Microstructure and Mechanical Properties of Austenitic Stainless Steel 310 Manufactured via WAAM

Aline Cipriano [1], Célia de Fraga Malfatti [1], Henrique Cechinel Casagrande [2,*], Anderson Daleffe [3], Jovani Castelan [3] and Pedro Henrique Menegaro Possamai [3]

[1] Department Metallurgical Engineering, Universidade Federal do Rio Grande do Sul (UFRGS), Porto Alegre 90010-150, Brazil; aline.cipriano@satc.edu.br (A.C.); materiaisxenergia@gmail.com (C.d.F.M.)

[2] Department Metallurgical Engineering, Universidade Federal de Santa Catarina (UFSC), Florianópolis 88040-900, Brazil

[3] Department Additive Manufacturing, Centro Universitário SATC (UNISATC), Criciúma 88805-380, Brazil; anderson.daleffe@satc.edu.br (A.D.); jovani.castelan@satc.edu.br (J.C.); pedro.possamai@satc.edu.br (P.H.M.P.)

* Correspondence: henrique_cechinel@hotmail.com

Abstract

The objective of this study was to characterize austenitic stainless steel 310 produced by Wire and Arc Additive Manufacturing (WAAM), addressing a gap in the literature regarding this alloy. Microstructural, chemical, and mechanical analyses were performed. Optical and electron microscopy revealed a predominantly columnar grain structure with characteristic tracks along the deposition direction. Point and mapping EDS analyses indicated a homogeneous distribution of iron, chromium, and nickel; however, point measurements suggested a possible underestimation of nickel, likely due to high relative error. Tensile tests demonstrated anisotropic mechanical behavior, with yield strength meeting standards at 45° and 90°, but lower at 0°. Ultimate tensile strength and elongation were below conventional requirements, with a maximum elongation of 15% at 90°. Additionally, the sample exhibited a total porosity of approximately 0.89%, which contributes to the reduction in mechanical properties, especially in the direction parallel to the deposition tracks. Overall, the WAAM-produced 310 stainless steel presented a microstructure similar to hot-rolled and annealed AISI 310 steel, but with distinctive features related to the additive process, such as mechanical anisotropy and microstructural directionality. These limitations highlight the need for process optimization to improve mechanical performance but reinforce the alloy's structural potential in additive manufacturing.

Keywords: Wire and Arc Additive Manufacturing (WAAM); AISI 310 austenitic stainless steel; tensile strength and ductility; microstructural characterization; optical emission spectrometry (OES)

1. Introduction

Additive manufacturing (AM) represents a modern approach to producing complex components through layer-by-layer construction [1–4]. AM has the potential to revolutionize the global manufacturing and logistics landscape [5] by offering significantly greater geometric freedom compared to conventional methods such as casting, forging, and machining [6]. A promising example is Wire and Arc Directed Energy Deposition (WA-DED) [7–9], also known as Wire and Arc Additive Manufacturing (WAAM) [10,11].

This process has gained attention due to its high deposition rates [12] and suitability for the fabrication of large-scale and geometrically complex components [13,14]. Its operation resembles multipass welding [15], particularly regarding solidification behavior and thermal cycles [16]. According to ASTM (American Society for Testing and Materials), additive manufacturing is defined as a process of material consolidation to create objects from three-dimensional models, typically through layer-by-layer construction, in contrast to subtractive manufacturing techniques [17,18]. The use of metallic materials in AM contributes to cost reduction and environmental impact mitigation by significantly saving material and energy during production [19].

The manufacturing of stainless steels remains a relevant industrial challenge, particularly due to the machining difficulty of these high-value austenitic alloys, caused by their pronounced work hardening and low thermal conductivity [20,21]. Austenitic stainless steels stand out for their unique combination of advantageous properties, such as high strength at both elevated and cryogenic temperatures [22–24], rapid strain hardening, excellent ductility, good energy absorption, and strong resistance to corrosion and environmental degradation [25,26]. As a result, they are widely employed across sectors such as petrochemical [27], gas turbines, automotive, medical, civil construction [28], marine, offshore [29], and aerospace industries [30].

These alloys contain higher levels of chromium and nickel compared to conventional ferrous alloys. Their specific chemical composition provides high tensile strength and excellent oxidation resistance, enabling efficient use in applications up to approximately 1323 K (1050 °C) [31]. Numerous studies have investigated the relationship between processing parameters, microstructure, and the properties of austenitic stainless steels, aiming to leverage the economic and productive advantages of WAAM. However, most investigations have focused on the 304, 308, and 316 series [32–47], whereas the austenitic stainless steel 310, especially concerning its specific microstructural features and mechanical anisotropy, remains underexplored.

The mechanical properties of austenitic stainless steel components produced via WAAM are critical to ensuring reliable performance in safety-sensitive applications. Previous studies have highlighted several factors affecting these properties, including grain structure [48–51], dendrite geometries [52], δ-ferrite phase presence [53], dislocation density [54,55], inclusions [56], and cracking [57–59]. According to Siddiqui et al. [60], other relevant aspects include delamination caused by segregation between adjacent deposited tracks [61] and geometric accuracy [62].

Although some studies have addressed the microstructure and defects such as microcracks and segregations [63], significant gaps remain in the detailed understanding of the correlation between specific deposition conditions, cooling rates, and the anisotropic mechanical behavior of WAAM-fabricated 310 stainless steel. This study differentiates itself by providing an integrated and in-depth investigation of microstructural characterization using both optical and electron microscopy, combined with quantitative analysis of mechanical anisotropy in multiple deposition orientations—areas still scarcely explored in the literature.

Therefore, this study expands the knowledge on the behavior of WAAM-produced 310 stainless steel, providing quantitative data and comparative analyses that support future process improvements and industrial applications of this material, contributing to technological advancement in the additive manufacturing of high-performance stainless steel alloys.

2. Experimental Procedure

This study employed the WAAM process to fabricate austenitic stainless steel 310 components. A wall measuring approximately 250 × 130 × 10 (mm) was produced. The feedstock wire, supplied by Superfine, had a diameter of 1.2 mm. The manufacturing system comprised a YASKAWA industrial robot (YASKAWA, São Paulo, Brazil) coupled with a Lincoln Electric welding power source (Lincoln Electric, São Paulo, Brazil). Figure 1 illustrates the equipment setup used in the additive manufacturing process.

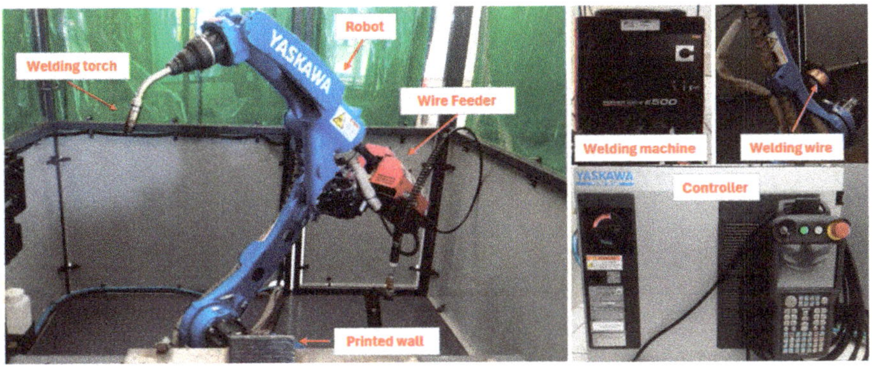

Figure 1. Industrial robot for additive manufacturing.

After the additive manufacturing of the material, no post-processing treatments or pre-processing steps were performed on the austenitic 310 steel. Optical emission spectrometry testing was conducted for chemical characterization. During deposition, each bead was laid sequentially to form continuous tracks, ensuring proper overlap and consistent layer height throughout the structure.

Wire and Arc Additive Manufacturing (WAAM) of austenitic stainless steel 310 was successfully carried out using carefully selected operating parameters, namely, a voltage of 19 V, a current of 165 A, a wire feed speed of 5.5 m/min, a torch travel speed of 300 mm/min, a shielding gas flow rate of 16 L/min, and a wire diameter of 1.2 mm. These parameters resulted in an approximate power of 3.14 kW and a heat input (Equation (1)) of 627 J/mm, a value considered adequate to ensure proper interlayer fusion and a high deposition rate of approximately 49 g/min (Equation (2)), without compromising the geometric integrity of the built part.

$$\text{Heat Input (J/mm)} = \frac{\text{V} \times \text{A} \times 60}{\text{Torch travel speed}} \quad (1)$$

where

- V = voltage (V).
- A = current (A).
- Travel speed (mm/min).
- Result in joules per millimeter (J/mm).

$$\text{Heat Input} = \frac{19 \times 165 \times 60}{300} = 627\,\frac{\text{J}}{\text{mm}}$$

$$\text{Deposition Rate (g/min): m} = \text{Area vw}\cdot\rho \quad (2)$$

where:

- Area = $\frac{\pi d^2}{4}$ = cross-sectional area of the wire (mm^2).

- d = wire diameter (mm).
- vw = wire feed speed (mm/min).
- ρ = density of stainless steel 310 = (7.9 g/cm^3 or 0.0079 g/mm^3) [26].
- m = deposition rate (g/min).

m = 49.1 g/min.

The process parameters used for fabricating the austenitic stainless steel 310 wall are presented in Table 1.

Table 1. Parameters used for the additive manufacturing of austenitic stainless steel 310.

Variables	Parameters
Voltage (V)	19
Current (A)	165
Gas flow rate (L/min)	16
Wire feed speed (m/min)	5.5
Power (kW)	3.135
CNC travel speed (mm/min)	300
Wire diameter (mm)	1.2

As shown in Table 2, the metallic wall was successfully fabricated using Wire and Arc Additive Manufacturing (WAAM), exhibiting an average thickness of 10 mm and a layer height of approximately 2.5 mm, which indicates good dimensional stability and process control throughout the build. A total of 100 successive layers were deposited, resulting in a wall height of approximately 250 mm, consistent with the vertical spacing between layers and the intended geometry. This vertical growth rate per layer (2.5 mm) falls within the expected range for high-deposition-rate processes such as WAAM and reinforces the effectiveness of the selected parameters, particularly in terms of heat input and interlayer fusion.

Table 2. Dimensions of the wall fabricated by additive manufacturing.

Variables	Values
Average thickness (mm)	10
Average layer height (mm)	2.5
Number of deposited layers	100
Deposition time per bead (S)	26
Total active fabrication time (min)	43
Shielding gas argon (%)	100

The average deposition time per bead was 26 s, reflecting a relatively fast deposition rate, a desirable feature in industrial processes aimed at high productivity. The total active fabrication time was only 43 min, demonstrating the potential of the WAAM process for the rapid and efficient production of medium-sized metallic components.

It is worth noting that the relationship between the total manufacturing time and the number of layers suggests that the interpass time was minimal or even nonexistent. This may warrant further microstructural evaluation, as consecutive thermal cycles can influence columnar grain growth and the formation of undesirable phases in highly alloyed materials such as 310 stainless steel.

Overall, the results show that the applied parameters enabled efficient construction, with adequate dimensional control, high deposition rate, and low production time—characteristics that highlight WAAM as a viable alternative for fabricating structural components using corrosion- and heat-resistant stainless steels.

Figure 2 depicts the wall fabricated by additive manufacturing using a zig-zag bead deposition strategy. The image also presents the tensile test specimens prepared for anisotropy analysis, along with a sample designated for metallographic examination. The sample detailed in Figure 2 for mechanical tensile testing follows the ASTM E8 standard [64].

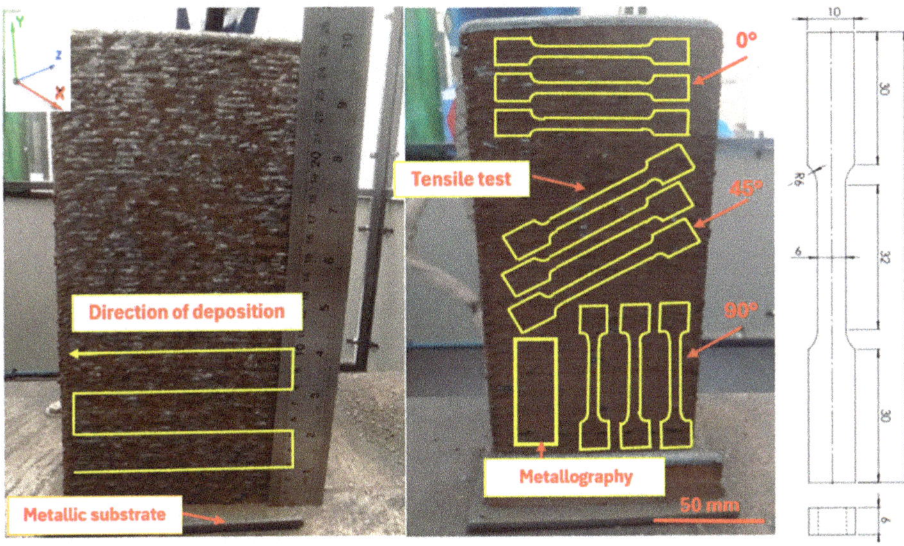

Figure 2. Wall printed using the Wire and Arc Additive Manufacturing (WAAM) process with 310 stainless steel.

Subsequently, the edges of the wall were symmetrically trimmed, leaving a remaining structure measuring 120 mm in length. This was performed to remove the inherent surface waviness caused by the WAAM bead step effect [65].

The material used as the substrate was a 310 stainless steel plate with a thickness of 6.14 mm, 75 mm in width, and 150 mm in length. During the additive manufacturing process, the deposition torch was positioned at 90° relative to the track deposition direction (Y-axis).

2.1. Microstructural Characterization

Scanning electron microscopy (SEM) and energy dispersive spectroscopy (EDS) analyses were performed using a Zeiss EVO MA 10 microscope (São Paulo, Brazil) operated at 10 kV with a tungsten filament. To reveal the phases present in the samples, Vilella's reagent was applied for 180 s for both SEM and optical microscopy (OM) analyses. Optical microscopy (OM) was conducted using an Olympus SC30 microscope (São Paulo, Brazil). The samples were sectioned using a metallographic cutter, brand Struers (Milton, Australia), model MESOTOM. Subsequently, the samples were extracted along the deposition track and underwent grinding with different grit sizes (80, 120, 200, 320, 400, 600, and 1200). After grinding, the samples were polished using a polisher, brand Fortel (Willenhall, UK), model PFL, with Fortel alumina polishing suspension of 1-micron particle size diluted in water.

2.2. Tensile Testing

Tensile tests were conducted at a crosshead speed of 5 mm/min using an InterMetric GR173 testing machine (São Paulo, Brazil) with a maximum capacity of 200 kN. The specimens to be tested were machined using a ROMI U30 milling machine (São Paulo, Brazil). After milling, to reduce the surface roughness, the specimens were subjected to grinding using a Mello P36 surface grinder (São Paulo, Brazil). Following machining, the specimens were cut according to the ASTM E8 standard using an AMADA FLC 3015 AJ laser cutting machine (Kanagawa, Japan). To evaluate the anisotropy of the WAAM-processed material, tensile specimens were extracted in three orientations relative to the deposition track direction: 0°, 45°, and 90°. For each orientation, the tests were performed in triplicate to ensure statistical reliability of the mechanical behavior assessment.

2.3. Optical Emission Spectrometry (OES)

For the optical emission spectrometry (OES) analysis, a BRUKER Q2 ION model spectrometer (Billerica, MA, USA) was used, operating at a power of 400 Watts for 30 s. The sample was initially sectioned using a metallographic cutter, brand Struers, model MESOTOM. Subsequently, it was machined using a Romi U30 milling machine (São Paulo, Brazil) and then ground with a Mello P36 surface grinder (São Paulo, Brazil) in order to reduce surface roughness.

2.4. Characterization of Pore Distribution

The sample was initially prepared using a metallographic cutter from Struers, model MESOTOM. The samples underwent a progressive grinding process using abrasive papers with grit sizes of 80, 120, 200, 320, 400, 600, and 1200. Subsequently, the samples were polished using a polisher from Fortel, model PFL, employing a Fortel alumina polishing suspension with 1-micron particle size diluted in water. After the surface preparation, the sample was analyzed using an Olympus SC30 optical microscope (São Paulo, Brazil), without chemical etching, aiming exclusively to reveal the presence of pores. The quantification of porosity was performed using the software PRECiV Pro 1.2, which enabled the determination of the distribution and amount of pores in the sample.

3. Results and Discussion

3.1. Metallographic Analysis Using an Optical Microscope on 310 Austenitic Steel

The columnar grain growth observed in Figures 3 and 4 is typical of arc-based deposition processes, due to the strong thermal gradient and the preferential solidification direction [1]. This microstructure can induce anisotropy in the mechanical properties, as supported by the tensile test results presented. The presence of grain boundary segregations, suggested by contrast variations and EDS analyses, may negatively affect the corrosion resistance and ductility of the material, as reported by [66,67].

Compared to other studies involving austenitic stainless steel fabricated by the WAAM process [68–70], our results confirm that the thermal and cooling conditions during deposition are decisive for the resulting microstructure, highlighting the importance of precise control of welding parameters to optimize component properties. In particular, recent studies estimate that the cooling rates during WAAM of AISI 310 stainless steel can range from approximately 200 to 1130 K/s, depending on the process parameters and geometry, which is consistent with the refined and columnar dendritic structures observed in the present study [1].

Figure 3. Metallographic analysis of 310 austenitic stainless steel produced by Wire and Arc Additive Manufacturing (WAAM) at 200× magnification.

Figure 4. Metallographic analysis of 310 austenitic stainless steel produced by Wire and Arc Additive Manufacturing (WAAM) at 1000× magnification.

More specifically, microstructural analyses using optical microscopy and scanning electron microscopy, focusing on the interlayer regions, revealed features that contribute to the observed anisotropic mechanical behavior. These include localized porosity and micropores along the interfaces between successive layers, partial lack of fusion between adjacent weld beads, interlayer oxide films detected by compositional contrast and EDS, localized microsegregation of elements, especially nickel depletion, and discontinuities in dendritic grain boundaries formed during directional solidification. Additionally, the relatively uniform size of the columnar grains directly influences the observed ductile fracture mode, promoting energy absorption and the formation of dimples on the fracture surface. These interfacial and microstructural features act as critical points for crack initiation and propagation, weakening mechanical cohesion and explaining the reduction in tensile strength and ductility in samples tested parallel to the deposition direction.

3.2. Scanning Microscope Metallographic Analysis of 310 Austenitic Steel

The SEM micrograph of AISI 310 stainless steel produced by WAAM, presented in Figure 5, reveals a relatively uniform matrix with the presence of microporosity and localized granular agglomerates. These features are typical of the overlapping thermal cycles and rapid solidification conditions inherent to arc-based additive manufacturing processes. The presence of porosity and microstructural heterogeneity may compromise the mechanical performance and durability of the component, particularly in corrosive environments. Previous studies have shown that reducing porosity is closely associated with the optimization of welding parameters, such as current, deposition speed, and bead overlap [71].

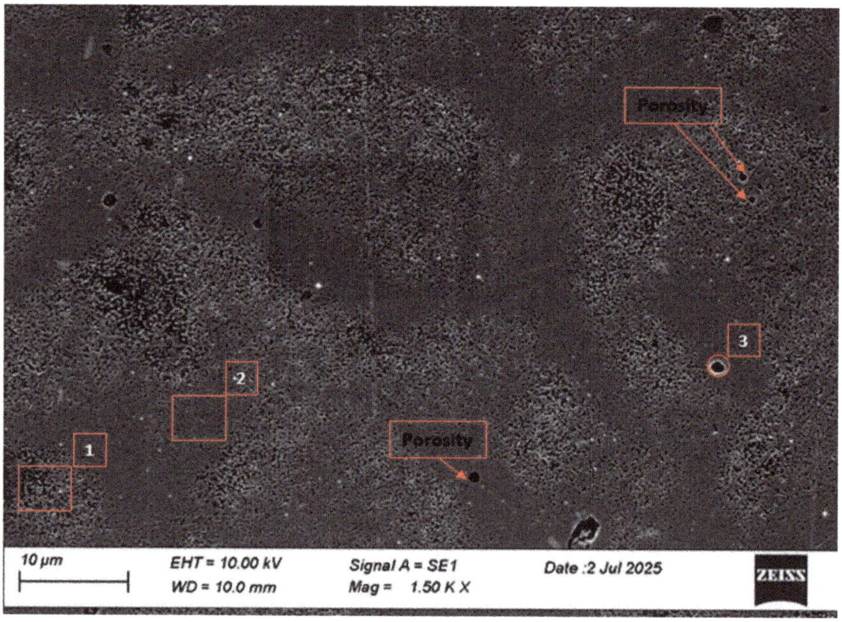

Figure 5. SEM metallographic analysis of 310 austenitic stainless steel fabricated by Wire and Arc Additive Manufacturing (WAAM).

The EDS analysis performed at Point 1 (Figure 6), located in a visually homogeneous region of the matrix, identified the elements Fe (iron), Cr (chromium), Ni (nickel), Mn (manganese), and Si (silicon), all consistent with the nominal composition of 310 stainless steel. The prominent peaks of Fe, Cr, and Ni confirm that the analyzed matrix is predominantly an austenitic solid solution. However, the presence of Si at detectable levels, although expected (as silicon acts as a deoxidizer and ferrite stabilizer), may also suggest a certain degree of segregation during solidification. This behavior has been previously reported in studies on stainless steels produced by WAAM, where thermal gradients and constant reheating promote element partitioning between dendritic regions or along grain boundaries [63,72–74]. Although the analyzed point was not directly located within such segregated zones, the slight increase in silicon and manganese suggests that even apparently homogeneous areas may exhibit chemical composition fluctuations, which can affect localized corrosion resistance and mechanical performance.

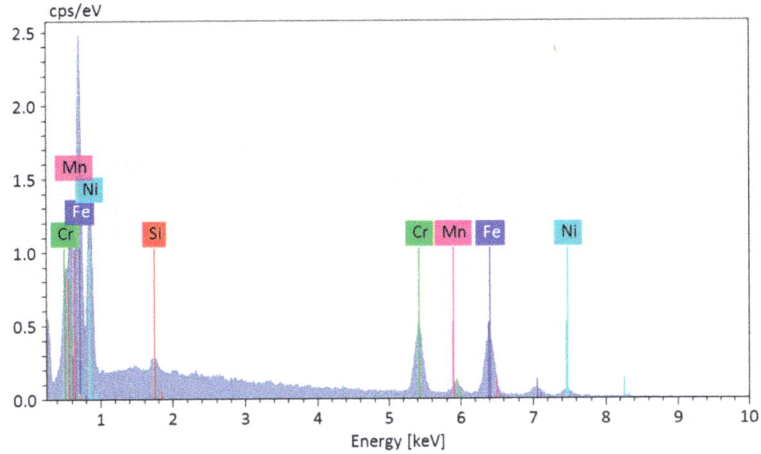

Figure 6. EDS analysis at point 1 for 310 austenitic stainless steel produced by Wire and Arc Additive Manufacturing (WAAM).

The point EDS analysis conducted at Region 2 (Figure 7) revealed the presence of key alloying elements characteristic of 310 stainless steel, such as iron (Fe), chromium (Cr), and nickel (Ni), as well as manganese (Mn) and silicon (Si). The high Cr and Ni contents are essential for stabilizing the austenitic phase and ensuring oxidation and corrosion resistance at elevated temperatures [26,53]. The detection of Si, albeit at lower intensity, may be associated with the use of deoxidizing elements during wire production or with the formation of surface silicon oxide inclusions, which may be confirmed through elemental mapping analyses. The observed surface texture, with contrast variations, indicates potential microsegregation of alloying elements, especially Cr and Ni, among different solidification regions. According to Scudino et al. [75] and Jin et al. [76], this behavior is typical of WAAM processes, where multiple thermal cycles promote elemental redistribution and the formation of zones with distinct metallographic morphologies.

Figure 8 presents the EDS analysis corresponding to Point 3. While Fe and Cr were again identified as the predominant elements consistent with the austenitic matrix of the alloy, a significant reduction in Ni concentration was observed. This discrepancy may be attributed to the low excitation efficiency of Ni peaks under the analysis conditions, the local surface topography, or more likely, to microsegregation or local nickel depletion during solidification. Such behavior has been reported by Keller et al. [77] and Nedjad et al. [78], who observed localized Ni depletion in specific regions of austenitic stainless steels deposited by WAAM, with direct implications for phase stability. Despite the uncertainty in Ni quantification at Point 3, the detection of Mn and Si in typical proportions suggests that the austenitic matrix was generally preserved. Nevertheless, the chemical heterogeneity observed, particularly in Ni distribution, may compromise intergranular corrosion resistance and negatively affect local mechanical properties. These variations highlight the need for complementary strategies, such as post-deposition heat treatments and optimization of welding parameters, to reduce interlayer chemical gradients and promote greater microstructural uniformity [79,80]. Therefore, the results obtained are consistent with several studies in the literature that describe the metallurgical challenges associated with the fabrication of stainless steels via the WAAM process. The observations confirm the presence of a typical austenitic microstructure; however, local compositional variations and process-inherent defects such as porosity and segregation were also identified. These factors must be carefully considered during the design and qualification of critical components manufactured through Wire and Arc Additive Manufacturing.

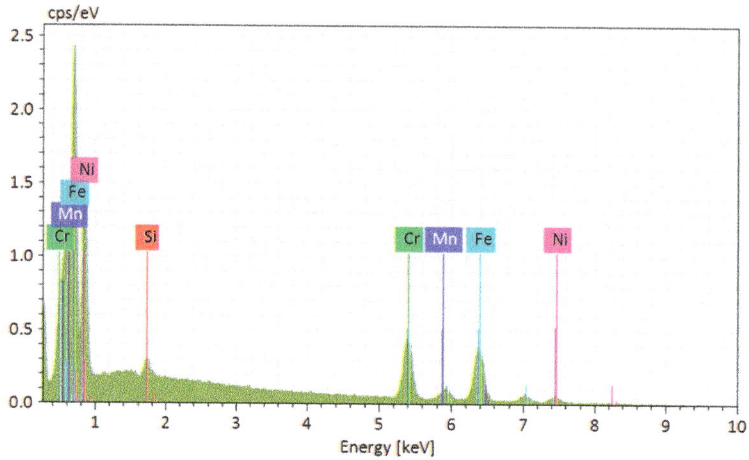

Figure 7. EDS analysis at point 2 for austenitic 310 stainless steel fabricated by Wire and Arc Additive Manufacturing (WAAM).

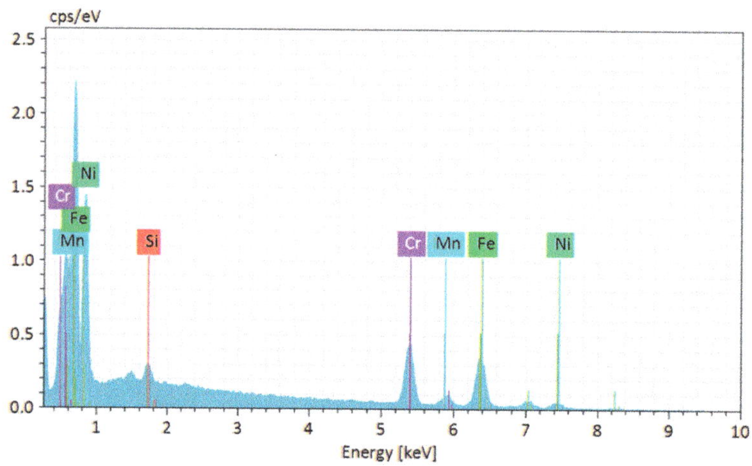

Figure 8. EDS analysis at point 3 for austenitic 310 stainless steel produced by Wire and Arc Additive Manufacturing (WAAM).

Figure 9 presents the elemental mapping obtained by energy-dispersive X-ray spectroscopy (EDS) of the surface of AISI 310 austenitic stainless steel produced by Wire and Arc Additive Manufacturing (WAAM), highlighting the distribution of chromium (Cr), iron (Fe), and nickel (Ni). The analysis reveals a homogeneous distribution of the main alloying elements within the mapped region, indicating good chemical uniformity at the micrometric scale. Chromium, a key element in the formation of the passive film and, consequently, in corrosion resistance, was uniformly dispersed across the analyzed area. This suggests the absence of chromium-depleted zones, which could otherwise compromise resistance to localized corrosion. Iron, the primary matrix element, also displayed a uniform distribution, supporting the structural stability of the alloy. Although the nickel mapping showed slight intensity variations, no significant segregation or localized accumulation of the element was observed. This observation aligns with the point EDS results, in which nickel quantification yielded slightly underestimated values in some regions; however, when viewed along with the mapping data, these deviations appear to result from local analytical limitations rather than actual microsegregation.

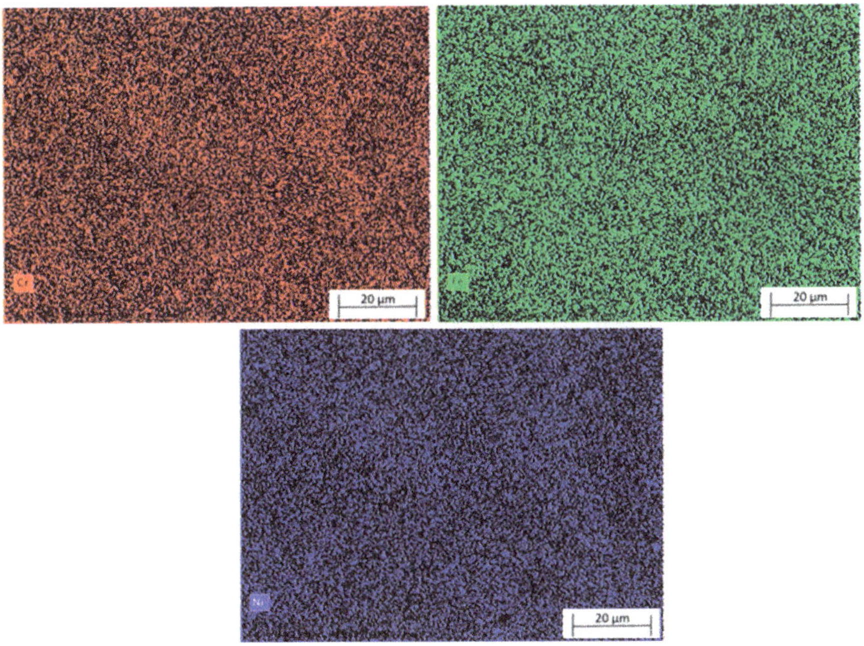

Figure 9. Scanning electron microscopy analysis of austenitic 310 stainless steel fabricated by Wire and Arc Additive Manufacturing (WAAM).

The observed compositional homogeneity indicates that, despite the intrinsic anisotropies of the WAAM process, such as columnar textures or interlayer thermal gradients, the deposition process ensured effective fusion and mixing between adjacent weld beads. Similar findings were reported by Rahimi et al. [63,76], who demonstrated that, despite the thermal complexities inherent to WAAM, the distribution of Cr and Ni remains stable when process parameters are properly optimized. Therefore, the EDS mapping results confirm the preservation of the austenitic matrix and the overall chemical uniformity of the material, both of which are critical for the mechanical performance and corrosion resistance of WAAM-deposited 310 stainless steel.

3.3. Mapping of Porosity in the Sample

Quantitative analysis of the micrograph revealed a total porosity of approximately 0.89% in the AISI 310 stainless steel sample produced by WAAM. Although this value falls within the range commonly reported for Wire and Arc Additive Manufacturing processes, its presence is not negligible and can significantly influence the mechanical behavior of the material. The pores act as stress concentrators and potential crack initiation sites, especially under tensile loading, contributing to the reduction in ductility and tensile strength. This influence is particularly relevant in the longitudinal direction (0°), where tensile test results showed the lowest elongation (12.5%) and lowest yield strength (198 MPa). The non-uniform distribution of pores along the deposited layers can also intensify the anisotropic characteristics observed, as pore morphology, orientation, and position relative to interlayer boundaries affect each loading direction differently. Figure 10 illustrates the distribution of pores in the analyzed region, reinforcing the microstructural heterogeneity introduced by the WAAM process. Therefore, the porosity identified in this study can be considered a contributing factor to the localized degradation of mechanical properties and the typical anisotropy found in WAAM-produced components.

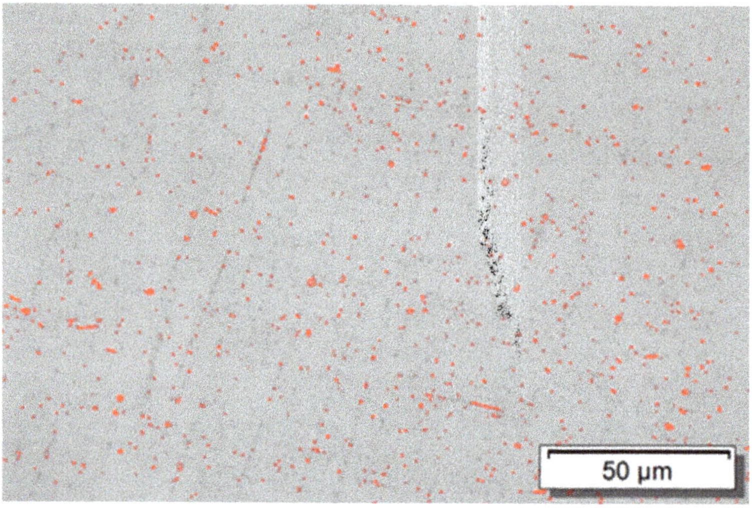

Figure 10. Porosity distribution in the sample produced by Wire and Arc Additive Manufacturing (WAAM).

3.4. Optical Emission Spectrometry for the Chemical Characterization of AISI 310 Stainless Steel

The chemical composition of the AISI 310 stainless steel sample was determined by optical emission spectrometry (OES), as presented in Table 3. The results confirm typical contents for this type of steel, with approximately 26.70 wt.% Cr, 20.10 wt.% Ni, 1.32 wt.% Mn, and the remainder predominantly composed of Fe (51.14 wt.%), in addition to minor fractions of Si (0.476 wt.%), Mo (0.14 wt.%), and C (0.12 wt.%). This analysis was carried out to quantitatively and reliably validate the data obtained from energy-dispersive X-ray spectroscopy (EDS), which, due to its intrinsic limitations, may lack quantitative accuracy, particularly for light or low-concentration elements such as nickel (Ni) and carbon (C).

Table 3. Chemical composition of the part made of AISI 310 stainless steel.

Elements	Values
Carbon (C)	0.120
Silicon (Si)	0.476
Manganese (Mn)	1.320
Chromium (Cr)	26.70
Nickel (Ni)	20.10
Molybdenum (Mo)	0.140
Iron (Fe)	51.144

In the specific case of Point 3 in the EDS analysis (Figure 8), although the austenitic matrix was confirmed by the predominance of Fe and Cr, a significant reduction in Ni content was observed. This discrepancy may be attributed to inefficient excitation of Ni peaks under EDS conditions, to local surface topography, or more likely, to microsegregation or localized nickel depletion during solidification, as previously reported in the literature (Keller et al. [77]; Nedjad et al. [78]). When these EDS results are compared to those obtained via OES, which indicate a bulk Ni content of approximately 20.10 wt.%, in line with AISI 310 specifications, it is possible to infer that the overall chemical matrix is preserved, while the local variations detected by EDS reflect the layer-wise deposition characteristics of the WAAM process, such as thermal gradients and partial remelting between adjacent tracks.

3.5. Mechanical Tensile Test on 310 Austenitic Steel

Figure 11 presents the tensile test results for AISI 310 austenitic stainless steel specimens fabricated via WAAM, tested in three orientations relative to the deposition direction: 0°, 45°, and 90°. The values shown correspond to the average of three specimens per condition, highlighting the anisotropic mechanical behavior typically observed in components produced by layer-by-layer additive manufacturing processes [81,82]. Samples tested at 0°, parallel to the deposition direction, exhibited the lowest tensile strength, with average values around 320 MPa and reduced final elongation, suggesting that the interfaces between successive layers act as weak zones.

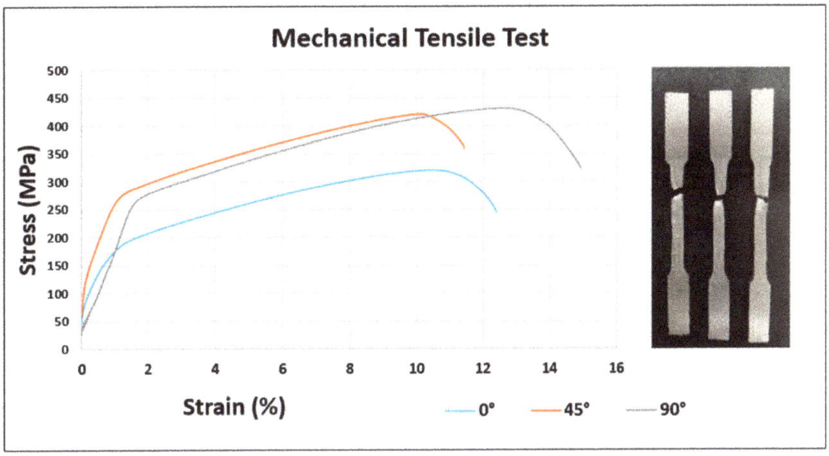

Figure 11. Mechanical tensile test at 0°, 45°, and 90° of specimens fabricated by Wire and Arc Additive Manufacturing (WAAM).

Microstructural analyses performed by optical microscopy (OM) and scanning electron microscopy (SEM), focused on the interlayer regions, revealed features that explain this fragility, including localized porosity and micropores, partial lack of fusion between adjacent beads, and localized elemental microsegregation, especially of nickel. Additionally, discontinuities were observed in the grain boundaries of the dendritic columns formed during directional solidification, constituting preferential sites for crack initiation and propagation. These microstructural particularities act as critical points that compromise mechanical cohesion and explain the lower strength and ductility in specimens tested parallel to the deposition direction.

This behavior is widely reported in the literature for metallic alloys produced by WAAM, where lack of fusion or interlayer oxidation can compromise longitudinal cohesion [83,84]. Conversely, specimens tested at 90°, perpendicular to the deposition direction, reached higher tensile strength values (up to 430 MPa) and exhibited elongations exceeding 15%, indicating greater structural integrity in the transverse direction. This improved performance is attributed to the morphology of dendritic columns formed during directional solidification, which, in this orientation, provides greater resistance to crack propagation [85].

Specimens oriented at 45° showed intermediate mechanical behavior in terms of both ultimate tensile strength and elongation. This gradient of properties reinforces the importance of loading orientation in WAAM-fabricated structural components, as mechanical anisotropy can impact the predictability of in-service performance. Despite the inherent ductility of 310 stainless steel, the results clearly demonstrate that the final mechanical performance is significantly influenced by the manufacturing direction. Similar findings have been reported by Wang et al. [83], Rodriguez et al. [86], and Laghi et al. [87], who showed that tensile strength and toughness can vary depending on the testing direction in WAAM-produced parts. Therefore, for critical applications, especially those subjected to multiaxial loading, proper control of component orientation relative to the load direction is essential.

Table 4 summarizes the mechanical behavior of the WAAM-produced AISI 310 stainless steel in three different specimen orientations (0°, 45°, and 90°) relative to the deposition direction. A significant anisotropy is observed in the mechanical properties. The highest yield strength (280 ± 5.13 MPa) was obtained at 45°, while the lowest (198 ± 7.20 MPa) occurred at 0°, which is below the minimum standard requirement of 205 MPa for AISI 310 [88]. Ultimate tensile strength also varied with orientation, reaching a maximum of 430 ± 12.06 MPa at 90°, but remaining below the nominal requirement of 520 MPa in all positions. Elongation followed a similar trend, with the highest ductility (15 ± 1.69%) in the 90° direction, although still significantly lower than the expected 40%. The standard deviations observed, particularly in tensile strength and elongation, reflect the inherent variability of the WAAM process and further support the evidence of anisotropy. These results highlight the directional dependence of the mechanical performance in WAAM-produced components, likely due to thermal cycling and microstructural variations inherent to the process.

Table 4. Tensile test results.

Variables: Position	**0°**	**45°**	**90°**
Yield strength (MPa)	198 ± 7.20	280 ± 5.13	269 ± 10.21
Ultimate tensile strength (MPa)	320 ± 4.16	419 ± 12.66	430 ± 12.06
Strain (%)	12.5 ± 0.98	11.5 ± 2.72	15 ± 1.69

To provide additional context and support a broader comparison, Table 5 presents tensile test results from previous studies involving different austenitic stainless steels fabricated by WAAM, including alloys 304, 308L, and 316L. These data help frame the performance of WAAM AISI 310 within the existing literature and emphasize the need to understand alloy-specific responses and anisotropy patterns.

Table 5. Tensile test results for different austenitic stainless steels produced by WAAM.

Reference	Alloy	Tested Orientations	Ys (MPa)	UTS (MPa)	Elongation (%)
Gowthaman et al., 2023 [89] (WAAM)	316L	Horizontal (0°) Vertical (90°)	302 295	576 469	35 39
Anand et al., 2023 [90] (WAAM)	308L	Horizontal (0°)	364	548	38
Rani et al., 2022 [91] (WAAM)	304	Horizontal (0°) Vertical (90°)	- -	600 597	- -
ASM Handbook 1990 [92] (Sheet)	310	Horizontal (0°) Vertical (90°)	240 240	570 600	40 46
ASM Handbook 1990 [92] (Forging)	310L	Vertical (90°)	260	585	54

Compared to the literature summarized in Table 5, the mechanical properties of the WAAM-produced AISI 310 stainless steel in this study are generally lower, especially in terms of elongation and ultimate tensile strength (UTS). While other austenitic stainless steels such as 316L and 308L exhibit elongation values above 35% and UTS exceeding 540 MPa, the 310 alloy produced via WAAM in this study achieved a maximum elongation of only 15 $\pm$ 1.69% and a UTS of 430 $\pm$ 12.06 MPa (in the vertical orientation).

Additionally, when comparing these results to the standard values reported for laminated sheet and forged AISI 310 (ASM Handbook, 1990) [92], a similar trend is observed. The sheet-processed AISI 310 exhibits yield strength (YS) of 240 MPa, UTS of 570–600 MPa, and elongation of 40–46%, while the forged version of 310L shows even higher ductility, with elongation reaching 54%. In contrast, the WAAM-fabricated samples presented YS ranging from 198 $\pm$ 7.20 MPa (0°) to 280 $\pm$ 5.13 MPa (45°) and elongation between 11.5 $\pm$ 2.72% (45°) and 15 $\pm$ 1.69% (90°).

This discrepancy may be related to several microstructural and processing factors, including the presence of pores detected in the microstructural analysis, interlayer interfaces, microstructural heterogeneities, and potential elemental segregation during solidification. Furthermore, the limited and discontinuous presence of δ-ferrite, combined with clean γ-dendritic grain boundaries, may contribute to localized ductility loss in specific orientations. These results highlight the influence of alloy-specific solidification behavior and thermal cycling in the WAAM process on the anisotropic mechanical response of the components, indicating the need for further process optimization in the case of AISI 310.

4. Conclusions

Based on the results obtained in this study, it is concluded that the austenitic stainless steel 310 manufactured via Wire and Arc Additive Manufacturing (WAAM) exhibits a microstructure characteristic of directional solidification, with prominent columnar grain morphology, dispersed porosity, and visible layer overlap marks. Energy-dispersive X-ray spectroscopy (EDS), through point and mapping analyses, confirmed a global chemical composition consistent with the standard alloy, showing a homogeneous distribution of Fe, Cr, and Ni. However, a slight underestimation of nickel content was observed in specific regions, likely due to local limitations of the point analysis, which highlights the importance of mapping for a more representative and accurate compositional assessment. Optical emission spectrometry (OES) was employed for quantitative validation of the overall chemical composition, confirming a nickel content around 20.10 wt.%, in agreement with the nominal composition of AISI 310 stainless steel. This complementary analysis reinforces the reliability of the chemical characterization and the preservation of the austenitic matrix, despite the local fluctuations detected by EDS.

Tensile tests revealed anisotropic mechanical behavior, with lower strength and elongation in samples extracted parallel to the deposition direction (0°), attributed to the presence of interlayer interfaces, which may be weakened by lack of fusion or porosity. Specifically, specimens tested at 0° presented the lowest yield strength (198 ± 7.20 MPa), ultimate tensile strength (320 ± 4.16 MPa), and elongation (12.5 ± 0.98%). In contrast, the best mechanical performance was observed at 90°, with higher values of yield strength (269 ± 10.21 MPa), tensile strength (430 ± 12.06 MPa), and ductility (15 ± 1.69%). These values, along with the obtained standard deviations, indicate not only the directional dependence of mechanical properties but also the variability inherent to the WAAM process. The highest yield strength (280 ± 5.13 MPa) was unexpectedly observed at 45°, suggesting a complex interaction between grain orientation and stress direction. This behavior is consistent with the literature reports describing the influence of columnar grain orientation and the nature of layer interfaces on the final mechanical response of components produced by Wire and Arc Additive Manufacturing (WAAM).

Furthermore, the metallographic analysis performed without chemical etching allowed the exclusive revelation of pores in the sample. The porosity quantification showed a total value of approximately 0.89%. This porosity directly impacts the tensile test results, especially in samples with orientation parallel to the deposition, contributing to the reduction in mechanical properties due to the presence of internal discontinuities.

Overall, the results confirm that AISI 310 stainless steel manufactured by Wire and Arc Additive Manufacturing (WAAM) maintains chemical composition and mechanical performance suitable for industrial applications, provided that its specific microstructural features and mechanical anisotropy are properly considered in the design and structural assessment of components subjected to multidirectional loading. The potential development of residual stresses inherent to the WAAM process and their implications for dimensional stability are also acknowledged and warrant further investigation. It is recommended to perform X-ray diffraction (XRD) analyses for precise determination of the phases and crystallographic structures present, as well as accelerated corrosion tests to evaluate the material's resistance under severe service conditions, aiming for a more comprehensive understanding of the long-term performance of AISI 310 stainless steel components produced by WAAM.

When comparing the results obtained in this study with the values presented in Table 5, it is observed that the mechanical properties of the 310 stainless steel produced by WAAM were lower than those reported for similar austenitic alloys fabricated by the same process, such as 316L and 308L, as well as lower than the properties of 310 steels processed by conventional routes, such as rolling and forging. In particular, the ultimate tensile strength and ductility showed reduced values, with a maximum elongation of only 15% in the 90° orientation. This difference can be attributed to the presence of interlayer porosity, weak interfaces between deposited layers, microstructural heterogeneities, and possible elemental segregation during solidification, which negatively affect mechanical performance. A limitation of the present study is the absence of corrosion resistance and high-temperature performance analyses, which are essential aspects for the industrial application of the material. Therefore, future investigations should prioritize these topics to broaden the understanding of the durability and reliability of 310 stainless steel produced by Wire and Arc Additive Manufacturing.

Author Contributions: Conceptualization, A.C. and C.d.F.M.; methodology, H.C.C. and A.C.; software, P.H.M.P.; validation, A.D. and C.d.F.M.; formal analysis, C.d.F.M. and J.C.; investigation, A.C. and H.C.C.; resources A.D.; data curation, P.H.M.P.; writing—original draft preparation, A.C. and H.C.C.; writing—review and editing, C.d.F.M. and J.C.; visualization, A.D.; supervision, C.d.F.M.;

project administration, A.D. and A.C.; funding acquisition A.D. and J.C. All authors have read and agreed to the published version of the manuscript.

Funding: This research was funded by Fundação de Amparo à Pesquisa e Inovação do Estado de Santa Catarina grant number Termo de Outorga Nº: 2021TR2276.

Data Availability Statement: The original contributions presented in this study are included in the article. Further inquiries can be directed to the corresponding author.

Acknowledgments: The authors gratefully acknowledge the Centro Universitário UNISATC for providing access to laboratory facilities that supported this research. The authors also thank the Fundação de Amparo à Pesquisa e Inovação do Estado de Santa Catarina (FAPESC) for financial support in the acquisition of equipment and materials essential to this study.

Conflicts of Interest: The authors declare no conflict of interest.

References

1. Rahimi, A.; Yazdizadeh, M.; Ara, M.V.; Pouranvari, M. CMT wire-arc additive manufacturing of 310 austenitic stainless steel: Microstructure-properties relationships. *J. Mater. Res. Technol.* **2025**, *35*, 881–891. [CrossRef]
2. Moreira, É.A.; Bomfim, M.H.S.; Liberato, F.M. A evolução da fabricação: Comparação e sinergia entre manufatura aditiva e subtrativa. *Res. Soc. Dev.* **2005**, *14*, e9114548869. [CrossRef]
3. Cunningham, C.R.; Flynn, J.M.; Shokrani, A.; Dhokia, V.; Newman, S.T. Invited review article: Strategies and processes for high quality wire arc additive manufacturing. *Addit. Manuf.* **2018**, *22*, 672–686. [CrossRef]
4. Zhang, T.; Li, H.; Gong, H.; Wu, H.; Ahmad, A.S.; Chen, X. Effect of rolling force on tensile properties of additively manufactured Inconel 718 at ambient and elevated temperatures. *J. Alloys Compd.* **2021**, *884*, 161050. [CrossRef]
5. Frazier, W.E. Metal Additive Manufacturing: A Review. *J. Mater. Eng. Perform.* **2014**, *23*, 1917–1928. [CrossRef]
6. Lee, J.; Lee, C.; Kim, D. Repair of damaged parts using wire arc additive manufacturing in machine tools. *J. Mater. Res. Technol.* **2022**, *16*, 13–24. [CrossRef]
7. Jeffus, L.F. *Welding and Applications*; Cengage Learning: Boston, MA, USA, 1997.
8. Zhang, C.; Li, Y.; Gao, M.; Zeng, X. Wire arc additive manufacturing of Al-6Mg alloy using variable polarity cold metal transfer arc as power source. *Mater. Sci. Eng. A* **2018**, *711*, 415–423. [CrossRef]
9. Rodrigues, T.A.; Duarte, V.; Miranda, R.M.; Santos, T.G.; Oliveira, J. Current Status and Perspectives on Wire and Arc Additive Manufacturing (WAAM). *Materials* **2019**, *12*, 1121. [CrossRef]
10. Campatelli, G.; Montevecchi, F.; Venturini, G.; Ingarao, G.; Priarone, P.C. Integrated WAAM-subtractive versus pure subtractive manufacturing approaches: An energy efficiency comparison. *Int. J. Precis. Eng. Manuf. Technol.* **2020**, *7*, 1–11. [CrossRef]
11. Tabernero, I.; Paskuala, A.; Álvarezb, P.; Suárezcb, A. Study on Arc Welding processes for High Deposition Rate Additive Manufacturin. In Proceedings of the CIRP Conference on Electro Physical and Chemical Machinin, Bilbao, Spain, 20 April 2018; Volume 68, pp. 358–362.
12. Yan, L.; Chen, Y.; Liou, F. Additive manufacturing of functionally graded metallic materials using laser metal deposition. *Addit. Manuf.* **2020**, *31*, 100901. [CrossRef]
13. Ferreira, C.; Schaeffer, L.; Daleffe, A.; Castelan, J.; Casagrande, H.; March, G.M.A. Hybrid Manufacturing Process and Hot Forging: Preform Manufacturing Through Localized Melting Material Deposition Using UTP AF DUR 600 Wire. *RGSA J. Soc. Environ. Manag.* **2025**, *19*, e012427. [CrossRef]
14. Casagrande, H.C.; Daleffe, A.; Ferreira, C.A.; Milanez, A.; March, G.; Schaeffer, L. Additive Manufacturing in the Forging Process: Preform Manufacturing Through Localized Melting Material Deposition Using Low Carbon Wire. *RGSA J. Soc. Environ. Manag.* **2025**, *19*, e012422. [CrossRef]
15. Casagrande, H.C.; Daleffe, A.; Ferreira, C.; Fritzen, D.; March, G.; Castelan, J. Processo de fabricação de peças metálicas por manufatura aditiva com fusão localizada de aços baixa liga. *Open Sci. Res.* **2023**, *11*, 997–1008.
16. Li, K.; Chen, W.; Gong, N.; Pu, H.; Luo, J.; Zhang, D.Z.; Murr, L.E. A critical review on wire-arc directed energy deposition of high-performance steels. *J. Mater. Res. Tecnol.* **2023**, *24*, 9369–9412. [CrossRef]
17. Kumar, S.A.; Prasad, R.V.S. Chapter 2—Basic Principles of Additive Manufacturing: Different Additive Manufacturing Technologies. In *Additive Manufacturing. A Tool for Industrial Revolution 4.0*; Woodhead Publishing Reviews: Mechanical Engineering Series; Elsevier: Amsterdam, The Netherlands, 2021; pp. 17–35.
18. *ASTM international F2792-12A*; Standard Terminology for Aditive Manufacturing Technologies. ASTM International: West Conshohocken, PA, USA, 2012; pp. 918–928.
19. Uralde, V.; Veiga, F.; Aldalur, E.; Suarez, A.; Ballesteros, T. Symmetry and Its Application in Metal Additive Manufacturing (MAM). *Symmetry* **2022**, *14*, 1810. [CrossRef]

20. Korkut, I.; Kasap, M.; Ciftci, I.; Seker, U. Determination of optimum cutting parameters during machining of AISI 304 austenitic stainless steel. *Mater. Des.* **2004**, *25*, 303–305. [CrossRef]
21. Pixer, F.; Arnoldt, A.; Unger, M.; Scneider-Broskamp, C.; Bharadwaj, K.; Mayrhofer, F.; Gradinger, R.; Klein, T. On the viability of in-situ alloying via process gas mixtures in wire arc directed energy deposition of austenitic stainless steel. *J. Mech. Work. Technol.* **2025**, *337*, 118738. [CrossRef]
22. Kou, S. *Welding Metallurgy*, 2nd ed.; John Wiley & Sons Incorporated: Hoboken, NJ, USA, 2003.
23. Lippold, J.C. *Welding Metallurgy and Weldability*; Ohio State University: Columbus, OH, USA, 2015.
24. Li, J.; Yi, M.; Wu, H.; Fang, Q.; Liu, Y.; Liu, B.; Zhou, K.; Liaw, P.K. Fine-grain-embedded dislocation-cell structures for high strength and ductility in additively manufactured steels. *Mater. Sci. Eng. A* **2020**, *790*, 139736. [CrossRef]
25. Gardner, L.; Talja, A.; Baddoo, N.R. Structural design of high-strength austenitic stainless steel. *Thin-Walled Struct.* **2006**, *44*, 517–528. [CrossRef]
26. *ASM Specialty Handbook: Stainless Steels*; ASM International: Materials Park, OH, USA, 1994.
27. Sousa, J.M.B.A. Influência dos elementos de liga nos aços inoxidáveis duplex e superduplex utilizados na indústria do petróleo: Uma revisão. Trabalho de Conclusão de Curso apresentado ao curso de Engenharia Metalúrgica do Departamento de Engenharia Metalúrgica e de Materiais da Universidade Federal do Ceará. 2021. Available online: https://repositorio.ufc.br/bitstream/riufc/60634/3/2021_tcc_jmbasousa.pdf (accessed on 27 June 2025).
28. Souza, V.P.; Labiapari, W.S.; Lins, V.F.C. Stainless steels as a solution for corrosion and erosion problems involving grains in agribusiness sector applications. *J. Mater. Res. Technol.* **2024**, *30*, 5605–5621. [CrossRef]
29. Yadav, A.; Srivastava, M.; Jain, K. Design and fabrication of wire arc additive manufacturing setup and enhanced tailored properties of dissimilar steel additively deposited by WAAM process. *Estruturas* **2025**, *72*, 108228. [CrossRef]
30. Blakey-Milner, B.; Gradl, P.; Snedden, G.; Brooks, M.; Pitot, J.; Lopez, E.; Leary, M.; Berto, F.; Plessis, A. Metal additive manufacturing in aerospace: A review. *Mater. Des.* **2021**, *209*, 110008. [CrossRef]
31. Kanni, R.A. On High-Temperature Matersials: A Case on Creep and Oxidation of a Fully Austenitic Heat-Resistant Super-alloy Stainless Steel Sheet. *J. Mater.* **2013**, *1*, 124649.
32. Liao, H.; Wang, Z.; Chi, P.; Zhang, B.; Ding, T.; Zhangc, Q. Evolutions of microstructure and mechanical property of 308L stainless steel repaired by the local dry underwater wire arc additive manufacturing. *Mater. Sci. Eng. A* **2024**, *898*, 146365. [CrossRef]
33. Chamim, M.; Darmadi, D.B.; Purnowidodo, A.; Widodo, T.D.; Ismail, Z. Influence of the welding thermal cycle on δ-ferrite evolution in the first layer of austenitic stainless steel (ASS) 308L produced by WAAM-GTAW. *Case Stud. Therm. Eng.* **2024**, *64*, 105489, Erratum in *Case Stud. Therm. Eng.* **2024**, *64*, 105575. [CrossRef]
34. Niu, F.; Bi, W.; Zhang, K.; Sun, X.; Ma, G.; Wu, D. Additive manufacturing of 304 stainless steel integrated component by hybrid WAAM and LDED. *Mater. Today Commun.* **2023**, *35*, 106227. [CrossRef]
35. Andrade, D.G.; Tankova, T.; Zhu, C.; Silva, L.S.; Rodrigues, D.M. Mechanical properties of 3D printed CMT-WAAM 316LSi stainless steel walls. *J. Constr. Steel Res.* **2024**, *215*, 108527. [CrossRef]
36. Vinoth, V.; Sathiyamurthy, S.; Natarajan, U.; Venkatkumar, D.; Prabhakaran, J.; Prakash, S.K. Examination of microstructure properties of AISI 316L stainless steel fabricated by wire arc additive manufacturing. *Mater. Proc.* **2022**, *66*, 702–706. [CrossRef]
37. Chen, W.; Guo, S.; Xuan, Y.; Xu, J.; Li, S.; Fang, W.; Zhou, Q.; Wang, K. Influence of different deposition strategies on the microstructure and mechanical properties of the laminated heterostructured material with ER130S-G HSS and 316 L SS fabricated by WAAM. *Mater. Charact.* **2024**, *210*, 113838. [CrossRef]
38. Rezende, R.F.; Árias, A.R.; Lima, E.J.; Coelho, F.G.F. Pulsed GMAW-based WAAM–Influence of droplet detachment mode on the geometry and mechanical properties of 308 L stainless steel. *J. Adv. Join. Process.* **2025**, *11*, 100286. [CrossRef]
39. Yadav, A.; Srivastava, M.; Jain, P.K.; Rathee, S. Microstructure transformations and improving wear resistance of austenitic stainless steel additively fabricated by arc-based DED process. *Def. Technol.* **2024**, *38*, 194–204. [CrossRef]
40. Chen, Y.; Zhao, X.; Yang, B.; Liu, Y.; Liang, Y.; Li, Z.; Chen, C. Study on properties of 304 wire arc additive manufacturing stainless steel TIG welded joints. *Mater. Lett.* **2024**, *361*, 136107. [CrossRef]
41. Sandeep, O.S.; Kuriachen, B.; Anantharam, G.S.; Eldose, K.K. Influence of microstructure on the anisotropic tribocorrosion behaviour of WAAM-fabricated SS304L in marine environments. *Tribol. Int.* **2025**, *209*, 110668. [CrossRef]
42. Wang, L.; Xue, J.; Wang, Q. Correlation between arc mode, microstructure, and mechanical properties during wire arc additive manufacturing of 316L stainless steel. *Mater. Sci. Eng. A* **2019**, *751*, 183–190. [CrossRef]
43. Ayan, Y.; Kahraman, N. Investigation of tensile and fatigue properties of an austenitic stainless steel part fabricated by WAAM. *Mater. Chem. Phys.* **2024**, *315*, 128937. [CrossRef]
44. Yao, P.; Lin, H.; Wu, W.; Tang, H. Influence of Duty Ratio and Current Mode on Robot 316L Stainless Steel Arc Additive Manufacturing. *Metals* **2021**, *11*, 508. [CrossRef]
45. Xiao, X.; Yin, Y.; Li, C.; LI, C.; Zhang, K. Influence of thermal behavior on microstructure evolution of TIG arc additive manufacture for 316L stainless steel. *Trans. Mater. Heat. Treat.* **2020**, *41*, 97–105.

46. Zhang, H.; Huang, J.; Liu, C.; Ma, Y.; Han, Y.; Xu, T.; Lu, J.; Fang, H. Fabricating Pyramidal Lattice Structures of 304 L Stainless Steel by Wire Arc Additive Manufacturing. *Materials* **2020**, *13*, 3482. [CrossRef] [PubMed]
47. Guan, K.; Wang, Z.; Gao, M.; Li, X.; Zeng, X. Effects of processing parameters on tensile properties of selective laser melted 304 stainless steel. *Materials* **2013**, *50*, 581–586. [CrossRef]
48. Zhang, D.; Prasad, A.; Bermingham, M.J.; Todaro, C.J.; Benoit, M.J.; Patel, M.N.; Qiu, D.; StJohn, D.; Qian, M.; Easton, M.A. Grain Refinement of Alloys in Fusion-Based Additive Manufacturing Processes. *Metall. Mater. Trans.* **2020**, *51*, 4341–4359. [CrossRef]
49. Liao, H.; Kuang, Y.; Zhou, D.; Zhang, Q.; Wu, X.; Xu, L.; Wang, X.; Wang, Z. Achieving fine-grained structure with larger ferrite fraction via local dry underwater WAAM to enhance corrosion behavior of 308 L stainless steel. *Constr. Build. Mater.* **2024**, *457*, 139510. [CrossRef]
50. Spinasa, A.; Weber, B.; Meng, X.; Zhang, R.; Nitawaki, M.; Gardner, L. Influence of process parameters on the physical and material properties of WAAM steels. *Constr. Build. Mater.* **2025**, *479*, 141078. [CrossRef]
51. Belhadj, M.; Werda, S.; Kromer, R.; Darnis, P. Influence of operating parameters on the mechanical and geometric properties of 316L Stainless Steel structures fabricated by WAAM-CMT. *Procedia CIRP* **2025**, *133*, 585–590. [CrossRef]
52. Wang, F.; Chen, W.; Wang, D.; Hou, H.; Zhao, Y. Phase-field modeling and Experimental investigation for rapid solidification in wire and arc additive manufacturing. *J. Mater. Res. Technol.* **2024**, *28*, 4585–4599. [CrossRef]
53. Gurol, U.; Kocaman, E.; Dilibal, S.; Koçak, M. A comparative study on the microstructure, mechanical properties, wear and corrosion behaviors of SS 316 austenitic stainless steels manufactured by casting and WAAM Technologies. *CIRP J. Manuf. Sci. Technol.* **2023**, *47*, 215–227. [CrossRef]
54. Bertsch, K.M.; Bellefon, G.M.; Kuehl, B.; Thoma, D.J. Origin of dislocation structures in an additively manufactured austenitic stainless steel 316L. *Acta Mater.* **2020**, *199*, 19–33. [CrossRef]
55. Wang, G.; Ouyang, H.; Fan, C.; Guo, Q.; Li, Z.; Yan, W.; Li, Z. The origino f high-density dislocations in additively manufactured metals. *Mater. Res. Lett.* **2020**, *8*, 283–290. [CrossRef]
56. Derekar, K.S. A review of wire arc additive manufacturing and advances in wire arc additive manufacturing of aluminium. *Mater. Sci. Technol.* **2018**, *34*, 895–916. [CrossRef]
57. Mendez-Morales, M.; Jesus, J.S.; Branco, R.; Tankova, T.; Rebelo, C. Fatigue crack growth of untreated and heat-treated WAAM ER70S-6 carbon steel. *Int. J. Fatigue* **2025**, *198*, 109008. [CrossRef]
58. Kun, L.; Ping, Y.; Sindo, K. Solidification Cracking Susceptibility of Stainless Steels: New Test and Explanation. *Weld. Res.* **2020**, *99*, 225–270.
59. Zhang, C.; Gao, M.; Zeng, X. Workpiece vibration augmented wire arc additive manufacturing of high strength aluminum alloy. *J. Mater. Process. Technol.* **2019**, *217*, 85–92. [CrossRef]
60. Siddiqui, N.A.; Muzamil, M.; Jamil, T.; Hussain, G. Heat sources in wire arc additive manufacturing and their impact on macro-microstructural characteristics and mechanical properties—An overview. *Smart Mater. Manuf.* **2025**, *3*, 100059. [CrossRef]
61. Vimal, K.E.K.; Srinivas, M.N.; Rajak, S. Wire arc additive manufacturing of aluminium alloys: A review. *Mater. Today Proc.* **2021**, *41*, 100059. [CrossRef]
62. Strano, G.; Hao, L.; Everson, R.M.; Evans, K.E. Surface roughness analysis, modelling and prediction in selective laser melting. *J. Mater. Process. Technol.* **2013**, *213*, 589–597. [CrossRef]
63. Rahimi, A.; Yazdizadeh, M.; Ara, M.V.; Pouranvari, M. Strength and ductility of additively manufactured 310 austenitic stainless steel via wire-arc directed energy deposition: The role of columnar grain growth and ductility-dip cracking. *Mater. Sci. Eng. A* **2025**, *920*, 147554. [CrossRef]
64. *ASTM E8*; Standard Test Methods for Tension Testing of Metallic Materials. ASTM International: West Conshohocken, PA, USA, 2024.
65. Xiong, J.; Tzempelikos, A.; Bilionis, I.; Awalgaonkar, N.; Lee, S.; Konstantzos, I.; Sadeghi, S.; Karava, P. Inferring personalized visual satisfaction profiles in daylit offices from comparative preferences using a Bayesian approach. *Build. Environ.* **2018**, *138*, 74–88. [CrossRef]
66. Felice, I.; Shen, J.; Barragan, A.F.; Moura, I.A.; Li, B.; Wang, B.; Khodaverdi, H.; Mohri, M.; Schell, N.; Ghafoori, E.; et al. Wire and arc additive manufacturing of Fe-based shape memory alloys: Microstructure, mechanical and functional behavior. *Mater. Des.* **2023**, *231*, 112004. [CrossRef]
67. Alagha, A.; Hussain, S.; Zaki, W. Additive manufacturing of shape memory alloys: A review with emphasis on powder bed systems. *Mater. Des.* **2021**, *204*, 109654. [CrossRef]
68. Ji, L.; Lu, J.; Liua, C.; Jing, C.; Fan, H.; Ma, S. Microstructure and mechanical properties of 304L steel fabricated by arc additive manufacturing. In Proceedings of the 2017 International Conference on Electronic Information Technology and Computer Engineering (EITCE 2017), Sibiu, Romania, 7–9 June 2017.
69. Sasikumar, C.; Oyyaravelu, R. Mechanical properties and microstructure of SS 316 L created by WAAM based on GMAW. *Mater. Commun.* **2024**, *38*, 107807. [CrossRef]

70. Long, P.; Wen, D.; Min, J.; Zheng, Z.; Li, J.; Liu, Y. Microstructure Evolution and Mechanical Properties of a Wire-Arc Additive Manufactured Austenitic Stainless Steel: Effect of Processing Parameter. *Materials* **2021**, *14*, 1681. [CrossRef]
71. Omiyale, B.; Ogedengbe, I.; Olugbade, T.; Farayibi, P. Corrosion Performance of Wire Arc Additive Manufacturing of Stainless Steel: A Brief Critical Assessment. *3D Print. Addit. Manuf.* **2024**, *11*, e572–e585. [CrossRef]
72. de Brito Neto, F.M.; Ferreira, M.O.A.; dos Santos, S.A.C.; Pereira, J.N.; Meza, D.L.C.; Ahmed, W.; Nossa, T.d.S.; Moreto, J.A.; Pinto, H.C.; Arantes, V.L. Mechanical and corrosion behaviour of 316L stainless steel processed by wire arc additive manufacturing (WAAM). *Prog. Addit. Manuf.* **2025**, *10*, 6453–6469. [CrossRef]
73. Chen, M.; Gong, Z.; Zhang, T.; Zuo, W.; Zhao, Y.; Zhao, O.; Zhang, G.; Wang, Z. Mechanical behavior of austenitic stainless steels produced by wire arc additive manufacturing. *Thin-Walled Struct.* **2024**, *196*, 111455. [CrossRef]
74. Nagasai, B.; Malarvizhi, S.; Balasubramanian, V. Mechanical Properties and microstructural characteristics of wire arc aditive manufactured 308L stainless steel cylindrical componentes made by gas metal arc and colda metal transfer arc welding processes. *J. Mater. Process. Technol.* **2022**, *307*, 117655. [CrossRef]
75. Karlina, A.I.; Kondratyev, V.V.; Balanovskiy, A.E.; Astafyeva, N.A.; Yamshchikova, E.A. Porosity reduction in metal with hybrid wire and arc additive manufacturing technology (WAAM). *CIS Iron and Steel Review.* **2024**, *27*, 91–95. [CrossRef]
76. Jin, W.; Zhang, C.; Jin, S.; Tian, Y.; Wellmann, D.; Liu, W. Wire Arc Additive Manufacturing os Stainless Steels: A Review. *Materials* **2020**, *10*, 1563.
77. Keller, T.; Lindwall, G.; Ghosh, S.; Ma, L.; Lane, B.M.; Zhang, F.; Kattner, U.R.; Lass, E.A.; Heigel, J.C.; Idell, Y.; et al. Application of Finite Element, Phase-field, and CALPHAD-based Methods to Additive Manufacturing of Ni-based Superalloys. *Mater. Sci.* **2017**, *139*, 244–253. [CrossRef] [PubMed]
78. Nedjad, S.; Yildiz, M.; Saboori, A. Solidification behaviour os austenitic stainless steels during welding and directed energy deposition. *Sci. Technol. Weld. Join.* **2023**, *28*, 1–17. [CrossRef]
79. Kotecki, D.J.; Lippold, J.C. *Welding Metallurgy and Weldability of Stainless Steels*; Wiley: Hoboken, NJ, USA, 2005.
80. Ettefagh, A.H.; Guo, S.; Raush, J. Corrosion performance of additively manufactured stainless steel parts: A review. *Addit. Manuf.* **2021**, *37*, 101689. [CrossRef]
81. Charmi, A.; Falkenberg, R.; Ávila, L.; Mohr, G.; Sommer, K.; Ulbricht, A.; Sprengel, M.; Neumann, R.S.; Skrotzki, B.; Evans, A. Mechanical anisotropy of additively manufactured stainless steel 316L: An experimental and numerical study. *Mater. Sci. Eng. A* **2021**, *799*, 140154. [CrossRef]
82. Wang, H.; Jiang, P.; Yang, G.; Yan, Y. An Investigation of the Anisotropic Mechanical Properties of Additive-Manufactured 316L SS with SLM. *Materials* **2024**, *17*, 2017. [CrossRef]
83. Wang, C.; Liu, T.G.; Zhu, P.; Lu, Y.H.; Shoji, T. Study on microstructure and tensile properties of 316L stainless steel fabricated by CMT wire and arc additive manufacturing. *Mater. Sci. Eng. A* **2020**, *796*, 140006. [CrossRef]
84. Ajay, V.; Nakrani, J.; Mishra, N.K.; Shrivastava, A. Anisotropic fatigue crack propagation in wire arc additively manufactured 316L stainless steel. *Int. J. Fatigue* **2023**, *177*, 107976. [CrossRef]
85. Becker, T.H.; Kumar, P.; Ramamurty, U. Fracture and fatigue in additively manufactured metals. *Acta Mater.* **2021**, *219*, 117240. [CrossRef]
86. Rodriguez, N.; Vázquez, L.; Huarte, I.; Arruti, E.; Tabernero, I.; Alvarez, P. Wire and arc additive manufacturing: A comparison between CMT and TopTIG processes applied to stainless steel. *Weld. World* **2018**, *62*, 1083–1096. [CrossRef]
87. Laghi, V.; Palermo, M.; Tonelli, L.; Gasparini, G.; Ceschini, L.; Trombetti, T. Tensile properties and microstructural features of 304L austenitic stainless steel produced by wire-and-arc additive manufacturing. *Int. J. Adv. Manuf. Technol.* **2020**, *106*, 3693–3705. [CrossRef]
88. *ASTM A276-13a;* Standard Specification for Stainless Steel Bars and Shapes. ASTM International: West Conshohocken, PA, USA, 2013.
89. Gowthaman, P.S.; Jeyakumar, S.; Sarathchandra, D.T. Experimental characteristics of 316L stainless steel thin wall processed in wire-based arc additive manufacturing process. *Inst. Mech. Eng.* **2023**, *239*, 1712–1721. [CrossRef]
90. Anand, S.; Haldar, N.; Datta, S.; Das, A. Experimental investigation on microstructure and mechanical property of wire arc additively manufactured SS308L built part. *Sadhana* **2023**, *48*, 1–20. [CrossRef]
91. Rani, K.U.; Kumar, R.; Mahapatra, M.M.; Mulik, R.S.; Swierczynska, A.; Fydrych, D.; Pandey, C. Wire Arc Additive Manufactured Mild Steel and Austenitic Stainless Steel Components: Microstructure, Mechanical Properties and Residual Stresses. *Materials* **2022**, *15*, 7094. [CrossRef] [PubMed]
92. *ASM Handbook, Volume 1. Properties and Selection: Irons, Steels, and High Performance Alloys Section: Publication Information and Contributors*; ASM international: Materials Park, OH, USA, 1990.

materials

MDPI

Article

Effects of Homogenization Heat Treatment on Microstructure of Inconel 718 Lattice Structures Manufactured by Selective Laser Melting

Lucia-Antoneta Chicos [1], Camil Lancea [1,*], Sebastian-Marian Zaharia [1], Grzegorz Cempura [2], Adam Kruk [2] and Mihai Alin Pop [3]

[1] Department of Manufacturing Engineering, Transilvania University of Brasov, 500036 Brasov, Romania; l.chicos@unitbv.ro (L.-A.C.); zaharia_sebastian@unitbv.ro (S.-M.Z.)
[2] Faculty of Metals Engineering and Industrial Computer Science, AGH University of Krakow, Al. Mickiewicza 30, PL-30 059 Krakow, Poland; cempura@agh.edu.pl (G.C.); kruczek@agh.edu.pl (A.K.)
[3] Department of Materials Science, Transilvania University of Brasov, 500036 Brasov, Romania; mihai.pop@unitbv.ro
* Correspondence: camil@unitbv.ro

Abstract

Inconel 718 is a nickel-based superalloy that has a wide range of applications in the industries that require corrosion resistance or high-temperature resistance. It is well known that parts display internal stresses, anisotropy, and alloying element segregation after the selective laser melting (SLM) process. A homogenization heat treatment, which reduces internal stresses and homogenizes the material structure, can resolve these shortcomings. The present study focuses on the impact of this heat treatment on the microstructure of the Inconel 718 material produced by SLM. The research results indicate that this heat treatment improves both the material microstructure and mechanical performance by lessening the microstructural inhomogeneities, dissolving the Laves phases, and promoting grain coarsening. The findings of this study can contribute to the optimization of post-fabrication strategies for Inconel 718 parts fabricated by SLM.

Keywords: Inconel 718; homogenization heat treatment; microstructure; selective laser melting; Laves phase; δ phase; γ' and γ'' phases; post-fabrication optimization

1. Introduction

Inconel 718 (IN718) is a nickel-based alloy that was developed by Huntington Alloys (which became part of Special Metals Corporation) for jet engines and gas turbines, because of its excellent high temperature properties and corrosion resistance [1]. With the advent of additive manufacturing technologies for metals, such as selective laser melting (SLM), the manufacture of parts with complex geometries made of IN718 has become increasingly popular. The use of SLM technology in manufacturing IN718 parts has led to the production of high-performance materials while also allowing for the production of parts with complex geometric configurations coupled with good mechanical properties [2–4]. IN718 is characterized by particularly good mechanical properties (high strength, good ductility, large fatigue resistance), high corrosion resistance, and good thermal resistance [5–7]. These special properties, coupled with moderate manufacturing costs, have led to the widespread use of this material in various industries where resistance to high temperatures and oxidation is crucial, such as the following: the aerospace industry (turbine elements, combustion

 https://doi.org/10.3390/ma18174149

chambers, and engine parts) [8–10], automotive (engine parts and exhaust systems) [11,12], the oil and gas industry (valves and piping systems that operate in highly corrosive environments) [13], and nuclear industries [14]. Despite its advantages, the selective laser melting (SLM) process has a notable drawback: the layer-by-layer deposition of molten material generates extremely high temperatures followed by rapid cooling, resulting in significant internal stresses and microstructural heterogeneities. This necessarily requires a stress-relieving and homogenizing heat treatment aimed at strengthening the mechanical characteristics of the material.

Compared to other metal alloys produced by SLM, such as alloys with high titanium concentrations, IN718 presents numerous advantages, among which the most important are its superior mechanical properties at high temperatures, lower cost, excellent weldability, and welded joints that have a high resistance to cracking. This last attribute regarding weldability makes it very suitable for additive manufacturing, which is quite comparable to the welding process [15]. Furthermore, its excellent corrosion resistance, even in particularly aggressive environments, increases the durability and longevity of the components compared to conventional stainless steels or titanium alloys in similar conditions of use. All these qualities make IN718 an extremely useful material for engineering applications that require resistance to corrosion or high temperatures, which also confers long-term reliability [13].

Despite the advantages offered by SLM, the rapid melting and solidification process can lead to a columnar grain structure aligned along the build direction, along with micro-segregation and Laves phase formation [16]. These non-equilibrium phases lead to anisotropic mechanical properties and can significantly degrade mechanical properties, reducing the material's ductility and creep resistance, reducing overall performance.

Due to the specific SLM conditions, which include a series of high temperatures in the material deposition area, followed by abrupt cooling, the appearance of microstructural defects and internal stresses is unavoidable. For this reason, the parts produced by this method must undergo homogenization and stress relief heat treatment to improve the microstructure and material properties. This heat treatment serves to mitigate the difficulties outlined above by facilitating the diffusion of alloying elements and dissolving unwanted precipitates [17]. Thus, after the heat treatment, a more uniform microstructure, as well as better mechanical properties such as yield strength, tensile strength, and fatigue resistance, is obtained. By reducing microstructural flaws and eliminating unstable phases, homogenization heat treatment increases the long-term durability and mechanical performance of IN718 parts. Recent studies have shown that homogenization heat treatment contributes to grain recrystallization, lowers micro-segregation, and improves mechanical isotropy, resulting in good tensile strength, significantly increased ductility, and high thermal stability [18]. All of these enhancements will enable parts made of this material to function properly in a variety of key applications across industries.

This study extends prior research [19] and aims to demonstrate, through detailed analyses and both qualitative and quantitative evaluations, the impact of homogenization heat treatment on the microstructural evolution of the IN718 alloy. Accordingly, high-resolution characterization techniques were employed to investigate the microstructure and chemical composition of the as-fabricated and heat-treated samples, including scanning electron microscopy (SEM), Transmission and Scanning Transmission Electron Microscopy (TEM and STEM, respectively), and energy-dispersive X-ray spectroscopy (EDS).

2. Materials and Methods

2.1. SLM Fabrication of Samples

All samples feature a lattice structure with dimensions of 250 × 250 × 250 mm. The material of this study was gas-atomized IN718 alloy powder with spherical particles in a diameter range from 10 to 45 μm [5,20]. The manufacturing process of the samples was carried out at Nikon SLM Solutions AG in Lübeck, Germany.

The SLM process was carried out using an SLM 280HL system [21], equipped with two 400 W yttrium fibre lasers and a build volume measuring 280 × 280 × 350 mm^3. The process parameters included a scanning speed of 900 mm/s, with a hatch distance of 120 μm and a layer thickness of 30 μm. For the directed energy deposition, the samples were vertically oriented, and a continuous cross-snake hatch strategy was used on the skin surface (Figure 1). The axis of deposition was parallel to the Z-direction according to ISO/ASTM52921-13 [22]. The manufacturing process was conducted in an argon inert gas environment, with a gas consumption rate of 2.5 L/min. The chemical composition of this metal powder is listed in Table 1.

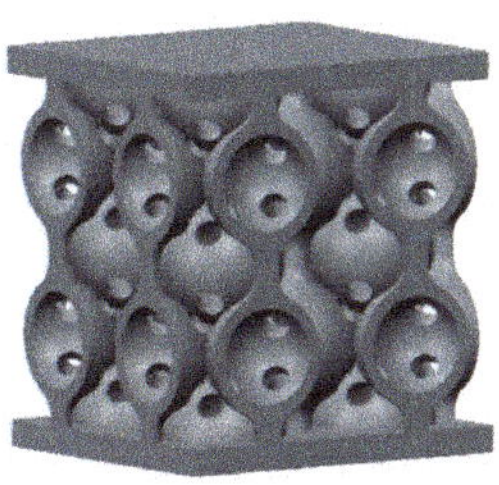

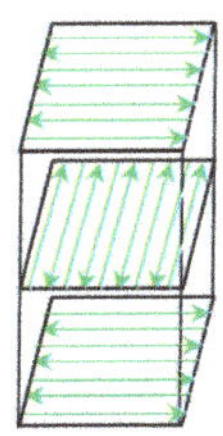

Figure 1. The geometry of the samples and the cross-snake manufacturing strategy [19].

Table 1. Chemical composition (in mass percent, wt%) of IN718 SLM ASTM F3055 powders used in this research [5].

	ASTM F3055/ASTM B637/AMS5664													
	Fe	**Ni**	**Cr**	**Ta+Nb**	**Mo**	**Ti**	**Co**	**Al**	**Si**	**Mn**	**Cu**	**C**	**P**	**S**	**B**
Min.	Bal.	50.00	17.00	4.75	2.80	0.65		0.20							
Max.		55.00	21.00	5.50	3.30	1.15	1.00	0.80	0.35	0.35	0.30	0.08	0.015	0.015	0.006

2.2. The Heat Treatment of SAMPLES

IN718 samples subjected to SLM were heat-treated in the Nabertherm™ L-245K2ANL Muffle Furnace with Flap Door (Nabertherm™, Nabertherm GmbH, Lilienthal (Germany)), max temperature: 1200 °C. The samples were heat-treated according to the following schedule, consistent with the industrial standard for casting IN718 [23,24]: homogenization treatment at 1080 °C for 1.5 h followed by air cooling (AC); solution treatment at 980 °C for 1 h with subsequent air cooling; and double ageing consisting of 8 h at 720 °C with furnace cooling (FC) at a rate of 55 °C/h down to 620 °C, followed by 8 h of air cooling. Figure 2 provides a graphical representation of the homogenization heat treatment schedule applied to the specimens.

Cutting; mounting in resin; grinding with SiC abrasive papers of 600, 1200, and 2000 grit sizes; and polishing in a diamond suspension on a LaboPol (Struers, Ballerup, Denmark) machine were all steps that were taken to get the samples ready for optical microscopic and SEM examinations. In order to reveal the microstructure, the specimens' surfaces were etched using Kroll's reagent (90 mL H_2O, 6 mL HNO_3, and 3 mL HF) [25].

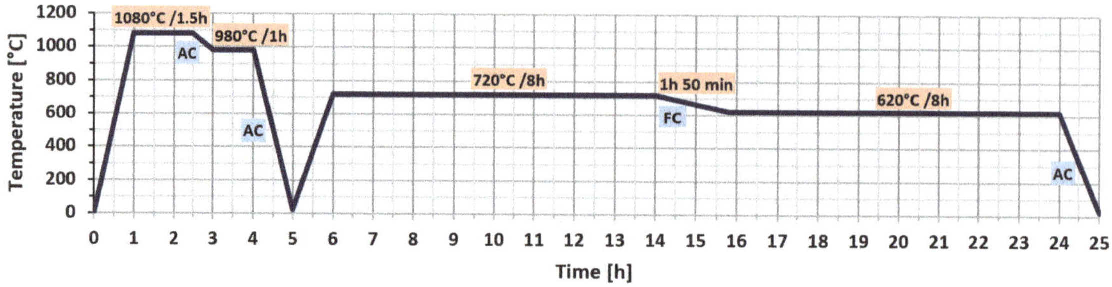

Figure 2. Schematic diagram of homogenization heat treatment process [23].

A ZEISS Axio Imager optical microscope was used for optical microscopy (AIOM), and a scanning electron microscope (SEM, Merlin Gemini II of ZEISS, Oberkochen, Germany) equipped with an electron-dispersive spectrometer (EDS, Bruker Quantax 800, Ettlingen, Germany) was utilized for SEM analysis. The elemental composition was measured by energy-dispersive X-ray spectroscopy (EDS) [26]. The microstructure of the samples was characterized using a ChemiSTEM system-equipped scanning transmission electron microscope (STEM, Titan Cubed G2 60-300, FEI, Hillsboro, OR, USA) and a transmission electron microscope (TEM, Tecnai G2 20 TWIN, FEI) [26]. Energy-dispersive X-ray spectroscopy (EDS) mapping and STEM imaging with high-angle annular dark-field (HAADF) contrast were employed to investigate structural features down to the atomic scale. Phase identification was conducted using selected area electron diffraction (SAED) and further supported by EDS microanalysis, which was also utilized for elemental composition analysis.

Java Electron Microscopy Software (JEMS) (https://www.jems-swiss.ch/, accessed on 20 July 2025) was used to evaluate the electron diffraction pattern [27].

The NEON CrossBeam 40EsB (ZEISS) was utilized in the focused ion beam (FIB) technique to prepare the lamellae for TEM and STEM investigations. The procedure described in reference [28] was employed to prepare the region of interest (ROI) and to fabricate thin sections with a thickness ranging from 10 to 100 nm, as illustrated in Figure 3. The cross-section of the samples included all of the lamellae needed for TEM and STEM investigation (Figure 3).

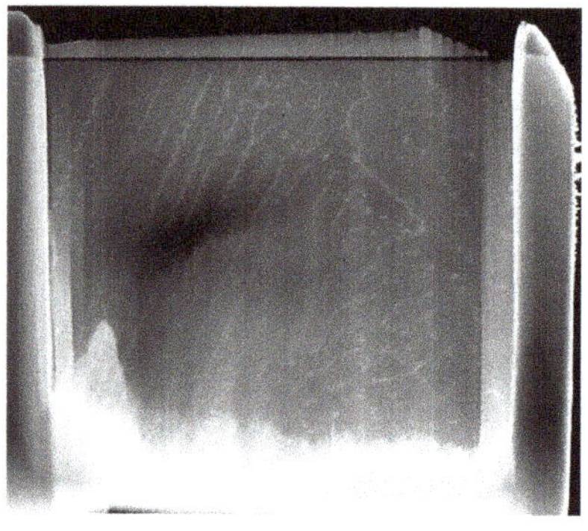

Figure 3. Lamella for TEM examination.

3. Results and Discussion

To analyze the microstructure of the samples, before and after the heat treatment, they were sectioned longitudinally and transversely (Figure 4) and then prepared for microscopic investigation, using the above-described procedures.

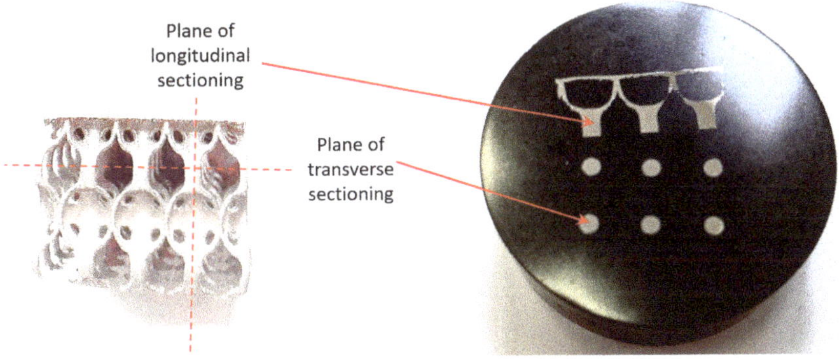

Figure 4. The sectioning planes of the IN718 samples subjected to SLM.

The microstructure of the as-fabricated IN718 subjected to SLM following examination with the ZEISS Axio Imager optical microscope is shown in Figures 5a, 6a and 7a, where a mixed-grain texture can be seen, which includes columnar grains, mainly oriented toward the building direction, and fine grains having varied orientation, due to the temperature gradient. In this regard, the columnar grains resulted from epitaxial growth in the centre of molten pools, while the fine grains resulted from the material's heating history, due to the overlap of laser scanning at the edges of the molten pools [1]. Following the applied heat treatment, the grains tended to grow up, and the scanning stripes on plane XY disappeared after the heat treatment. The grain boundaries are clear now, and additional precipitations were observed in the grains and along the boundaries, as shown in Figures 5b and 6b, and Figure 7b. At a magnification scale of 50×, the δ phase cannot be seen because of its small size.

Two sets of heat-treated and untreated samples were also examined through TEM analysis to analyze the microstructure in detail. A lamella for TEM analysis is shown in Figure 3 and was extracted from the cross-sections of both samples sets. When examining the microstructure of the as-built samples, the same orientation of all dendrites can be observed in Figure 8a. The particles with high Nb content, having a flower shape morphology, can be seen in the γ' phase. The particles with a fine granulation morphology, with a high Fe content, make up the Laves phase, as shown in Figure 8b. The Laves phase and γ' phase precipitates in the cellular and columnar dendrites are a result of the SLM manufacturing process.

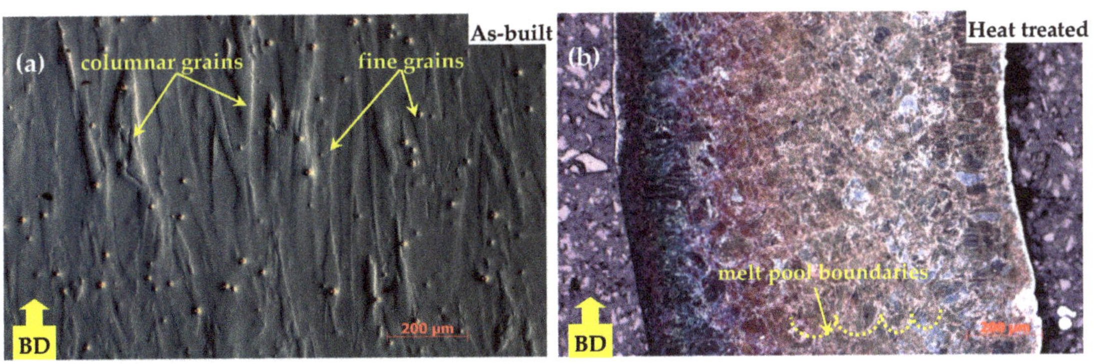

Figure 5. Microstructure of IN718 samples subjected to SLM in longitudinal section: as-built (**a**) and heat-treated (**b**).

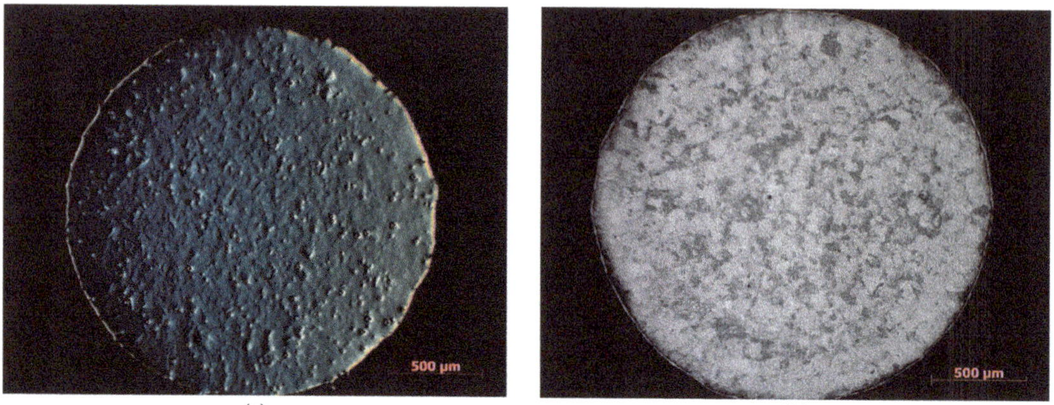

(**a**) (**b**)

Figure 6. Microstructure of IN718 samples subjected to SLM in transversal section: as-built (**a**) and heat-treated (**b**).

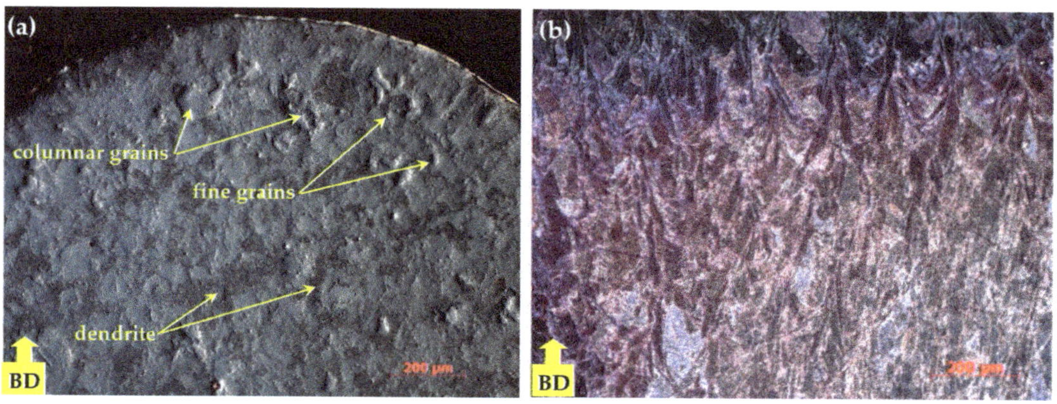

Figure 7. Microstructure of IN718 samples subjected to SLM in transversal section: as-built (**a**) and heat-treated (**b**) at higher magnification factor.

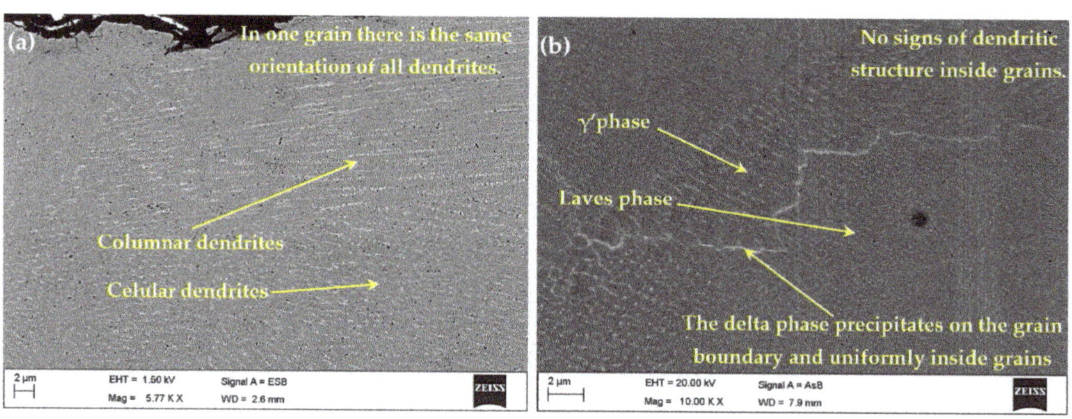

Figure 8. The microstructure of the IN718 as-built samples subjected to SLM in the longitudinal section at a lower (**a**) and higher (**b**) magnification factor.

A higher magnification of the microstructure, given by the SEM images, also reveals the presence of elements that contribute to a decrease in the mechanical strength of the as-built material, as shown in Figure 9. In this regard, besides the Laves phase (Figure 10a), a significant dislocation density can also be observed (Figure 10b).

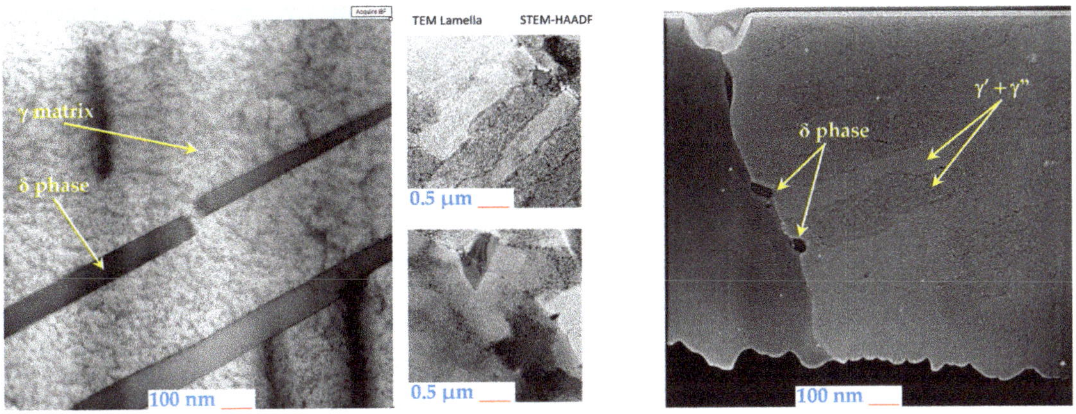

Figure 9. TEM microstructure analyses of the heat-treated IN718 samples subjected to SLM in the longitudinal section.

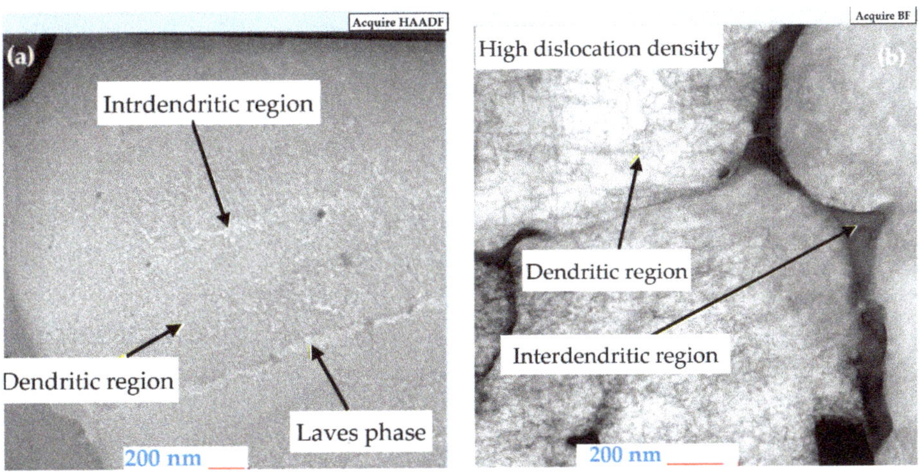

Figure 10. The microstructure of the IN718 as-built samples subjected to SLM in the longitudinal section at 200 nm magnification, Laves Phases (**a**) and dislocation density (**b**).

The Laves phase is harmful because it reduces the ductility of the material [28,29]. Therefore, the heat treatment is very important because it leads to the dissolution of this phase. The homogenization heat treatment process increases the percentage of the γ' and γ'' phases and also enhances the mechanical performance of the Inconel 718 superalloy because these phases are strengthening phases.

After the homogenization heat treatment process, the Laves and γ phases transform into γ', γ'', and δ phases. In this regard, γ' and γ'' phases were detected within the TEM analysis (Figure 9). In the same figure, the precipitation of the δ phase particles at the grain boundaries can also be observed. To confirm the nature and type of these precipitates [30],

EDX analysis for the as-built and heat-treated samples was performed, and the results are shown in Figure 11.

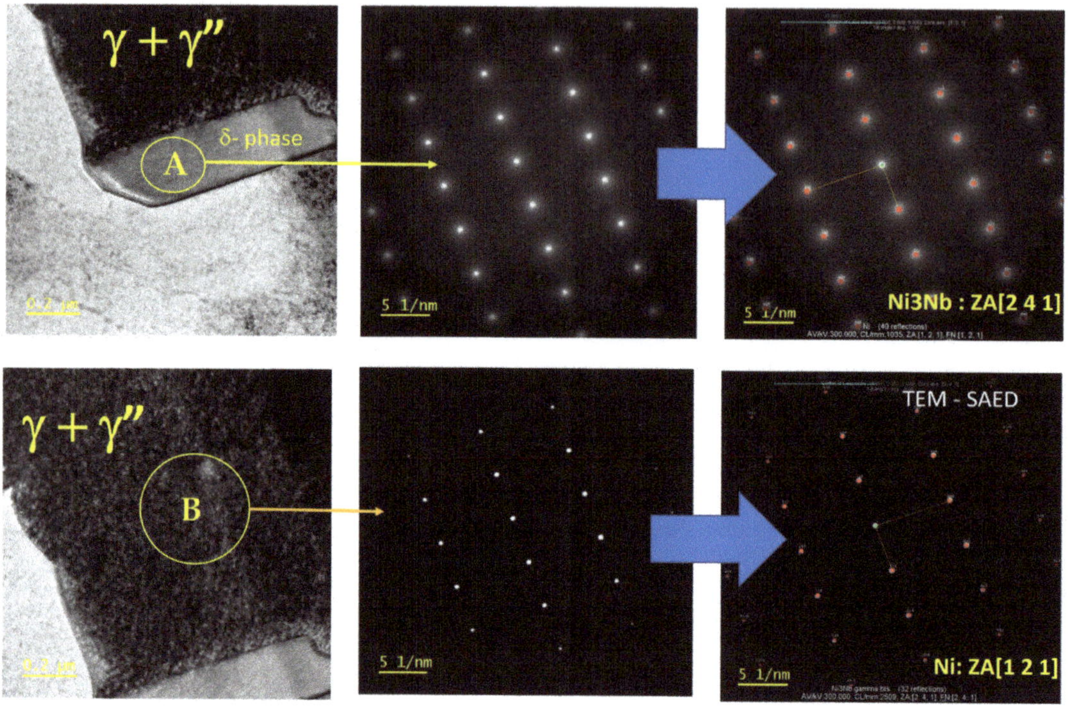

Figure 11. TEM- SAED diffraction pattern of the heat-treated IN718 samples subjected to SLM in the longitudinal section.

The diffraction results (Figure 11) show that the light grey phase in region A is Ni3Nb, and the dark grey phase in region B is Ni, which verifies the conclusion of the above analysis. In addition to the above-presented aspects, it should also be mentioned that the structure obtained after the heat treatment is very similar to the Inconel 718 equilibrium structure [30].

In the as-deposited state (as illustrated in Figure 8), the solidification process leads to an enrichment in titanium (Ti), molybdenum (Mo), and niobium (Nb), leading to the formation of TiN, NbC, and the Laves phase. The presence of a Nb-rich intermetallic Laves phase harms the toughness and strength of the material [31]. After the heat treatment (as shown in Figure 7), elemental redistribution occurs, resulting in the development of δ phase needle-shaped formations, particularly in proximity to the partially dissolved Laves phase (Figures 12b and 13). Additionally, the formation of the δ phase is associated with a reduction in chromium (Cr) and iron (Fe) concentrations within the dendritic interphase region [32].

Figure 13a shows the STEM-HAADF elemental analysis results for the major phases of the as-built material, and Figure 12b shows the elemental analysis results for the major phases after homogenization heat treatment.

Figure 13, Tables 2 and 3 present the quantitative results of EDX analysis, before homogenization heat treatment, which shows a maximum $\approx$21% Cr and $\approx$21% Fe segregation at the selected spots (data collected at nineteen different spots). Table 3 shows the quantitative results of EDX analysis, after homogenization heat treatment, which shows a maximum ≈19% Cr and ≈17% Fe segregation at the selected spots (data collected at nineteen different spots).

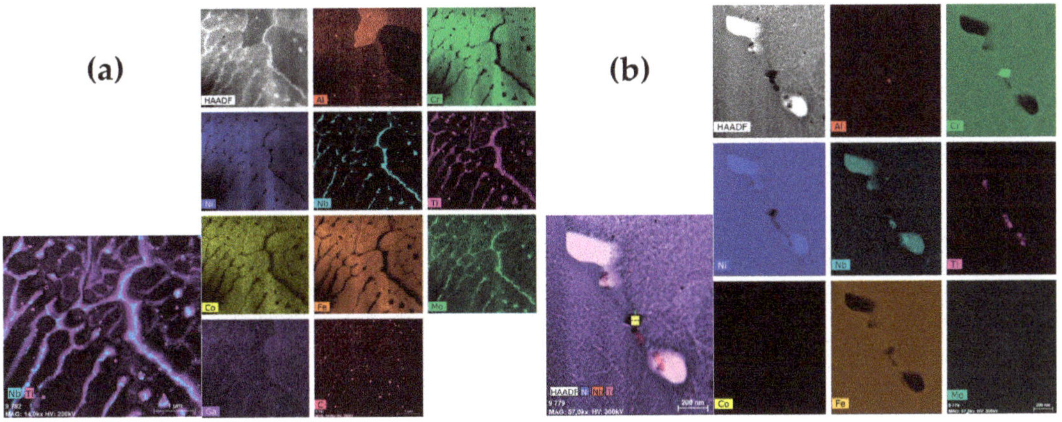

Figure 12. STEM-HAADF image of microstructure before (**a**) and after homogenization heat treatment (**b**) and corresponding EDS spectral images showing chemical element distribution maps.

Table 2. STEM-EDX elemental analysis results for major phases before homogenization heat treatment—mass percentages.

Zone No.	Al-K	Cr-KA	Ni-KA	Nb-KA	Ti-KA	Co-KA	Fe-KA	Mo-KA
1	1.35	22.14	50.08	3.72	1.23	0.38	18.38	2.72
2	1.35	20.18	52.86	2.84	1.02	0.37	18.84	2.55
3	2.47	20.01	51.83	3.28	1.14	0.37	18.36	2.55
4	1.55	20.18	52.77	2.72	0.9	0.37	18.98	2.54
5	1.54	20.48	51.96	2.59	0.81	0.37	19.72	2.52
6	1.33	21.24	50.62	2.15	0.7	0.35	21.19	2.42
7	1.5	20.4	52.28	2.92	0.97	0.37	18.94	2.62
8	1.96	20.15	52.22	2.95	1.07	0.38	18.72	2.56
9	2.55	20.06	51.67	3.32	1.18	0.37	18.25	2.59
10	1.37	20.53	52.88	2.3	0.75	0.36	19.45	2.36
11	1.2	16.13	46.82	16.33	2.18	0.34	12.74	4.26
12	1.42	15.96	47.24	15.59	2.32	0.35	12.97	4.16
13	1.27	15.98	46.02	17.03	2.35	0.35	12.76	4.23
14	1.4	18.88	51.1	7.21	1.74	0.3	16.27	3.09
15	1.05	15.87	45.05	18.38	2.11	0.32	12.75	4.47
16	1.08	15.53	45.94	18.13	2.12	0.37	12.7	4.03
17	1.34	15.36	46.39	17.86	2.17	0.37	12.52	4
18	1.22	15.96	45.36	17.49	2.61	0.32	12.8	4.23
19	1.4	17.83	49.46	10.68	1.98	0.36	14.56	3.73

Table 3. STEM-EDX elemental analysis results for major phases after homogenization heat treatment—mass percentages.

Zone No.	Al-K	Cr-KA	Ni-KA	Nb-KA	Ti-KA	Co-KA	Fe-KA	Mo-KA
1	5.93	19.28	50.48	12.44	0.81	0.51	5.71	4.84
2	3.92	19.18	48.74	4.27	1.63	0.99	17.5	3.78
3	3.99	19.04	48.69	4.26	1.71	1.02	17.45	3.84
4	4.01	19.14	48.81	4.23	1.6	0.99	17.45	3.78
5	3.92	19.1	48.81	4.21	1.64	1.02	17.52	3.79
6	4.08	19.33	48.79	4.03	1.57	0.99	17.55	3.66
7	4.1	6.19	52.36	14.17	1.72	1.03	17.62	3.82
8	3.57	2.24	54.09	20.39	3.45	0.96	12.28	3.03
9	4.62	10.6	53.74	19.91	3.46	0.96	3.67	3.03
10	4.95	9.97	53.41	20.81	3.63	0.88	3.34	3.02
11	4.55	9.98	51.25	20.21	14,01	0.78	9.22	4.01
12	4.14	9.14	42.32	17.44	13.31	0.88	8.65	4.12

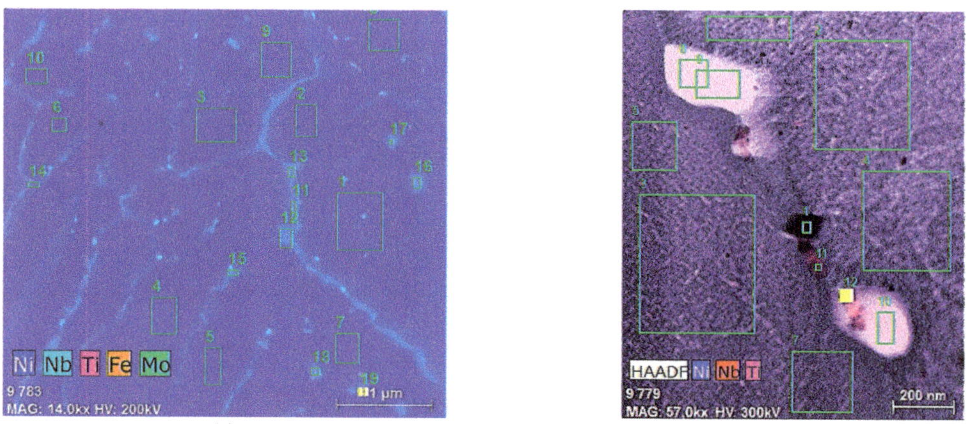

(a) (b)

Figure 13. STEM-EDX elemental analysis image for major phases before (**a**) and after homogenization heat treatment (**b**).

After calculating the arithmetic averages of all mass percentages before and after heat treatment for Cr and Fe, we obtain 18.57 vs. 13.6% Cr and 16.36 vs. 12.33% Fe. An analysis of the mass percentages reveals a decrease in the concentrations of both iron and chromium in the heat-treated samples. This observation confirms the formation of the δ phase [33–35].

4. Conclusions

This study demonstrates that homogenization heat treatment significantly enhances both the mechanical properties and microstructure of the Inconel 718 material produced by selective laser melting. To investigate this, the microstructures and precipitates of the as-built and heat-treated Inconel 718 samples, fabricated via the SLM process, were analyzed using optical microscopy (OM), scanning electron microscopy (SEM), and transmission electron microscopy (TEM). The key findings and conclusions can be summarized as follows:

- After heat treatment, the Laves phase generates a δ phase in the vicinity due to the diffusion effect, and γ′ and γ″ phases are also generated in the material structure. These phases contribute to the enhanced mechanical performance of the Inconel 718 superalloy due to their strengthening effect.
- Optical microscopy confirms the need for heat treatment by revealing a columnar dendritic structure caused by the rapid thermal cycles and radiative heat transfer in the SLM process.
- The quantitative results of EDX analysis allow for a simultaneous and quantitative original position statistical distribution analysis of different elements in the Inconel 718 alloy, before and after homogenization heat treatment. This analysis shows a maximum ≈21% Cr and ≈21% Fe segregation, at the selected spots. After homogenization heat treatment, the results of EDX analysis indicate a reduction in the mass percentages of these elements, with a maximum segregation of approximately 19% Cr and 17% Fe at the selected spots. These findings confirm the formation of the δ phase.
- A reduction in the mass percentages of these elements is further evidenced by the calculation of the arithmetic averages of all mass percentages before and after heat treatment. Specifically, the average mass percentages of Cr and Fe decrease from 18.57% to 13.6% and from 16.36% to 12.33%, respectively.
- Ageing heat treatment significantly enhances the microstructure and mechanical properties of additively manufactured IN718 by promoting grain growth, dissolving microstructural inhomogeneities, and reducing internal stresses.

The results presented in this study provide useful information for optimizing post-fabrication strategies applied to Inconel 718 material fabricated by SLM. By analyzing both the microstructural evolution and mechanical behaviour after heat treatment, the study provides useful information regarding the selection of thermal treatments that can improve the properties of the alloy. This is of particular importance because Inconel 718 is widely used in applications where mechanical strength, thermal stability, and structural homogeneity are of particular importance. Consequently, the results of this study can contribute to the development of more efficient and reliable manufacturing workflows in the field of additive manufacturing.

Author Contributions: Conceptualization, C.L., L.-A.C., and S.-M.Z.; methodology, C.L. and L.-A.C.; software, C.L., L.-A.C., and S.-M.Z.; validation, C.L., L.-A.C., and M.A.P.; formal analysis, L.-A.C., A.K., and G.C.; investigation, C.L., L.-A.C., A.K., G.C., S.-M.Z., and M.A.P.; resources, C.L., L.-A.C., A.K., G.C., S.-M.Z., and M.A.P.; data curation, C.L., L.-A.C., and S.-M.Z.; writing—original draft preparation, C.L., L.-A.C., and S.-M.Z.; writing—review and editing, C.L., L.-A.C., and S.-M.Z.; visualization, C.L., L.-A.C., and S.-M.Z.; supervision, C.L., L.-A.C., A.K., G.C., S.-M.Z., and M.A.P.; project administration, C.L., L.-A.C., A.K., G.C., S.-M.Z., and M.A.P.; funding acquisition, C.L., L.-A.C., A.K., G.C., S.-M.Z., and M.A.P. All authors have read and agreed to the published version of the manuscript.

Funding: This project received funding from the European Union's Horizon 2020 research and innovation programme under Grant Agreement No. 823717-ESTEEM3. This study was also supported by the Polish Ministry of Science and Higher Education: subvention AGH University of Krakow No. 16.16.110.663.

Institutional Review Board Statement: Not applicable.

Informed Consent Statement: Not applicable.

Data Availability Statement: The original contributions presented in this study are included in the article. Further inquiries can be directed to the corresponding author.

Acknowledgments: We also acknowledge the European Union's Horizon 2020 research and innovation programme under Grant Agreement No. 823717-ESTEEM3 and the Research and Development Institute of Transylvania University of Brasov for providing the necessary infrastructure for a certain amount of this work.

Conflicts of Interest: The authors declare no conflicts of interest.

References

1. Qi, H. Review of INCONEL 718 alloy: Its history, properties, processing and developing substitutes. *J. Mater. Eng.* **2012**, 92–100.
2. Adzhamsky, S.V. Implementation of SLM—Technologies for the Manufacture of Samples from Inconel 718 Heat-Resistant Alloy, Used in Aviation-Space Technology. *Aerosp. Tech. Technol.* **2019**, *2*, 69–75.
3. Borisov, E.V.; Popovich, V.A.; Popovich, A.A.; Sufiiarov, V.S.; Zhu, J.-N.; Starikov, K.A. Selective laser melting of Inconel 718 under high laser power. *Mater. Today Proc.* **2020**, *30*, 784–788. [CrossRef]
4. Kladovasilakis, N.; Charalampous, P.; Tsongas, K.; Kostavelis, I.; Tzovaras, D.; Tzetzis, D. Influence of Selective Laser Melting Additive Manufacturing Parameters in Inconel 718 Superalloy. *Materials* **2022**, *15*, 1362. [CrossRef]
5. Material Data Sheet IN718, Nikon SLM Solutions AG. Available online: https://nikon-slm-solutions.com/wp-content/uploads/2025/02/mds5128.pdf (accessed on 13 April 2025).
6. Blackwell, P.L. The mechanical and microstructural characteristics of laser-deposited IN718. *J. Mech. Work. Technol.* **2005**, *170*, 240–246. [CrossRef]
7. Smith, D.H.; Bicknell, J.; Jorgensen, L.; Patterson, B.M.; Cordes, N.L.; Tsukrov, I.; Knezevic, M. Microstructure and mechanical behavior of direct metal laser sintered Inconel alloy 718. *Mater. Charact.* **2016**, *113*, 1–9. [CrossRef]
8. Pollock, T.M.; Tin, S. Nickel-Based Superalloys for Advanced Turbine Engines: Chemistry, Microstructure and Properties. *J. Propuls. Power.* **2006**, *22*, 361–374. [CrossRef]
9. Sullivan, T.N.; Wang, B.; Espinosa, H.D.; Meyers, M.A. Extreme lightweight structures: Avian feathers and bones. *Mater. Today* **2017**, *20*, 377–391. [CrossRef]
10. Niesłony, P.; Grzesik, W.; Jarosz, K.; Laskowski, P. FEM-based optimization of machining operations of aerospace parts made of Inconel 718 superalloy. *Proc. CIRP* **2018**, *77*, 570–573. [CrossRef]
11. Iturbe, A.; Giraud, E.; Hormaetxe, E.; Garay, A.; Germain, G.; Ostolaza, K.; Arrazola, P.J. Mechanical characterization and modelling of Inconel 718 material behavior for machining process assessment. *Mater. Sci. Eng. B* **2017**, *682*, 441–453. [CrossRef]
12. Muralidharan, B.G.; Shankar, V.; Gill, T.P.S. *Inconel 718 Alloy—A Review*; Indira Gandhi Centre for Atomic Research: Kalpakkam, India, 1996.
13. Tang, Z.; Yang, C.; Duan, Y.; Ma, L.; Zheng, S.; Li, M. Corrosion and wear behaviors of Inconel 718 nickel-based alloy by boroaluminizing. *Surf. Coatings Technol.* **2024**, *478*, 130500. [CrossRef]
14. Arva, E.R.U.; Abrudeanu, M.; Negrea, D.A.; Galatanu, A.; Galatanu, M.; Rizea, A.-D.; Anghel, D.-C.; Branzei, M.; Jinga, A.I.; Petrescu, M.I. Experimental Research on the Influence of Repeated Overheating on the Thermal Diffusivity of the Inconel 718 Alloy. *Appl. Sci.* **2024**, *14*, 8574. [CrossRef]
15. Sonar, T.; Balasubramanian, V.; Malarvizhi, S.; Venkateswaran, T.; Sivakumar, D. An overview on welding of Inconel 718 alloy—Effect of welding processes on microstructural evolution and mechanical properties of joints. *Mater. Charact.* **2021**, *174*, 110997. [CrossRef]
16. Zhang, S.; Wang, L.; Lin, X.; Yang, H.; Huang, W. The formation and dissolution mechanisms of Laves phase in Inconel 718 fabricated by selective laser melting compared to directed energy deposition and cast. *Compos. B Eng.* **2022**, *239*, 109994. [CrossRef]
17. Ajay, P.; Dabhade, V.V. Heat treatments of Inconel 718 nickel-based superalloy: A Review. *Met. Mater. Int* **2024**, *31*, 1204–1231. [CrossRef]
18. Yao, X.; Moon, S.K.; Lee, B.Y.; Bi, G. Effects of heat treatment on microstructures and tensile properties of IN718/TiC nanocomposite fabricated by selective laser melting. *Int. J. Precis. Eng. Manuf.* **2017**, *18*, 1693–1701. [CrossRef]
19. Zaharia, S.-M.; Lancea, C.; Kruk, A.; Cempura, G.; Gruszczyński, A.; Chicos, L.-A.; Pop, M.A. Mechanical Properties and Microstructure of Inconel 718 Lattice Structures Produced by Selective Laser Melting Process. *Materials* **2024**, *17*, 622. [CrossRef]
20. Chang, L.; Sun, W.; Cui, Y.; Zhang, F.; Yang, R. Effect of heat treatment on microstructure and mechanical properties of thehot-isostatic-pressed Inconel 718 powder compact. *J. Alloys Compd.* **2014**, *590*, 227–232. [CrossRef]
21. SLM®280 Machine, by Nikon SLM Solutions AG. Available online: https://nikon-slm-solutions.com/slm-systems/slm280-2-0/ (accessed on 15 April 2025).
22. *ISO/ASTM52921-13*; Standard Terminology for Additive Manufacturing-Coordinate Systems and Test Methodologies. ASTM International: West Conshohocken, PA, USA, 2013.

23. Huang, W.; Yang, J.; Yang, H.; Jing, G.; Wang, Z.; Zeng, X. Heat treatment of Inconel 718 produced by selective laser melting: Microstructure and mechanical properties. *Mater. Sci. Eng. A.* **2019**, *750*, 98–107. [CrossRef]
24. Hosseini, E.; Popovich, V.A. A review of mechanical properties of additively manufactured Inconel 718. *Addit. Manuf.* **2019**, *30*, 100877. [CrossRef]
25. *ASTM E407-07 e1*; Standard Practice for Microetching Metals and Alloys. ASTM International: West Conshohocken, PA, USA, 2015.
26. Kruk, A.; Cempura, G.; Lech, S.; Wusatowska-Sarnek, A.M.; Czyrska-Filemonowicz, A. Application of analytical electron microscopy and tomographic techniques for metrology and 3D imaging of microstructural elements in Allvac ® 718Plus™. In Proceedings of the 9th International Symposium on Superalloy 718 & Derivatives: Energy, Aerospace, and Industrial Applications; Ott Al, E.E., Ed.; The Minerals, Metals & Materials Series. Springer: Cham, Switzerland, 2018.
27. Stadelmann, P. Simulation of Diffraction Patterns and High-Resolution Images Using JEMS. Available online: https://www.jems-swiss.ch/Home/Docs/PDF/Manual.pdf (accessed on 16 April 2025).
28. Chicos, L.A.; Zaharia, S.M.; Cempura, G.; Kruk, A.; Lech, S.; Kryshtal, O.; Ziętara, M.; Michta, G.; Rodríguez, J.; Cosnita, M.; et al. Effect of concentrated solar energy on microstructure evolution of selective laser melted Ti-6Al-4V alloy. *Int. J. Adv. Manuf. Technol.* **2022**, *118*, 3183–3207. [CrossRef]
29. Li, X.; Shi, J.J.; Wang, C.H.; Cao, G.H.; Russell, A.M.; Zhou, Z.J.; Li, C.P.; Chen, G.F. Effect of heat treatment on microstructure evolution of Inconel 718 alloy fabricated by selective laser melting. *J. Alloys Compd.* **2018**, *764*, 639–649. [CrossRef]
30. Cao, G.H.; Sun, T.Y.; Wang, C.H.; Li, X.; Liu, M.; Zhang, Z.X.; Hu, P.F.; Russell, A.M.; Schneider, R.; Gerthsen, D.J.M.C.; et al. Investigations of γ', γ'' and δ precipitates in heat-treated Inconel 718 alloy fabricated by selective laser melting. *Mater. Charact.* **2018**, *136*, 398–406. [CrossRef]
31. Gullipalli, C.; Burad, P.; Thawari, N.; Bhatt, J.; Gupta, T.V.K. Microstructure Evolution in Direct Energy Deposited Multilayer Inconel 718. *Arab. J. Sci. Eng.* **2021**, *47*, 7985–7994. [CrossRef]
32. Santhosh, V.; Babu, N. A Study on Microstructure and Mechanical Properties of Inconel 718 Superalloy Fabricated by Novel CMT-WAAM Process. *Mat. Res* **2024**, *27*, e2023025.
33. Li, Y.; Yang, S.; Peng, Z.; Wang, Z.; Gao, Z. Effects of Heat Treatment and Carbon Element Invasion on Microstructure and Mechanical Properties of Inconel 718 Alloys Fabricated by Selective Laser Melting. *J. Mater. Eng. Perform.* **2023**, *33*, 8015–8032. [CrossRef]
34. Zhou, X.; Li, D.; Zhou, Q.; Jiang, F.; Song, Y.; Liang, W.; Liu, M.; Shen, X.; Wang, H. Quantitative Characterization of Elemental Segregation in Inconel 718 Superalloy by Micro-Beam X-ray Fluorescence Spectroscopy and Its Correlation Study. *Materials* **2023**, *16*, 7163. [CrossRef]
35. Tajyar, A.; Brooks, N.; Holtham, N.; Rowe, R.; Newell, D.J.; Palazotto, A.N.; Davami, K. Effects of a modified heat-treatment on microstructure and mechanical properties of additively manufactured Inconel 718. *Mater. Sci. Eng. A* **2022**, *838*, 142770. [CrossRef]

materials

MDPI

Article

Influence of Laser Metal Deposition Process Parameters on the Structural Integrity of $CuSn_{11}Bi_3$ Coatings on C45

Federico Mazzucato [1,*], Edouard Baer [2], Samuel Rey-Mermet [2] and Anna Valente [1]

1 Department of Innovative Technologies, University of Applied Science and Arts of Southern Switzerland (SUPSI), Via La Santa 1, 6962 Lugano, Switzerland; anna.valente@supsi.ch
2 Institute of Systems Engineering, School of Engineering, University of Applied Sciences and Arts HES-SO Valais-Wallis, Systems Engineering Rue de l'Industrie 23, 1950 Sion, Switzerland; edouard.baer@hevs.ch (E.B.); samuel.rey-mermet@hevs.ch (S.R.-M.)
* Correspondence: federico.mazzucato@supsi.ch; Tel.: +41-58-666-67

Abstract

Bronze-steel bimetallic structures are structural components finding a growing application in industrial sectors such as aerospace, power generation, and machinery. Recent legislation on green economy and sustainable manufacturing is boosting industry to implement innovative manufacturing processes and new metal alloys capable of lowering environmental footprint by avoiding toxic substances. Laser Metal Deposition is a cost-effective Additive Manufacturing technique for producing bimetallic components by limiting material waste and reducing energy consumption. In this research work, the influence of the main LMD process parameters on the final quality of $CuSn_{11}Bi_3$ coatings on C45 surfaces is analyzed. The Cu-based powder is specifically designed and developed to reduce environmental pollution and increase worker safety by avoiding the use of hazardous chemical elements. The performed observations demonstrate that high-density (99.90%) and crack-free clads are feasible by preventing melt pool dilution zones. Cu diffusion into the C45 substrate deteriorates the structural integrity at the clad-substrate interface by inducing liquid metal embrittlement cracking, whereas steel diffusion into the as-deposited clad promotes crack propagation. High-density (up to 99.97%) and crack-free $CuSn_{11}Bi_3$ coatings are achieved by using low specific energies (from 17 J/mm^2 to 40 J/mm^2) and reducing the Oxygen content during sample manufacturing up to 0.02%.

Keywords: Laser Metal Deposition; Cu-based alloys; multi-material; chemical analysis; bimetallic structure

1. Introduction

Bimetallic structures consisting of a steel bulk coated with pure or pre-alloyed copper (Cu) are advanced structural components that are finding growing applications in industrial sectors such as aerospace, machinery, and power generation [1–3]. Specifically, copper-steel composites are conceived to take advantage of the joining of the dissimilar thermal, chemical, and mechanical properties of their heterogeneous constituents, exhibiting unprecedented performance not achievable by mono-material solutions in terms of high mechanical strength, improved corrosion and cavitation resistance, low dynamic friction coefficient, and high thermal conductivity [4–7]. Examples of copper-steel bimetallic components are bearing bushes, cooling staves, liquid pipelines, and hydraulic pumps [8]. Currently, the primary production method of such bimetallic components consists of casting pure Cu or Cu-based alloys on a pre-machined steel bulk, followed by heat treatments and a final step of machining to meet product requirements in terms of mechanical and tribological properties, dimensional accuracy, geometrical tolerances, and surface roughness. Despite the above fabrication method proving to be reliable in the production of high-quality structural components, it exhibits several drawbacks in terms of limited achievable part geometrical complexity, low process repeatability, high material waste, and high CO_2 emissions. Moreover, recent stringent legislations concerning circular economy and sustainable manufacturing (e.g., REACH Regulation [9]) are forcing the industry to renew its workshops by embracing cutting-edge manufacturing technologies capable of managing new metal alloys, ensuring high standards in the production of high-quality structural parts with reduced resource consumption and lower pollution and carbon emissions [10–12]. Direct Energy Deposition (DED) is a family of Additive Manufacturing (AM) techniques that has proven to be a cost-effective alternative to casting for the fabrication of both Functionally Graded Materials (FGMs) [13–15] and high-quality functional surface coatings [16–19] by decreasing the environmental impacts in terms of greenhouse gas emissions, scraps, and raw material management. Among DED processes, powder-based Laser Metal Deposition (LMD) is a recognized AM technique enabling an efficient optimization of the part design through an advanced functionalization of the material gradient structure [20]. Compared to Powder Bed Fusion (PBF), the LMD process fabricates metal parts layer upon layer by means of local deposition of metal powder injected into a melt pool, which is created by a constant interaction between a laser source and a metal surface [21]. When the laser moves, the melt pool solidifies, and a solid metal bonding between the as-deposited material and the underlying surface forms [22–27]. The capability to tailor the metal powder chemical composition on-the-fly and the limited Heat Affected Zone (HAZ) generated during the deposition process makes LMD a suitable and cost-effective manufacturing technique to realize high-quality graded structures [28–31].

Several authors have analyzed the influence of LMD process parameters on the manufacturing of defect-free Cu depositions. The scientific community agrees that the realization of high-density bulky features is feasible through LMD IR laser sources, but process optimization remains critical and resource-intensive due to the high reflectivity of the involved metal powder [32,33]. Moreover, an inert deposition chamber is required to prevent the deterioration of the tribological properties of the as-built Cu coating due to O_2 solubilization, which forms Cu_2O precipitates in the as-built microstructure [34]. Xiao et al. [35] were successful in repairing a motor commutator by depositing Cu-based metal powder ($Cu_{15}Sn$) on a pure Cu surface via a blue laser. Compared to the base material, they achieved a defect-free clad formation exhibiting a 2.22 times higher microhardness and 2.37 times wear resistance under the testing conditions of dry friction and wear at room temperature, proving that the final quality of the realized metallic components and the efficiency of the LMD process increase when using Cu-based alloys instead of pure Cu or blue laser sources instead of IR. Despite blue lasers ensuring enhanced deposition efficiency thanks to a higher coefficient of energy absorption [36–38], their service life, beam quality performance over time, and reliability are still one step behind IR systems, making blue thermal sources not yet mature for industrial production.

When dealing with bronze-steel bimetallic composites, the LMD of Cu-based alloys becomes even more challenging. Differences in the coefficients of thermal expansion, disparity in the melting temperatures and cooling rates, diverse crystal structures, and embrittlement phenomena undermine the structural integrity of the realized bimetallic structure by introducing structural cracks and delamination [39,40]. Recent research has focused on preventing the formation of detrimental defects, such as solidification cracking along the as-deposited Cu grain boundaries and liquid metal embrittlement (LME) cracks, by deploying a metallic interlayer between the two dissimilar metals [41]. Zhang et al. [42] demonstrated that defect-free bronze-SS304L bonding is achieved through a Deloro 22 interlayer deposition. Moreover, Chen et al. [43] and Wang et al. [44] confirmed that Nickel has an inhibitory effect on the formation of LME cracks in the steel substrate when Cu-based alloys are deposited. Despite the above-mentioned advantages, the implementation of an interlayer is not always feasible at the industrial level due to the diffusion of undesired chemical elements (e.g., Ni) into the as-deposited clad, which affects both the tribological behavior, such as friction coefficients and wear/corrosion resistance, and the thermal properties (e.g., thermal conductivity) of the Cu-based coating [45].

The objective of this research work is to characterize the influence of the powder-based LMD process parameters on the final quality of $CuSn_{11}Bi_3$ depositions performed on C45 surfaces to enable high-quality superficial coatings without the need for interlayers. The Cu-based powder used in this experimental activity is a new lead-free bronze alloy designed and developed to comply with the REACH regulation by ensuring high mechanical performance and enhanced tribological properties for hydraulic industrial applications [46]. The influence of laser power, axis speed, and plate preheating on the internal and chemical quality of as-built Cu-based clads is analyzed by assessing single track cross-sections. Acceptable combinations of LMD process parameters ensuring high-density and crack-free bronze depositions are used to realize single layers and assess the impact of the deposition environment on the final structural integrity of the performed coating. The achieved experimental results demonstrate that high-density $CuSn_{11}Bi_3$ depositions on C45 are achieved for low specific energies (i.e., ranging from 40 J/mm^2 to 17 J/mm^2), whereas keyhole phenomena and LME cracking occur at high laser power. Crack formation and propagation are avoided by preventing steel diffusion into the Cu-based clad and keeping the content of Oxygen (O_2) during material deposition below 0.02%. Section 2 deals with the material and equipment involved in the experimental campaign execution and sample analysis. Section 3 presents and discusses the achieved results. Finally, Sections 4 and 5, respectively, introduce the main conclusions and future works.

2. Materials and Methods

2.1. Metal Material and LMD Equipment

The metal powder involved in the experimental campaign was a $CuSn_{11}Bi_3$ spherical-shaped powder [46] designed and developed by Kugler Bimetal SA (Le Lignon, Switzerland) for the manufacturing of advanced coatings for the hydraulic, power generation, and aerospace sectors. The powder grain size ranged from 79.5 μm to 151.2 μm, and powder tap density was 4.9 g/cm^3 (see Figure 1).

The metal substrate was an 18 mm thick C45 steel disk (supplier: Brütsch/Rüegger Metals AG (Regensdorf, Switzerland)) with a nominal diameter of 115 mm. The material coupling was based on an industrial use case concerning a core component of a hydraulic pump and consisted of a bulk of medium carbon steel coated with a Cu-based alloy to ensure high mechanical strength in conjunction with enhanced superficial tribological properties. Table 1 shows the chemical composition of both the $CuSn_{11}Bi_3$ powder and the C45 disk, whereas Table 2 shows the main thermo-mechanical properties of the two metallic materials.

Table 1. $CuSn_{11}Bi_3$ and C45 chemical composition (wt. %).

Material	Chemical Composition				
$CuSn_{11}Bi_3$	Bi: 3.00	Sn: 11.00	Ni: 2.00 max	Cu: Bal.	
C45	C: 0.45	Si: 0.40	Cr: 0.40	Ni: 0.40	Fe: Bal.

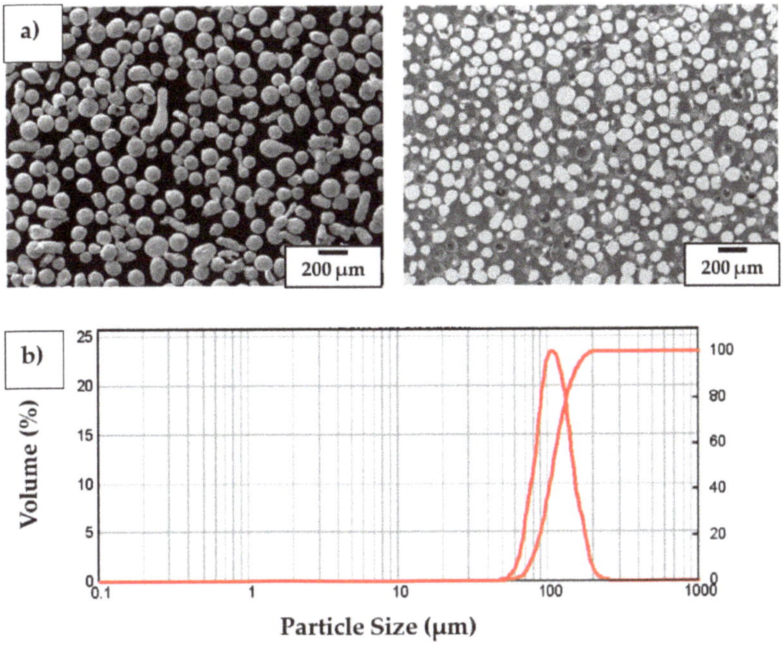

Dv_{10} [µm] 79.5	Dv_{50} [µm] 109.8	Dv_{90} [µm] 151.2

Figure 1. $CuSn_{11}Bi_3$ powder particles: (**a**) particle shape and cross-section; (**b**) particle grain size distribution.

Table 2. $CuSn_{11}Bi_3$ and C45 main thermo-mechanical properties.

Material	Bulk Density at 20° C [g/cm^3]	Yield Strength [MPa]	Young Modulus [GPa]	Thermal Expansion Coefficient [10^{-6} °C^{-1}]	Solidus/Liquidus Temperature [°C]
$CuSn_{11}Bi_3$	8.94	220	105	18	778/985
C45	7.8	300	210	12	1460/1420

The LMD system was a three-axis Laserdyne 430 machine, Prima Power, Turin, Italy, that consisted of a laser cutting machine retrofitted and adapted to perform the powder-based LMD process (see Figure 2a). The system was equipped with a Convergent Photonics CF1000 laser source, Convergent, France, featuring 1000 W maximum power, a nominal laser spot of 1 mm, and a wavelength of 1070 nm. A heating plate designed and developed by the Automation, Robotics, and Machines laboratory of SUPSI was implemented below the sample clamping system to enable a maximum preheating temperature of 450 °C on the substrate surface. Powder supply was ensured by a LENS™ Print Engine double-hopper powder feeding system, Optomec, NM, USA, connected to a commercial Optomec® multi-nozzle deposition head, Optomec, NM, USA (see Figure 2b). The inert gas used as a carrier and shielding gas was Argon 4.6. To reduce the O_2 content during the deposition process, the building volume was enclosed by a high-temperature resistant polymeric film (see Figure 2c), and the deposition chamber was filled with Argon to reach an O_2 content below 0.02% before executing $CuSn_{11}Bi_3$ depositions.

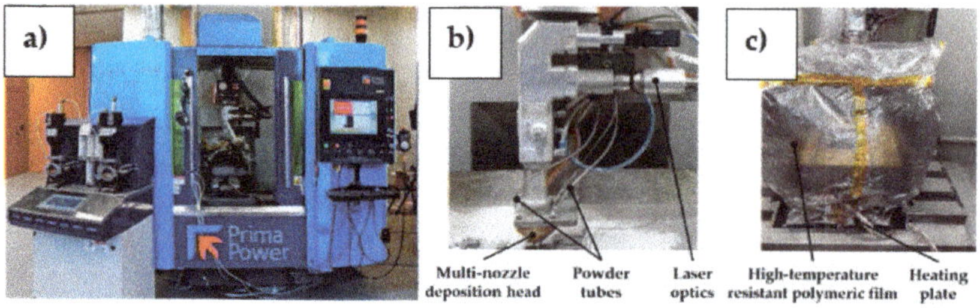

Figure 2. (**a**) Laserdyne 430 LMD system; (**b**) LMD deposition head; (**c**) enclosed deposition chamber.

2.2. Experimental Campaign

The experimental campaign consisted of two main phases: (1) the deposition and characterization of linear $CuSn_{11}Bi_3$ tracks on C45 disks (see Figure 3a); the realization and assessment of bidimensional coatings produced using suitable combinations of LMD process parameters, ensuring high-quality linear $CuSn_{11}Bi_3$ depositions (see Figure 3b).

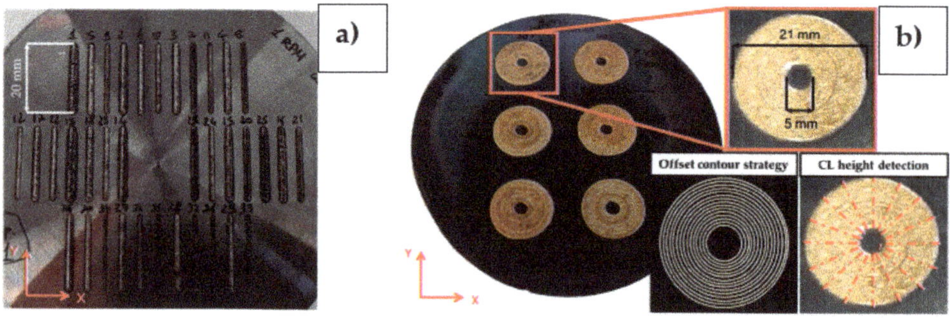

Figure 3. Performed $CuSn_{11}Bi_3$ depositions: (**a**) single tracks (STs); (**b**) circular layers (CLs).

The influence of the laser power (P), axis speed (F), and substrate preheating on the final structural quality of 20 mm single linear $CuSn_{11}Bi_3$ depositions, i.e., single tracks (STs), was analyzed by taking six levels for P, six levels for F, and two levels of substrate preheating temperatures (see Table 3). The reason for choosing P ranging from 300 W to 800 W and F ranging from 300 mm/min to 1050 mm/min was for the following:

- Based on the results discussed in previous studies involving the same LMD system [18,47], using a P lower than 300 W or an F higher than 1050 mm/min resulted in unstable depositions featuring a lack of fusion.
- A maximum P level of 800 W was chosen based on Zhang et al. [48] and Kim et al. [49], who suggested not exceeding an energy density threshold of 80 J/mm^2 to prevent keyhole formation. Given the high reflectivity of the Cu-based powder involved in this study and the corresponding energy dispersion, the authors decided to overcome the recommended threshold and set a maximum value of 160 J/mm^2 to have a larger overview of the iteration behavior between LMD parameters and material.

Three repetitions for each combination of process parameters were performed for a total of 216 realized STs. To enable internal quality assessment, each realized sample was cut by wire-EDM (see Figure 4a), and the resulting cross-sections were pre-polished with SiC-coated paper (FEPA 120), polished with diamond particles of 6, 3, and 1 μm, and finished with Oxide Polishing Suspension (OPS). A Zeiss Gemini 460 Scanning Electron Microscope (SEM), Zeiss, Oberkochen, Germany, was used to characterize the structural integrity of the clad cross-sections in terms of pores, cracks, and delamination. Specifically, porosity measurements were performed using the open-source ImageJ software v1.54p, which was used to both compute the pore size and the percentage in the analyzed area. After acquiring the ST cross-section with the SEM, the image was opened in ImageJ, the pixels were calibrated based on the scale bar of the SEM image, and the pore size was measured. The porosity percentage was calculated by determining the fraction of area occupied by pores relative to the total using the "Analyse Particles" tool in the software. Data analysis and representation were performed through Minitab Statistical Software v22 and Excel. The material diffusion and final chemical composition of the realized samples were evaluated through Energy Dispersive X-ray (EDX) analysis performed using an EDX Ultimax 100 system, Oxford Instruments, United Kingdom.

Table 3. LMD process parameters involved in ST depositions.

LMD Process Parameter	Min. Value	Max. Value	Number of Levels	Step Value
Laser power [W]	300	800	6	100
Axis speed [mm/min]	300	1050	6	150
Preheating [°C]	20	345	2	-
E_s [J/mm^2]	17	160	-	-
ρ [g/mm]	7×10^{-3}	23×10^{-3}	-	-
Laser spot [mm]	1		-	-
Powder feed [g/s]	0.116		-	-
Stand-off [mm]	10.5		-	-
Carrier [l/min]	4		-	-
Shielding [l/min]	16		-	-

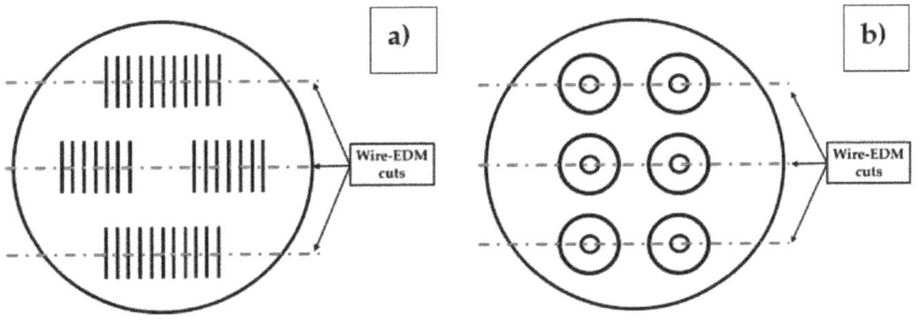

Figure 4. Performed EDM cut positions for (**a**) single tracks (STs); (**b**) circular layers (CLs).

The manufacturing of bidimensional coatings, i.e., Circular Layers (CLs), was implemented by using the combinations of process parameters resulting in crack-free and high-density $CuSn_{11}Bi_3$ linear clads (see Table 4). CL fabrication was performed with and without the implementation of a high-temperature resistant polymeric film (i.e., "enclosed" and "open" deposition chambers, ensuring an O_2 content of 0.02% and 21%, respectively) to analyze the influence of two different environmental printing conditions on the structural and dimensional quality of the performed coatings. Figure 3b shows the nominal size of the realized coatings and the implemented offset contour deposition strategy by using 60% ST overlapping. As performed for the linear $CuSn_{11}Bi_3$ depositions, each combination of process parameters was repeated three times for a total of 36 realized CLs. The final quality of the realized CLs was assessed in terms of layer final height, computed by averaging 16 equally spaced CL profiles traced radially through a Keyence ® VHX-6000 digital microscope, Keyence, Japan, that involved 1.47 μm planar resolution and 0.2 μm vertical pitch (see Figure 3b), and structural integrity by following the same methodology implemented for SL and performing the wire-EDM cuts shown in Figure 4b.

Table 4. LMD process parameters for CL printing.

P [W]	F [mm/min]	Preheating [°C]	O_2 Content [%]	E_s [J/mm^2]	ρ [g/mm]
300	450	-	0.02/21	40	0.015
400	750	-	0.02/21	32	0.009
400	900	-	0.02/21	27	0.008
300	750	345	0.02/21	24	0.009
300	900	345	0.02/21	20	0.008
300	1050	345	0.02/21	17	0.007

3. Results and Discussion

3.1. $CuSn_{11}Bi_3$ ST Analysis

The performed SEM optical analysis demonstrates that the influence of the three considered LMD input variables on the structural integrity of $CuSn_{11}Bi_3$ STs realized on the C45 substrate is significant (see Figures 5–14). To properly evaluate the effect of the LMD process parameters on the final quality of the as-deposited Cu-based material, the specific energy, E_s [J/mm^2], the linear mass density, ρ [g/mm], and the net laser power reaching the C45 substrate, P_{net} [W], must be introduced. E_s is defined as:

$$E_s = \frac{P}{F \times d} \quad (1)$$

where d is the laser spot diameter (see Table 3). E_s is a common LMD process parameter used to describe the energy balance during the deposition process, and it is directly proportional to laser power and inversely proportional to axis speed and laser spot. The linear mass density provides a measure of the melt pool powder supply, and it is defined as:

$$\varrho = \frac{f}{F} \quad (2)$$

where f [g/s] is the powder feeding rate. Finally, P_{net} is the amount of laser power that reaches the metallic substrate net of energy losses due to heat powder absorption (P_{powder}) and laser scattering and reflection (P_{loss}) [50]. It is defined as:

$$P_{net} = P_{nom} - P_{powder} - P_{loss} \quad (3)$$

where P_{nom} is the nominal laser power provided during the deposition process.

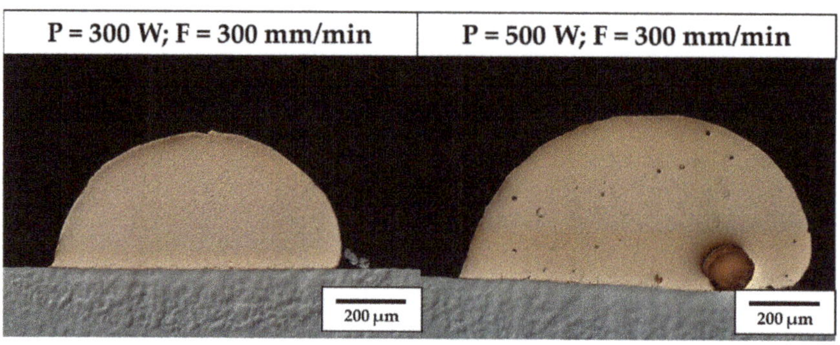

Figure 5. Influence of P on the structural integrity of $CuSn_{11}Bi_3$ STs.

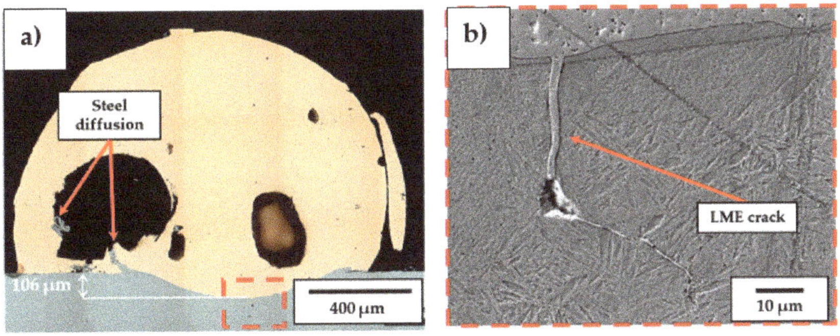

Figure 6. Steel diffusion on $CuSn_{11}Bi_3$ clad (**a**) and LME crack formation (**b**) for P = 800 W, F = 300 mm/min.

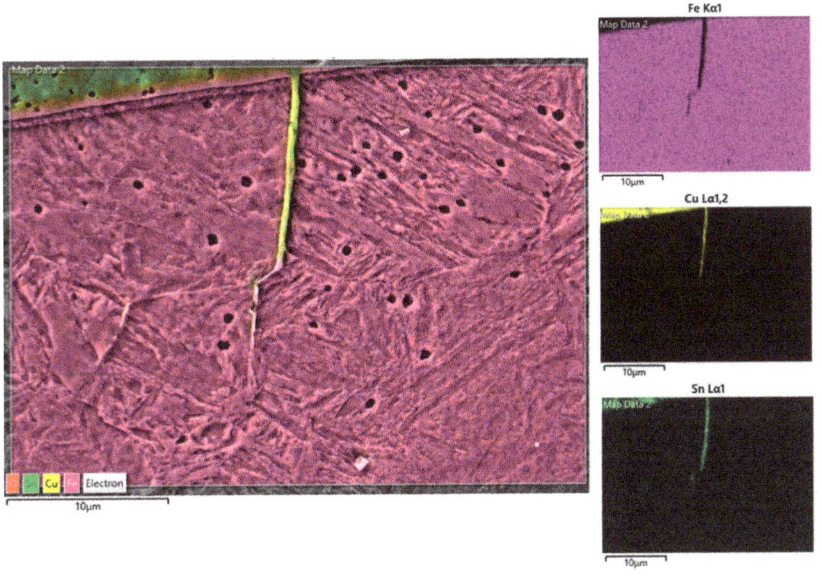

Figure 7. EDX mapping showing LME cracks filled by Cu and Sn.

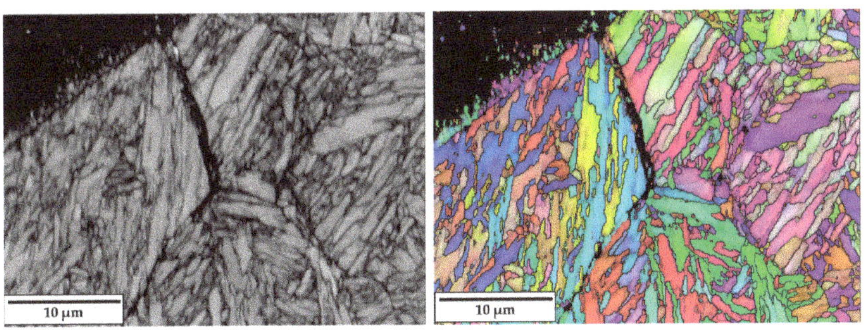

Figure 8. Band contrast image (**left**) and EBSD map (**right**) showing LME propagation along grain boundaries.

When no substrate preheating is involved, no melt pool penetration depth is observed for P ranging between 300 W and 600 W (see Figure 5), meaning that most of the P_{nom} is reflected and scattered by the powder flow exiting from the deposition nozzles. As a result, the P_{net} reaching the substrate is too low for melt pool formation, and the metal powder particles melt while still in suspension, dropping onto the C45 surface without forming any dilution zone. The observed phenomenon is in agreement with [51], where energy dispersion due to scattering was demonstrated to reach up to 97% of P_{nom} for the LMD of highly reflective spherical-shaped metal powder. By increasing P, substrate melting occurs, and melt pool penetration depths up to 106 µm are observed (see Figure 6a). The increased E_s promotes the formation of dilution zones and the mixing between $CuSn_{11}Bi_3$ and C45, inducing the nucleation of LME cracks, which propagate into the steel substrate (see Figure 6b). As discussed in [43], liquation cracks take place due to the diffusion of molten Cu into the steel grain boundaries, which deteriorates the cohesion between grains and entails a reduction in the minimum stress for intergranular cracking when high thermal loads are applied. Figures 7 and 8 show LME cracking detected on the ST cross-section realized with 800 W and 300 mm/min. At high P levels, the EDX map reveals Cu and Sn diffusion within the observed crack (see Figure 7), which propagates into the C45 substrate following grain boundary directions, as demonstrated by the EBSD map in Figure 8.

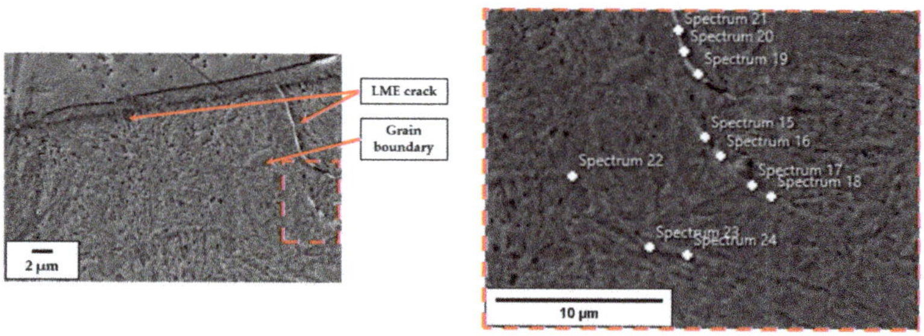

Spectrum label	Spectrum 15	Spectrum 16	Spectrum 17	Spectrum 18	Spectrum 19	Spectrum 20	Spectrum 21	Spectrum 22	Spectrum 23	Spectrum 24
O	2.13	2.84	2.52	2.93	3.36	4.54	4.42	2.29	2.02	2.61
Cr	0.13	0.01	0.03	0.13	0.08	0.00	0.12	0.08	0.30	0.15
Mn	0.31	0.42	0.61	0.81	0.27	0.23	0.20	0.47	0.55	0.61
Fe	50.21	73.11	86.83	93.99	29.10	31.00	34.27	95.88	95.16	95.24
Cu	42.29	19.98	9.28	1.78	59.91	52.98	46.97	1.21	1.88	0.97
Sn	4.55	2.08	0.33	0.16	6.04	5.21	4.33	0.07	0.00	0.16
Bi	0.38	1.56	0.41	0.21	1.24	6.04	9.70	0.00	0.09	0.26
Total	100.00	100.00	100.00	100.00	100.00	100.00	100.00	100.00	100.00	100.00

Figure 9. EDX point analysis performed on an LME crack and its surrounding area.

The diffusion of Cu into the steel substrate is also confirmed by the EDX point analysis when a medium-high P is used (i.e., 600 W and 700W), which demonstrates the presence of copper along C45 grain boundaries even when no LME cracks are observed on spectra 22, 23, and 24 of Figure 9.

The influence of P and F on the clad internal porosity is significant, as highlighted by the Pareto chart (see Figure 10a). By moving from 300 W to 800 W, internal porosity increases from 0.03% up to 6.50% on average (see Figure 10b), detecting pores with a maximum size of 460 µm (see Figure 6). The strong increase in the internal porosity with P values ranging from 600 W to 800 W is explained by the low melting point and the high coefficient of thermal conductivity of the involved bronze alloy (see Table 2). Indeed, assuming P_{loss} as a function of particle shape and optical material properties, for constant axis speeds, an increase in E_s induces an increment in the laser energy absorbed by the powder particles as they overheat and evaporate, forming large gaseous cavities. The high cooling rates of the Cu-based material do not allow the vaporized metal to escape from the melt pool, trapping the gas in the clad cross-section and forming keyholes, which deteriorate the structural integrity of the as-built material.

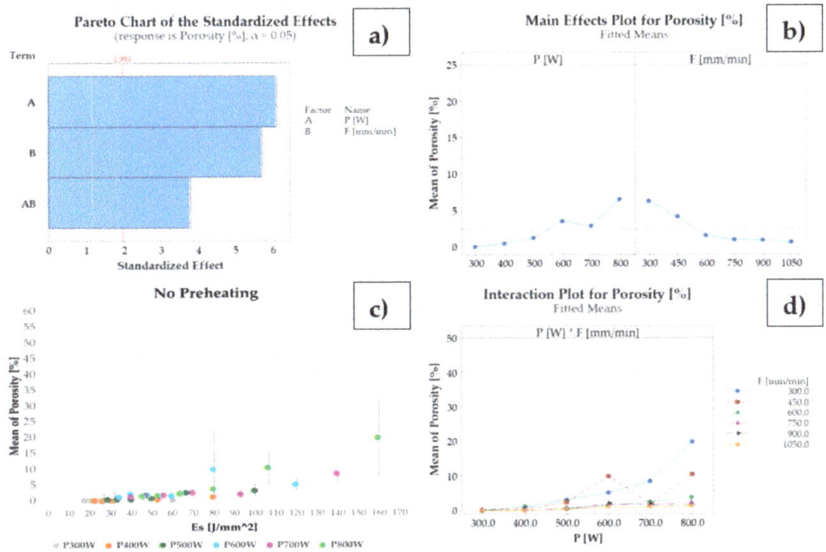

Figure 10. Influence of P, F, and E_s on the observed ST clad internal porosity for no substrate preheating: (**a**) Pareto chart, (**b**) P and F main effect plots, (**c**) ST clad internal porosity vs. E_s., (**d**) P and F interaction plot.

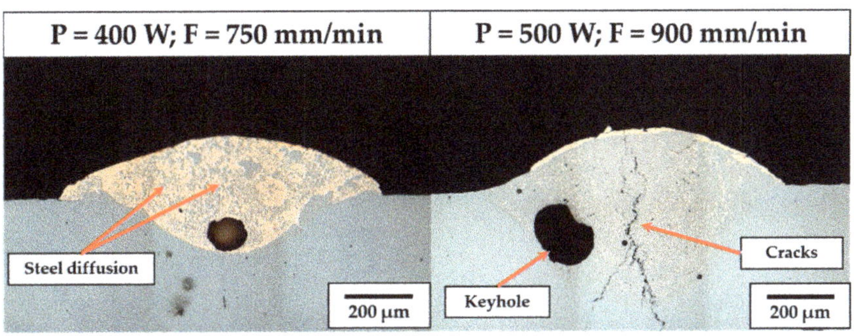

Figure 11. 345 °C substrate preheating: Steel diffusion into the Cu-based clad (**left**), keyhole and crack formation (**right**).

The pronounced impact of P on the internal porosity of the as-deposited clad and on keyhole formation is further highlighted by Figure 10c. For constant E_s (i.e., 40 J/mm^2, 60 J/mm^2, and 80 J/mm^2), an increase in laser power results in a deterioration of internal density, causing a strong keyhole formation for Es equal to or exceeding 80 J/mm^2, as confirmed by [48,49] and demonstrated by the high data variability shown in Figure 10c. Indeed, keyhole is an unstable phenomenon that causes the formation of large and irregular pores that significantly affect porosity measurement. As discussed in Section 2.2, having performed a discrete measurement of the internal density of the as-deposited clad based on EDM cuts and optical imaging, the same repetitions of process parameters can exhibit very different internal porosities depending on the presence or absence of keyholes. This fact leads to a wide variability in the observed results for those combinations of process parameters that involve medium-high laser power.

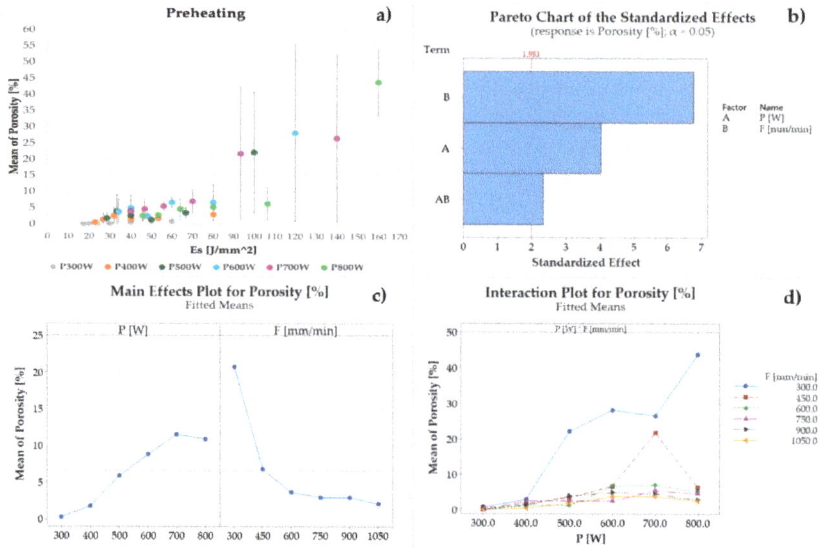

Figure 12. Influence of P, F, and E_s on the observed ST clad internal porosity for 345 °C substrate preheating: (**a**) ST clad internal porosity vs. E_s., (**b**) Pareto chart, (**c**) P and F main effect plots, (**d**) P and F interaction plot.

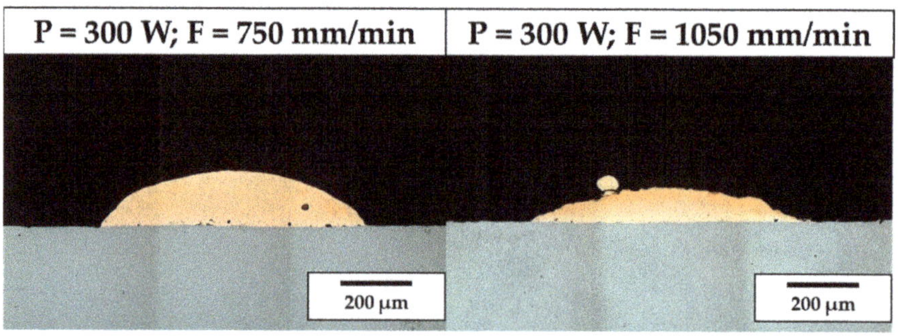

Figure 13. Substrate preheating of 345 °C: crack-free $CuSn_{11}Bi_3$ STs at 300 W.

Contrary to P, clad internal density increases with F (see Figure 10b,d). Moving from 300 mm/min to 1050 mm/min, internal porosity decreases below 1.5% on average. Moreover, by decreasing E_s through an increase in axis speed (i.e., F ranging between 750 mm/min and 1050 mm/min), no keyholes are observed. Nevertheless, for 300 W, lack of deposition occurs for ρ ranging between 7×10^{-3} g/mm and 9×10^{-3} g/mm as a result of the low specific energy and reduced melt pool powder supply induced by the high LMD axis speeds, degrading the structural integrity of the as-built clad.

In the case of no substrate preheating, high-quality and crack-free $CuSn_{11}Bi_3$ STs with an internal density of 99.97% are obtained using 300 W in combination with F ranging between 300 mm/min and 600 mm/min, and 400 W in combination with F ranging between 750 mm/min and 1050 mm/min (see Table 4).

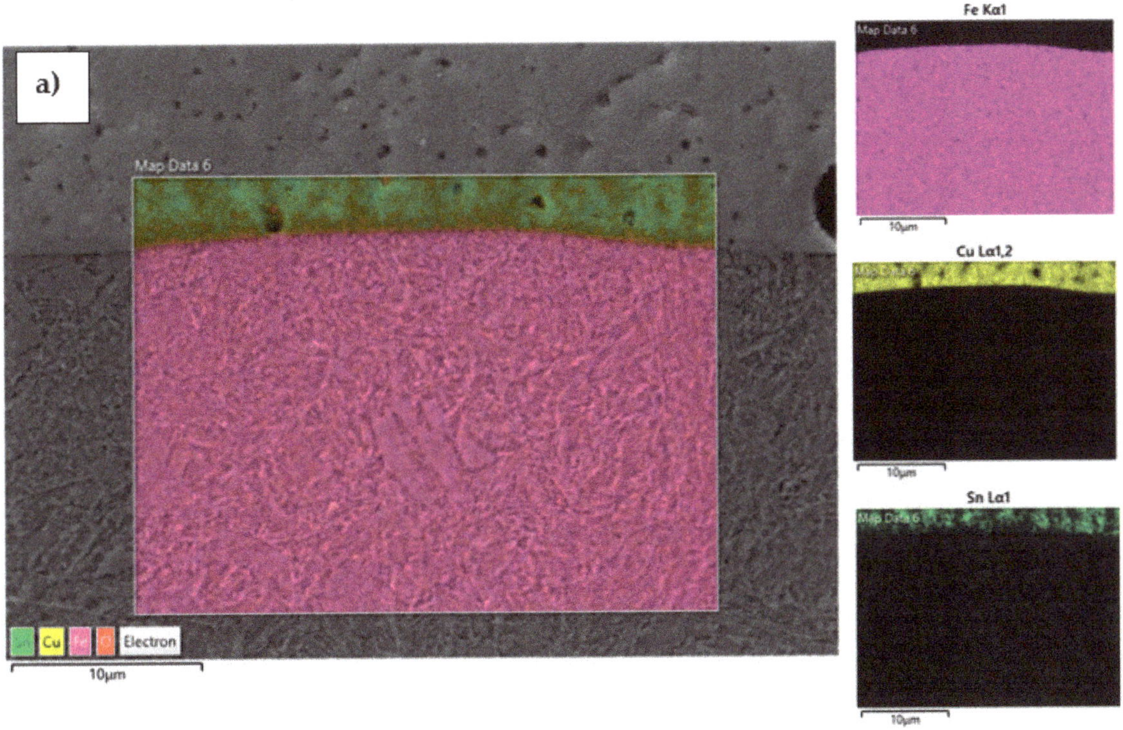

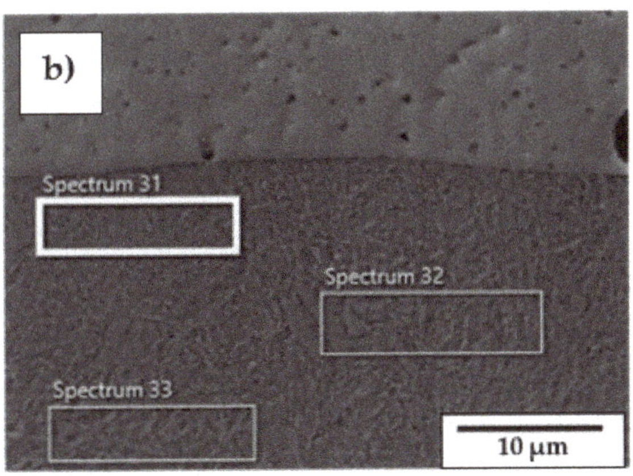

Spectrum label	Spectrum 31	Spectrum 32	Spectrum 33
O	2.58	2.40	2.37
Cr	0.08	0.24	0.05
Mn	0.95	0.79	0.80
Fe	94.68	95.79	95.81
Cu	1.49	0.78	0.85
Sn	0.00	0.00	0.11
Bi	0.22	0.00	0.00
Total	100.00	100.00	100.00

Figure 14. ST cross-section realized at P = 300 W and F = 750 mm/min: (**a**) EDX map; (**b**) EDX point analysis.

The use of 345 °C of substrate preheating promotes large melt pool penetration depths also for medium-low P levels, causing extensive steel diffusion into the as-built $CuSn_{11}Bi_3$ clad for P ranging between 400 W and 800 W (see Figure 11, left). As previously discussed, the presence of dilution zones induces the nucleation of liquidation cracks at the $CuSn_{11}Bi_3$-C45 interface, which, in this case, propagate along the clad cross-sections, significantly deteriorating the structural integrity of the bimetallic deposition (see Figure 11, right). The propagation of LME cracks is due to the diffusion of steel into the bronze alloy and the resulting reduced ductility of the as-built material, which breaks due to the residual stresses formed by volumetric shrinkage during melt pool solidification and the different thermal expansion coefficients between the substrate and the dilution zone during cooling. In contrast to the case of no substrate preheating, keyhole phenomena are also observed for 400 W and axis speeds up to 900 mm/min, as confirmed by the presence of potato-shaped pores with a size ranging between 80 μm and 270 μm (see Figure 11).

These findings indicate that substrate preheating promotes keyhole formation, extending the deterioration of internal clad integrity even at E_s below 80 J/mm^2. Notably, as shown in Figure 12a, a high result variability exists even at E_s equal to or exceeding 32 J/mm^2 when 400 W is involved. Although P and F are still significant in affecting pore formation and extension, for 345 °C of substrate preheating, the effect of axis speed is predominant on laser power, as demonstrated by the Pareto chart in Figure 12b. By increasing F up to 1050 mm/min, the formation of gas inclusions due to Cu vaporization is reduced, although not prevented, observing a reduction of internal porosity from 20.76% to 2.09% on average (see Figure 12c,d).

High-quality ST depositions are achieved for laser power of 300 W and axis speeds ranging between 600 mm/min and 1050 mm/min (see Figure 13 and Table 4). EDX chemical analysis confirms that for the LMD combinations of process parameters in which no melt pool penetration depth occurs, no Cu diffusion into the C45 substrate is observed (see Figure 14a), preventing LME crack formation and preserving the starting chemical composition of the raw metal powder. The small amount of Cu observed in Figure 14b is attributed to the secondary electron bounce back of the $CuSn_{11}Bi_3$ layer since a negligible presence of Sn and Bi is detected.

3.2. *$CuSn_{11}Bi_3$ Coating Analysis*

The CLs produced using the LMD parameter combinations listed in Table 4 exhibited no LME cracks at the $CuSn_{11}Bi_3$-C45 interface, achieving final actual heights ranging between 0.162 mm and 0.530 mm. Cross-section analysis reveals internal densities higher than 99.00%, achieving a minimum internal porosity when an inert deposition environment is implemented in conjunction with 300 W, 750 mm/min, and 345 °C of substrate preheating (see Table 5).

Table 5. $CuSn_{11}Bi_3$ CLs: detected internal porosity and actual height.

P [W]	F [mm/min]	ST Overlap [%]	Preheating [° C]	Internal Porosity [%]		CL Actual Height [mm]	
				$O_2 = 21\%$	$O_2 < 0.02\%$	$O_2 = 21\%$	$O_2 < 0.02\%$
300	450	60	-	0.45	0.15	0.530	0.336
400	750	60	-	0.51	0.99	0.307	0.215
400	900	60	-	0.66	0.53	0.262	0.162
300	750	60	345	0.82	0.03	0.432	0.315
300	900	60	345	0.83	0.2	0.315	0.245
300	1050	60	345	0.67	0.67	0.246	0.190

Nevertheless, structural cracks are observed in the Cu-based clad when CLs are realized in an open deposition chamber (see Table 6 and Figure 15a). The detected cracks initiate from the external surface of the realized sample and propagate into the bulk, degrading the structural integrity of the as-built material (see Figure 15b). On the contrary, samples realized with an enclosed deposition environment exhibit crack-free cross-sections, meaning that the reduction of O_2 content in the LMD process improves the structural integrity of $CuSn_{11}Bi_3$ coatings. The reason for sample degradation is related to the high chemical reactivity of Cu and Bi with O_2 at high temperatures. When LMD is performed in an open deposition chamber, the molten bronze alloy reacts with O_2, forming a surface oxide layer that, due to its brittle mechanical behavior, is inclined to break when subjected to the thermal cycles, promoting crack formation and propagation. On the contrary, by limiting the content of O_2 below 0.02%, oxidation is reduced, and crack formation is prevented. Figure 16 compares a coating realized with and without the use of an enclosed deposition environment, whereas Figure 17 shows both the EDX map and EDX point analysis performed on the structural cracks observed in CL realized at P 300 W, F 750 mm/min, and an open deposition environment. Chemical analysis reveals a high presence of O inside the observed crack and along the crack edges, confirming that significant oxidation forms when the Cu-based coating is realized with 21% O_2 content (see Figure 16a). On the contrary, for an O_2 content < 0.02%, superficial oxidation is strongly reduced, and the realized coating appears glossy (see Figure 16b).

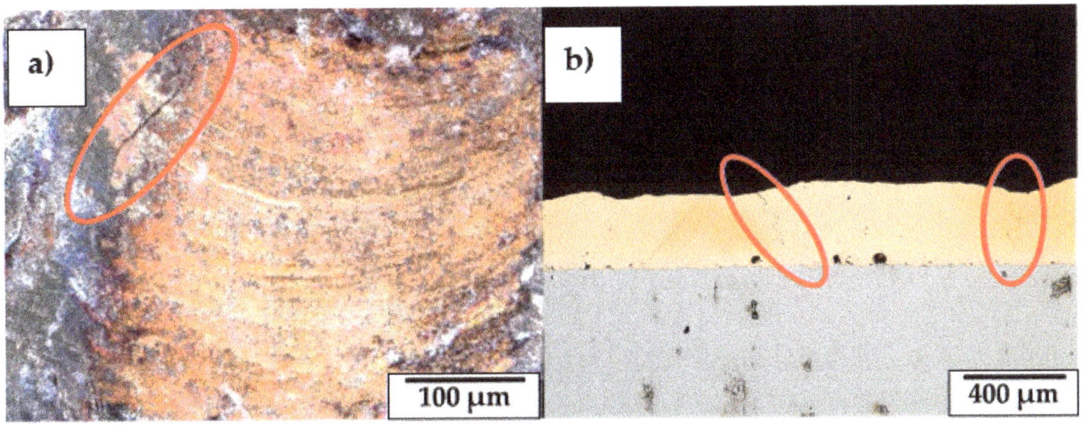

Figure 15. Structural cracks observed on $CuSn_{11}Bi_3$ CL realized at P300W, F900mm/min: (**a**) external surface; (**b**) cross-section.

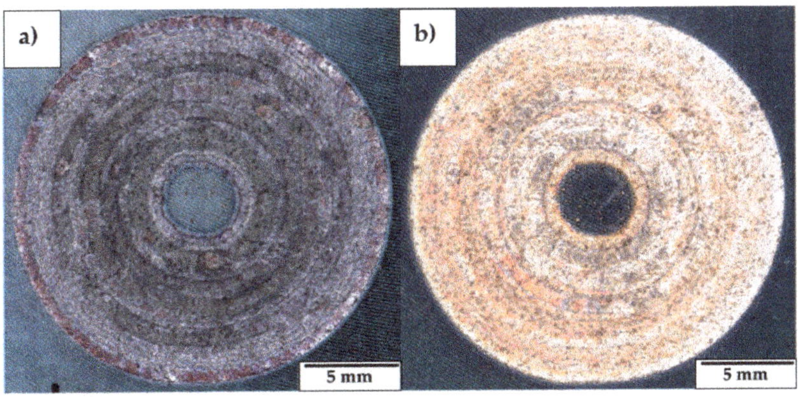

Figure 16. $CuSn_{11}Bi_3$ CL realized with P 300 W, F 900 mm/min, and 345 °C preheat: (**a**) 21% O_2 content; (**b**) 0.02% O_2 content.

Table 6. Structural cracks observed in the realized $CuSn_{11}Bi_3$ CLs.

P [W]	F [mm/min]	ST Overlap [%]	Preheating [°C]	Structural Cracks	
				O_2 = 21%	O_2 < 0.02%
300	450	60	-	Yes	No
400	750	60	-	No	No

Table 6. *Cont.*

P [W]	F [mm/min]	ST Overlap [%]	Preheating [°C]	Structural Cracks $O_2 = 21\%$	$O_2 < 0.02\%$
400	900	60	-	No	No
300	750	60	345	Yes	No
300	900	60	345	Yes	No
300	1050	60	345	Yes	No

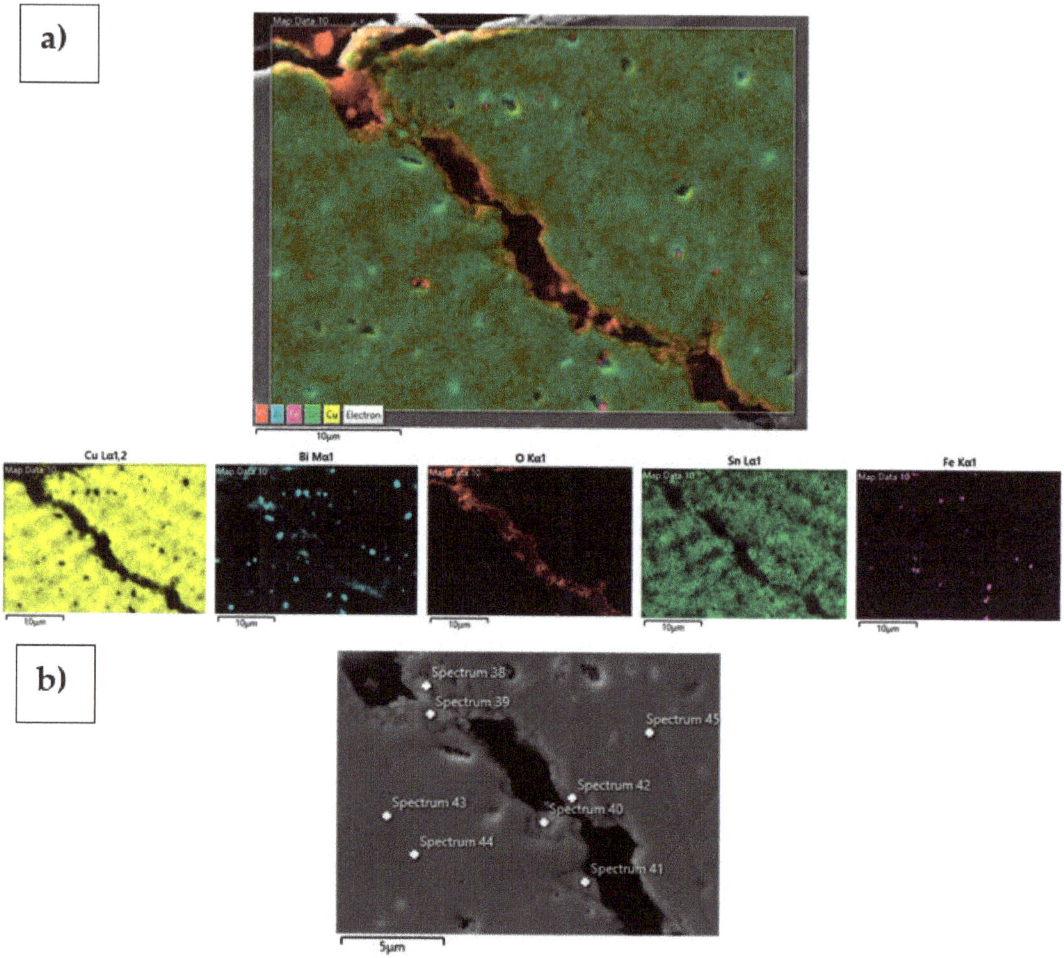

Spectrum label	Spectrum 38	Spectrum 39	Spectrum 40	Spectrum 41	Spectrum 42	Spectrum 43	Spectrum 44	Spectrum 45
O	11.70	9.14	13.65	13.80	12.63	0.82	0.69	0.56
Cr	0.03	0.00	0.05	0.09	0.10	0.11	0.07	0.00
Mn	0.14	0.07	0.09	0.04	0.00	0.00	0.03	0.00
Fe	0.46	0.62	1.14	0.37	0.42	0.73	0.33	0.36
Cu	34.37	66.66	53.55	50.03	61.87	91.03	91.01	88.73
Sn	11.02	7.14	12.36	8.78	8.48	7.03	7.75	9.92
Bi	42.28	16.37	19.16	26.88	16.51	0.28	0.13	0.43
Total	100.00	100.00	100.00	100.00	100.00	100.00	100.00	100.00

Figure 17. Structural cracks observed in the CL cross-section realized at P 300 W and F 750 mm/min in an open deposition environment: (**a**) EDX map; (**b**) EDX point analysis.

4. Conclusions

In this research work, the influence of the LMD process parameters on the manufacturing of high-density, crack-free $CuSn_{11}Bi_3$ depositions on C45 surfaces is analyzed by characterizing the influence of laser power, axis speed, substrate preheating, ST overlapping, and environmental printing conditions in terms of O_2 percentage inside the building chamber on the final structural integrity of single circular layers. Figure 18 summarizes the evolution of the observed internal defects in relation to the levels of E_s and substrate preheating involved in the experimental campaign. The main experimental outputs highlight:

- High-density (>99%) and crack-free $CuSn_{11}Bi_3$ clads on C45 are achieved through specific energy ranging between 17 J/mm^2 and 40 J/mm^2;
- The employment of high energy densities (E_s > 40 J/mm^2) degrades the structural integrity of the as-deposited material, inducing keyholes and LME cracks as a result of Cu diffusion along steel grain boundaries;
- The use of high laser power (ranging from 500 W to 800 W) and of substrate preheating produces large dilution zones and steel diffusion into the Cu-based clad, promoting brittle mechanical behavior of the as-deposited material and the propagation of structural cracks into as-built ST cross-sections;
- In crack-free depositions, no Cu diffusion into the C45 substrate is observed;
- In CL fabrication, no delamination is detected regardless of the use of substrate preheating;
- In Cu-based LMD, an enclosed deposition environment improves the final quality of the as-built coating by inhibiting material oxidation and preventing the formation of internal cracks that degrade the structural integrity of the deposition;
- Based on the achieved results, LMD demonstrates the capability of realizing thicker high-quality coatings compared to electroplating, ranging from 0.162 mm to 0.530 mm, which makes this AM technology more efficient in manufacturing bimetallic parts for the hydraulic, power generation, and aerospace sectors;

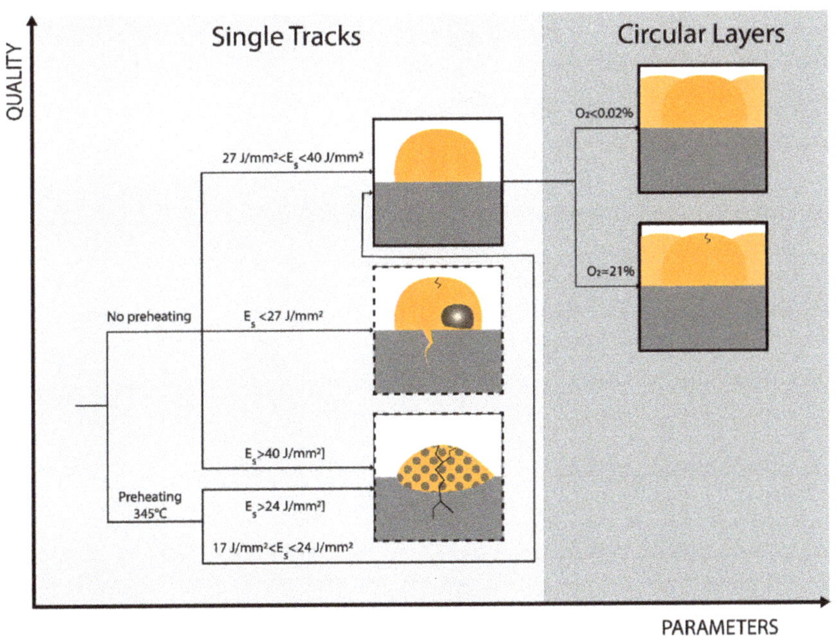

Figure 18. Diagram showing the evolution of the observed internal defects with regard to the involved Es and substrate preheating.

- In addition, despite Thermal Spraying (TS) being a competing technology for coating manufacturing, the high susceptibility of Cu-based alloys to form oxides and the moderate porosity typically featured by TS make process optimization challenging for the manufacture of defect-free Cu-Steel bimetallic components. On the contrary, the possibility of implementing inert deposition chambers and internal densities higher than 99% increases LMD attractiveness at the industrial level.

5. Future Works

Based on the main experimental findings, the next research activities will be related to (a) the characterization of the tribological properties of the as-built CL coating and (b) the optimization of the LMD process for the manufacturing of $CuSn_{11}Bi_3$ multi-layer depositions to enable thicker Cu-based depositions. Indeed, steel components coated with Cu-based materials generally exhibit a coating thickness exceeding 1 mm; therefore, it is essential to achieve coatings with a greater thickness than those measured on the CLs. To address this limitation, future efforts should focus on investigating how the alteration of the chemical and optical properties of the deposited material—resulting from the prior deposition of copper-based layers—affects the performance of the LMD process and the final coating quality. Structural characterizations, including porosity analysis, mechanical testing, and residual stress measurements, will be performed, and the follow-up activities will be extended to industrial use cases.

Author Contributions: F.M. is the main author of the manuscript, providing his contribution in Funding acquisition, Writing—original draft preparation, Writing—review and editing, Investigation, Methodology, and Project administration. E.B. carried out the internal and chemical assessment of the realized samples, giving Data curation, Writing—original draft preparation, Investigation, and Methodology. S.R.-M. contributed to Formal analysis, Funding acquisition, Methodology, Project administration, and Supervision. A.V. gave her contribution to Writing—review and editing and Supervision. All authors have read and agreed to the published version of the manuscript.

Funding: This work was supported by the innovation project 49013.1 IP-ENG "METIS" funded by the Innosuisse Swiss Innovation Agency.

Institutional Review Board Statement: Not applicable.

Informed Consent Statement: Not applicable.

Data Availability Statement: The original contributions presented in this study are included in the article. Further inquiries can be directed to the corresponding author.

Acknowledgments: Special recognition goes to the implementation partner Kugler Bimetal SA, owner of the experimental material used in this research work, as well as the main industrial partner of the innovation project, which supported the research activities by providing its expertise in the production of bimetallic structural components.

Conflicts of Interest: The authors declare no conflicts of interest.

References

1. Srivastava, M.; Rathee, S.; Maheshwari, S.; Kundra, T.K. Design and Processing of Functionally Graded Material: Review and Current Status of Research. In *3D Printing and Additive Manufacturing Technologies*; Kumar, L.J., Pandey, P.M., Wimpenny, D.I., Eds.; Springer: Singapore, 2019; pp. 243–255. [CrossRef]
2. Rodrigues, T.A.; Bairrão, N.; Farias, F.W.C.; Shamsolhodaei, A.; Shen, J.; Zhou, N.; Maawad, E.; Schell, N.; Santos, T.G.; Oliveira, J.P. Steel-copper functionally graded material produced by twin-wire and arc additive manufacturing (T-WAAM). *Mater. Des.* **2022**, *213*, 110270. [CrossRef]
3. Sanjeeviprakash, K.; Kannan, A.R.; Shanmugam, N.S. Additive manufacturing of metal-based functionally graded materials: Overview, recent advancements and challenges. *J. Braz. Soc. Mech. Sci. Eng.* **2023**, *45*, 1–35. [CrossRef]

4. Priyadarshi, P.; Yadav, A.; Maurya, R. Deposition of Cu-based composite coatings on mild steel substrate via electrodeposition method. *Mater. Today Proc.* **2022**, *62*, 7564–7569. [CrossRef]
5. Kumar, R.R.; Rajalakshmy, P.; Saranya, M.D.; Kirubakaran, S.; Elwin, J.G.R.; Marichamy, S. Process automation through internet of things on copper coating process of stainless steel. *Mater. Today Proc.* **2021**, *45*, 2354–2356. [CrossRef]
6. Kuksenova, L.I.; Kozlov, D.A.; Alekseeva, M.S. Operational Properties of Steel Friction Units with Copper-Alloy Antifriction Coating Formed via Electric Spark Alloying. *Inorg. Mater. Appl. Res.* **2022**, *13*, 745–753. [CrossRef]
7. Wan, L.; Cheng, M.; Fu, G.; Wei, C.; Shi, T.; Shi, S. Annular laser cladding of CuPb10Sn10 copper alloy for high-quality anti-friction coating on 42CrMo steel surface. *Opt. Laser Technol.* **2023**, *158*, 108878. [CrossRef]
8. Wang, Y.; Gao, Y.; Li, Y.; Zhai, W.; Sun, L.; Zhang, C. Review of preparation and application of copper—Steel bimetal composites. *Emerg. Mater. Res.* **2019**, *8*, 538–551. [CrossRef]
9. REACH Legislation—ECHA. Available online: https://echa.europa.eu/regulations/reach/legislation (accessed on 17 July 2020).
10. Jung, S.; Kara, L.B.; Nie, Z.; Simpson, T.W.; Whitefoot, K.S. Is Additive Manufacturing an Environmentally and Economically Preferred Alternative for Mass Production? *Environ. Sci. Technol.* **2023**, *57*, 6373–6386. [CrossRef]
11. Puukko, P.; Hepo-oja, L.; Vatanen, S.; Kaipainen, J.; Mäkinen, M.L.; Lindqvist, M. Case study of the environmental impacts and benefits of a PBF-LB additively manufactured optimized filtrate nozzle. *Sci. Rep.* **2024**, *14*, 22183. [CrossRef]
12. Baroutaji, A.; Arjunan, A.; Robinson, J.; Vance, A.; Arafat, A. Critical Review on the Sustainability of Metal Additive Manufacturing: Environmental and Economic Perspectives. *Renew. Sustain. Energy Technol.* **2019**, *8*, 538–551. [CrossRef]
13. Yan, L.; Chen, Y.; Liou, F. Additive manufacturing of functionally graded metallic materials using laser metal deposition. *Addit. Manuf.* **2020**, *31*, 100901. [CrossRef]
14. Singh, D.D.; Arjula, S.; Reddy, A.R. Functionally Graded Materials Manufactured by Direct Energy Deposition: A review. *Mater. Today Proc.* **2021**, *47*, 2450–2456. [CrossRef]
15. Ghanavati, R.; Naffakh-Moosavy, H.; Moradi, M.; Mazzucato, F.; Valente, A.; Bagherifard, S.; Saboori, A. Design optimization for defect-free AISI 316 L/IN718 functionally graded materials produced by laser additive manufacturing. *Mater. Charact.* **2025**, *220*, 114697. [CrossRef]
16. Pan, T.; Zhang, X.; Yamazaki, T.; Sutton, A.; Cui, W.; Li, L.; Liou, F. Characteristics of Inconel 625—Copper bimetallic structure fabricated by directed energy deposition. *Int. J. Adv. Manuf. Technol.* **2020**, *109*, 1261–1274. [CrossRef]
17. Leino, M.; Pekkarinen, J.; Soukka, R. The role of laser additive manufacturing methods of metals in repair, refurbishment and remanufacturing—Enabling circular economy. *Phys. Procedia* **2016**, *83*, 752–760. [CrossRef]
18. Mazzucato, F.; Forni, D.; Valente, A.; Cadoni, E. Laser Metal Deposition of Inconel 718 Alloy and As-built Mechanical Properties Compared to Casting. *Materials* **2021**, *14*, 437. [CrossRef] [PubMed]
19. Ehmsen, S.; Klar, M.; Aurich, J.C. Evaluating the environmental impact of additive manufacturing: A methodology to determine the environmental impact of parts manufactured by high-speed laser directed energy deposition. *Procedia CIRP* **2024**, *122*, 91–96. [CrossRef]
20. Yadav, S.; Paul, C.P.; Rai, A.K.; Singh, R.; Dixit, S.K. Elucidating laser directed energy deposition based additive manufacturing of copper-stainless steel functionally graded material: Processing and material behavior. *J. Manuf. Process.* **2023**, *92*, 107–123. [CrossRef]
21. Piscopo, G.; Atzeni, E.; Biamino, S.; Iuliano, L.; Mazzucato, F.; Saboori, A.; Salmi, A.; Valente, A. Analysis of single tracks of IN718 produced by laser powder directed energy deposition process. *Procedia CIRP* **2022**, *112*, 340–345. [CrossRef]
22. Mazzucato, F.; Avram, O.; Valente, A.; Carpanzano, E. Recent Advances Toward the Industrialization of Metal Additive Manufacturing. In *Systems Engineering in the Fourth Industrial Revolution: Big Data, Novel Technologies, and Modern Systems Engineering*; John Wiley & Sons, Inc.: Hoboken, NJ, USA, 2020; pp. 273–319. [CrossRef]
23. Piscopo, G.; Atzeni, E.; Saboori, A.; Salmi, A. An Overview of the Process Mechanisms in the Laser Powder Directed Energy Deposition. *Appl. Sci.* **2022**, *13*, 117. [CrossRef]
24. Saboori, A.; Aversa, A.; Marchese, G.; Biamino, S.; Lombardi, M.; Fino, P. Application of Directed Energy Deposition-Based Additive Manufacturing in Repair. *Appl. Sci.* **2019**, *9*, 3316. [CrossRef]
25. Solomon, I.J.; Sevvel, P.; Gunasekaran, J.; Tanushkumaar, P. A review on additive manufacturing of alloys using laser metal deposition. *Mater. Today Proc.* **2022**, *64*, 44–50. [CrossRef]
26. Gopan, V.; Dev, L.; Surendran, A. Innovative potential of additive friction stir deposition among current laser based metal additive manufacturing processes: A review. *CIRP J. Manuf. Sci. Technol.* **2021**, *32*, 228–248. [CrossRef]
27. Fatoba, O.S.; Akinlabi, S.A.; Akinlabi, E.T. Laser Metal Deposition Influence on the Mechanical Properties of Steels and Stainless Steel Composites: A Review. *Mater. Today Proc.* **2018**, *5*, 18603–18620. [CrossRef]
28. Makarenko, K.; Dubinin, O.; Shornikov, P.; Shishkovsky, I. Specific aspects of the transitional layer forming in the aluminium bronze—Stainless steel functionally graded structures after laser metal deposition. *Procedia CIRP* **2020**, *94*, 346–351. [CrossRef]
29. Khodabakhshi, F.; Farshidianfar, M.H.; Bakhshivash, S.; Gerlich, A.P.; Khajepour, A. Dissimilar metals deposition by directed energy based on powder-fed laser additive manufacturing. *J. Manuf. Process.* **2019**, *43*, 83–97. [CrossRef]

30. Grandhi, M.; Nguyen, V.; Liu, Z.; Romo-De-La-cruz, C.O.; Song, X. Copper-nickel functionally magnetic gradient material fabricated via directed energy deposition. *J. Manuf. Process.* **2023**, *100*, 47–54. [CrossRef]
31. Liu, K.; Yan, Z.; Pan, R.; Wang, F.; Chen, S. Effect of deposition sequence on interfacial characteristics of Inconel—Copper functional bimetallic structures fabricated by directed energy deposition-arc. *Mater. Lett.* **2023**, *345*, 134487. [CrossRef]
32. Yadav, S.; Paul, C.P.; Jinoop, A.N.; Rai, A.K.; Bindra, K.S. Laser Directed Energy Deposition based Additive Manufacturing of Copper: Process Development and Material Characterizations. *J. Manuf. Process.* **2020**, *58*, 984–997. [CrossRef]
33. Schäfle, M.B.; Fett, M.; Gärtner, J.; Kirchner, E. Additive Manufacturing of Copper—A Survey on Current Needs and Challenges. *J. Manuf. Mater. Process.* **2025**, *9*, 109. [CrossRef]
34. Zhong, H.Z.; Li, C.G.; Zhang, X.Y.; Gu, J.F. The graded microstructures evolving with thermal cycles in pure copper processed by laser metal deposition. *Mater. Lett.* **2018**, *230*, 215–218. [CrossRef]
35. Xiao, Y.; Sun, C.; Wu, X.; Zhu, X.; Kan, Y.; Zhang, Y. Restoration of pure copper motor commutator for aviation by laser powder deposition. *J. Mater. Res. Technol.* **2023**, *23*, 5796–5806. [CrossRef]
36. Dhara, S.; Finuf, M.; Zediker, M.; Masters, I.; Barai, A.; Das, A. Utilising blue laser over infrared laser to enhance control of penetration depth and weld strength for producing electric vehicle battery interconnects. *J. Mech. Work. Technol.* **2023**, *317*, 117989. [CrossRef]
37. Yamaguchi, M.; Yamazaki, M.; Funada, Y.; Yachi, T.; Saikai, A.; Furumoto, T. Influence of process parameters on the geometry and surface defects of the single-line track in the directed energy deposition of pure copper onto 304 stainless steel using a blue laser. *Int. J. Adv. Manuf. Technol.* **2022**, *121*, 1091–1100. [CrossRef]
38. Wei, Y.; Wen, Y.; Du, J.; Tian, Y.; Zhou, H.; Dong, X.; Shen, Z.; Liu, L.; Xiong, G.; Zhou, W.; et al. Melt pool behavior of pure copper-stainless steel multi-materials fabricated by the coaxial hybrid blue-infrared lasers additive manufacturing. *J. Manuf. Process.* **2024**, *125*, 50–63. [CrossRef]
39. Imran, M.K.; Masood, S.H.; Brandt, M. Direct Metal Deposition of H13 Tool Steel on Copper Alloy Substrate: Parametric Investigation. *Lasers Manuf. Mater. Process.* **2015**, *2*, 242–260. [CrossRef]
40. Singh, S.; Kumar, M.; Sodhi, G.P.S.; Buddu, R.K.; Singh, H. Development of thick copper claddings on SS316L steel for In-vessel components of fusion reactors and copper-cast iron canisters. *Fusion Eng. Des.* **2018**, *128*, 126–137. [CrossRef]
41. Nemani, A.V.; Ghaffari, M.; Bokati, K.S.; Valizade, N.; Afshari, E.; Nasiri, A. Advancements in Additive Manufacturing for Copper-Based Alloys and Composites: A Comprehensive Review. *J. Manuf. Mater. Process.* **2024**, *8*, 54. [CrossRef]
42. Zhang, X.; Pan, T.; Flood, A.; Chen, Y.; Zhang, Y.; Liou, F. Investigation of copper/stainless steel multi-metallic materials fabricated by laser metal deposition. *Mater. Sci. Eng. A* **2021**, *811*, 141071. [CrossRef]
43. Chen, C.; Zhou, J.; Xue, F.; Wu, Q. Elimination of liquid metal embrittlement cracks during arc cladding of tin bronze on steel sheet. *Mater. Lett.* **2020**, *269*, 127646. [CrossRef]
44. Wang, B.; Ge, F.; Zhang, H.; Du, R.; Fan, Y.; Pan, J. Interface microstructure and mechanical properties of copper-steel joints with nickel transition layer prepared by ultrasonic additive manufacturing. *J. Mater. Res. Technol.* **2023**, *23*, 6150–6161. [CrossRef]
45. Zhang, X.; Pan, T.; Chen, Y.; Li, L.; Zhang, Y.; Liou, F. Additive manufacturing of copper-stainless steel hybrid components using laser-aided directed energy deposition. *J. Mater. Sci. Technol.* **2021**, *80*, 100–116. [CrossRef]
46. Bronze Tokat—Kugler Bimetal SA. Available online: https://www.bimetal.ch/bronze-tokat/ (accessed on 4 July 2023).
47. Mazzucato, F.; Aversa, A.; Doglione, R.; Biamino, S.; Valente, A.; Lombardi, M. Influence of Process Parameters and Deposition Strategy on Laser Metal Deposition of 316L Powder. *Metals* **2019**, *9*, 1160. [CrossRef]
48. Zhang, K.; Chen, Y.; Marussi, S.; Fan, X.; Fitzpatrick, M.; Bhagavath, S.; Majkut, M.; Lukic, B.; Jakata, K.; Rack, A.; et al. Pore evolution mechanisms during directed energy deposition additive manufacturing. *Nat. Commun.* **2024**, *15*, 1715. [CrossRef] [PubMed]
49. Kim, K.H.; Jung, C.H.; Jeong, D.Y.; Hyun, S.K. Preventing Evaporation Products for High-Quality Metal Film in Directed Energy Deposition: A Review. *Metals* **2021**, *11*, 353. [CrossRef]
50. Svetlizky, D.; Das, M.; Zheng, B.; Vyatskikh, A.L.; Bose, S.; Bandyopadhyay, A.; Schoenung, J.M.; Lavernia, E.J.; Eliaz, N. Directed energy deposition (DED) additive manufacturing: Physical characteristics, defects, challenges and applications. *Mater. Today* **2021**, *49*, 271–295. [CrossRef]
51. Marot, L.; De Temmerman, G.; Oelhafen, P.; Covarel, G.; Litnovsky, A. Rhodium coated mirrors deposited by magnetron sputtering for fusion applications. *Rev. Sci. Instrum.* **2007**, *78*, 103507. [CrossRef]

materials

MDPI

Article

Crack Arrest Effect of FeMnNiSi-Inconel625-Ni60 Laminated Structure Prepared by Laser Cladding Additive Manufacturing

Lihong Ding [1,2,*], Weining Lei [2,3] and Jufang Chen [3]

1 Engineering Training Center, Jiangsu University of Technology, Changzhou 213001, China
2 Jiangsu Key Laboratory of Advanced Materials Design and Additive Manufacturing, Changzhou 213001, China; leiweining@jsut.edu.cn
3 School of Mechanical Engineering, Jiangsu University of Technology, Changzhou 213001, China; jfchen@jstu.edu.cn
* Correspondence: dlh900819@163.com

Abstract

This study addresses the technical challenges of cracking and surface crack initiation in Ni60 alloy cladding layers fabricated by laser cladding additive manufacturing on FeMnNiSi alloys. An innovative FeMnNiSi-Inconel625-Ni60 laminate design was proposed, achieving metallurgical bonding of the dissimilar materials through an Inconel625 transition layer. This effectively addresses the interfacial stress concentration issue caused by differences in thermal expansion coefficients in conventional processes. The results demonstrate that the interfacial microstructure is regulated by synergistic Nb-Mo element segregation, promoting the precipitation of γ'' phase and the formation of a nanoscale Laves phase. This phase not only inhibits carbide aggregation and growth, refining grain size, but also deflects crack propagation paths by pinning dislocations, achieving a dual mechanism of stress reduction and crack arrest. The Ni60 cladding layer in the laminated structure exhibits an average surface microhardness of 641.31 $HV_{0.3}$, 3.88 times that of the substrate (165.22 $HV_{0.3}$), while the Inconel625 base layer shows 340.71 $HV_{0.3}$, 2.06 times the substrate's value. Wear testing reveals the laminated cladding layer has a wear volume of 0.086 mm^3 (0.243 mm^3 less than the substrate's 0.329 mm^3) and a wear rate of 0.86×10^{-2} $mm^3/(N{\cdot}m)$, 73.86% lower than the substrate's 3.29×10^{-2} $mm^3/(N{\cdot}m)$, indicating superior wear resistance. The electrochemical test results show that under the same corrosion conditions, the self-corrosion potential and polarization resistance of the FeMnNiSi-Inconel625-Ni60 cladding layer are significantly higher than those of the substrate, while the corrosion current density is significantly lower than that of the substrate. The frequency stability region at the highest impedance modulus $|Z|$ is wider than that of the substrate, and the corrosion rate is 71.86% slower than that of the substrate, demonstrating excellent wear resistance. This study not only reveals the mechanism by which Laves phases improve interfacial properties through microstructural regulation but also provides a scalable interface design strategy for heterogeneous material additive manufacturing, which has important engineering value in promoting the application of laser cladding technology in the field of high-end equipment repair.

Keywords: laser cladding; laminated structure; crack arrest; wear resistance; corrosion resistance

1. Introduction

FeMnNiSi mold steel, as a typical low-cost medium-carbon steel, has a carbon content of about 0.45%. Due to its excellent processing performance and good comprehensive mechanical properties, it is widely used in mold manufacturing, mechanical transmission components, and fixtures [1]. However, this material has significant deficiencies in hardness, wear resistance, and corrosion resistance. Especially under high-load and high-wear conditions, it is very prone to wear failure and plastic deformation.

Laser cladding technology uses high-energy laser beams to achieve surface modification of materials and has outstanding performance in improving tribological properties. Compared with the two-phase process (nitriding), which is only suitable for surface hardening, laser cladding can directly repair worn parts. Its core advantage lies not only in the depth of the nitriding layer but also in its ability to simultaneously achieve alloying, cladding, and strengthening, and it is suitable for complex geometric shapes and various material systems. The non-equilibrium structure formed through the rapid cooling of this process makes the deposited layer usually present a fine-grained structure, with low porosity and metallurgical bonding characteristics. Its heat-affected zone (HAZ) is narrow, and the energy input can be precisely controlled, thereby reducing the thermal deformation of the substrate [2,3]. Due to its unique advantages, laser cladding technology has been widely used in material surface strengthening in recent years. However, its fatigue life improvement effect under compressive stress is limited and is greatly affected by material and process parameters [4].

In response to this problem, domestic and foreign scholars have conducted in-depth research on improving the surface properties of steel using laser cladding technology. Li T et al. [5] conducted multi-objective optimization of the process parameters of laser cladding Ni60 alloy powder on 45 steel. The optimized process parameters can obtain a cladding structure with refined grains and improve the surface properties of 45 steel. Luo Xixi et al. [6] studied the preparation of an Al-Cr composite strengthening layer on the surface of 45 steel and verified that a composite passivation film was formed on the surface of the Al-Cr coating with excellent corrosion resistance. Liu Lilan et al. [7] proposed a multivariate regression prediction method for optimization targets. The results showed that in the laser cladding process, increasing the laser power and reducing the powder feeding rate and scanning speed can reduce the crack density and improve the cladding performance. Cao Jinlong et al. [8] conducted research on the Ni60-TiC gradient layer structure and studied the cladding of TiC coating on the surface of 45 steel using a CO_2 constant current laser. Wang Sheng et al. [9] prepared Fe35 on the surface of 45 steel material. An alloy layer, under laser cladding parameters of a laser power of 2100 W, scanning speed of 5 mm/s, and powder feeding speed of 15 g/min, can obtain fine and uniform grains, effectively reducing brittle carbides. Wang Yuhao et al. [10] used electromagnetic field-assisted laser cladding technology to prepare a dense Ni60 coating on the surface of 45 steel. As a result, the stirring effect of the surface electromagnetic field on the molten pool will accelerate the heat and mass transfer of the cladding layer, thereby reducing the generation of substances such as CrB and Ni_2B and enhancing the surface properties of 45 steel.

Although the above researchers have made some progress, there are still some shortcomings in the following aspects:

1. Most of the existing studies prepare a single coating on the substrate surface, and there is a lack of research on laminated structures.
2. Most of the existing studies prepare specific alloy materials Ni60 or coating $Fe_{35}A$ on the substrate surface and optimize the process parameters, but lack the universal verification of the material system.
3. Most of the existing research aims to refine grains, reduce cracks, and enhance the fusion between substrate and cladding by optimizing process parameters and increasing surface electromagnetic fields, but there is no targeted related research.

Ni60 has a wide range of application prospects in the field of surface engineering due to its unique performance advantages. Not only does this alloy have excellent corrosion resistance, high hardness, and good wear resistance, but also in injection mold applications, the Ni60 cladding layer can extend mold life by 3–5 times and reduce maintenance costs by more than 40% [11,12]. The advantages of Ni60 alloy effectively make up for the shortcomings of FeMnNiSi steel performance, making it an ideal material for FeMnNiSi steel laser cladding [13]. However, the Ni60 cladding layer is prone to generate high residual stress during rapid solidification, resulting in a significant cracking tendency, especially the initiation and expansion of microcracks, which seriously restricts its service reliability in harsh environments [14–17]. The surface modification of the Ni60 cladding layer by laser cladding on FeMnNiSi steel faces two key technical difficulties in the preparation process: one is the residual stress problem caused by the difference in the thermal expansion coefficient between Ni60 and the substrate, which can easily cause longitudinal cracks [18]; the other is the formation of brittle carbide Cr_7C_3 in the interface area, which reduces the bonding strength by 30–40% [19].

This study addresses the technical challenge of cracking by laser-cladding Ni60 alloy on FeMnNiSi alloy. Through the proposal of an innovative design using an Inconel625 cladding layer as an intermediate transition layer, the goal is to construct an FeMnNiSi-Inconel625-Ni60 laminate structure, mitigating the difference in thermal expansion coefficients between FeMnNiSi and Ni60 while simultaneously achieving a metallurgical interlock between the FeMnNiSi matrix and the Ni60 functional layer. Not only does this research provide a theoretical basis for the additive manufacturing of heterogeneous materials, but its technical approach can also be directly applied to the surface modification of key components in aircraft engines, marine equipment, and other applications requiring synergistic wear and corrosion resistance, possessing significant engineering application value.

2. Experimental Materials and Methods

2.1. Test Materials

The substrate material employed in this study was FeMnNiSi die steel with dimensions of 100 mm × 100 mm × 10 mm. Two nickel-based alloy powders were selected for laser cladding: Ni60 (supplied by AVIC Zhongmai Metal Materials Research Institute, Beijing, China) and Inconel625 (supplied by Höganäs AB, Höganäs, Sweden). To ensure suitability for laser cladding processing, the powders were sieved to a particle size range of 45~105 μm, with a purity exceeding 99.9% and spheroidal morphology. Prior to laser cladding, both Ni60 and Inconel625 powders were subjected to a drying process at 200 °C for 2 h in an FZG-8 vacuum drying oven to eliminate moisture and enhance experimental precision. Table 1 presents the chemical composition and elemental content of the substrate, Ni60, and Inconel625.

Table 1. Composition and content of alloy elements in substrate, cladding layer Ni60, and Inconel625 (mass fraction/%).

Material	C	Ni	Cr	Mo	Mn	Nb	B	Si	Fe
FeMnNiSi	0.44	0.20	0.20	—	0.8	—	—	0.30	Bal
Ni60	0.10	60	16	2	—	—	3.5	3.5	Bal
Inconel625	0.10	63	20	8	0.3	3.55	—	0.1	3.0

2.2. Laser Cladding Process

The laser cladding test is a laser cladding forming system composed of a 4 kW IPG laser, KR60HA KUKA robot (Tennessee Industrial Electronics, La Vergne, TN, USA), GTVadjustable dual-material bin, airborne powder feeder, and worktable. The YLS-4000 fiber laser (IPG Photonics, Oxford, MA, USA) was used for single-channel and single-layer laser cladding tests with four synchronous coaxial powder feeders. The laser power was 1.5 kW, the spot diameter was 3 mm, the scanning speed was 6 mm/s, and the powder feeding rate was 10 g/min. During the laser cladding process, high-purity inert argon gas was used for protection, and the gas flow rate was set to 7 L/min. Two different cladding layers were prepared in the experiment, namely FeMnNiSi-Ni60 single cladding layer and FeMnNiSi-Inconel625-Ni60 laminated structure cladding layer.

2.3. Preparation and Characterization of Test Samples

Three sample geometries ($10 \times 10 \times 10$ mm^3, $20 \times 10 \times 10$ mm^3, $10 \times 10 \times 2$ mm^3) were sectioned perpendicular to the laser scan direction using an SG3000 electric discharge wire cutter (Suzhou Sanguang Science & Technology Co., Suzhou, China). For microstructural analysis, the $10 \times 10 \times 10$ mm^3 samples underwent cross-sectional grinding/polishing, followed by ultrasonic cleaning in anhydrous ethanol and chemical etching with a mixed solution of 20 mL HNO_3 and 20 mL HCl. Microhardness measurements (300 gf load, 10 s dwell time) were performed on the cladding layer using an HXD-1000TMS tester (SDL Atlas, Singapore), with three readings averaged per horizontal line.

Wear resistance was evaluated on $20 \times 10 \times 10$ mm^3 samples polished to 2000 grit. Pre- and post-test mass measurements (0.1 mg precision balance) were conducted after 30 min sliding against G5-G100 Si_3N_4 ceramic balls (6 mm diameter, 20 N load, 5 mm stroke). Three-dimensional wear scar profiles were quantified via NANOPS50 profilometry (SCANTECH, Hangzhou, China).

Corrosion behavior was assessed via potentiodynamic polarization tests (1 cm^2 exposed area) using a CHI600E workstation (Shanghai Chenhua, Shanghai, China) on polished $10 \times 10 \times 10$ mm^3 samples. Phase analysis of the $10 \times 10 \times 2$ mm^3 samples employed XRD (HD XpertPRO (ZEISS, Oberkochen, Germany)), while cross-sectional microstructural examination utilized SEM (SIGMA500 (ZEISS, Oberkochen, Germany)) with EDS (INCA OXFORD (Oxford Instruments, Abingdon, UK)) for elemental mapping.

3. Test Results

3.1. Interface Micromorphology

The micromorphology of the interface region between the FeMnNiSi-Ni60 single cladding layer and the FeMnNiSi-Inconel625-Ni60 laminated structure was observed under a scanning electron microscope. Figure 1(a-1,a-2) show the micromorphology of the interface region of the single cladding layer; Figure 1(b-1) shows the micromorphology of the FeMnNiSi-Inconel625 laminated structure interface, and Figure 1(b-2) shows the micromorphology of the Inconel625-Ni60 laminated structure interface. Figure 1(c-1) shows the micromorphology of the Inconel625 transition layer of the laminated structure, and Figure 1(c-2) shows the micromorphology of the Ni60 laminated structure cladding layer.

As shown in Figure 1(a-1), there is a black gap defect parallel to the substrate in the fusion zone. This is a non-fusion phenomenon caused by insufficient fluidity of the molten pool. Its outline is clear and linearly distributed, which conforms to the morphological characteristics of typical non-fusion defects. As shown in Figure 1(a-2), there is a large difference in the thermal expansion coefficient between the substrate FeMnNiSi and the Ni60 cladding layer. The difference in contraction during cooling produces high tensile stress. Ni60 itself is brittle and has weak tensile stress resistance, which leads to cracks initiating in the fusion zone and extending to the cladding layer area, presenting a transgranular fracture micromorphology.

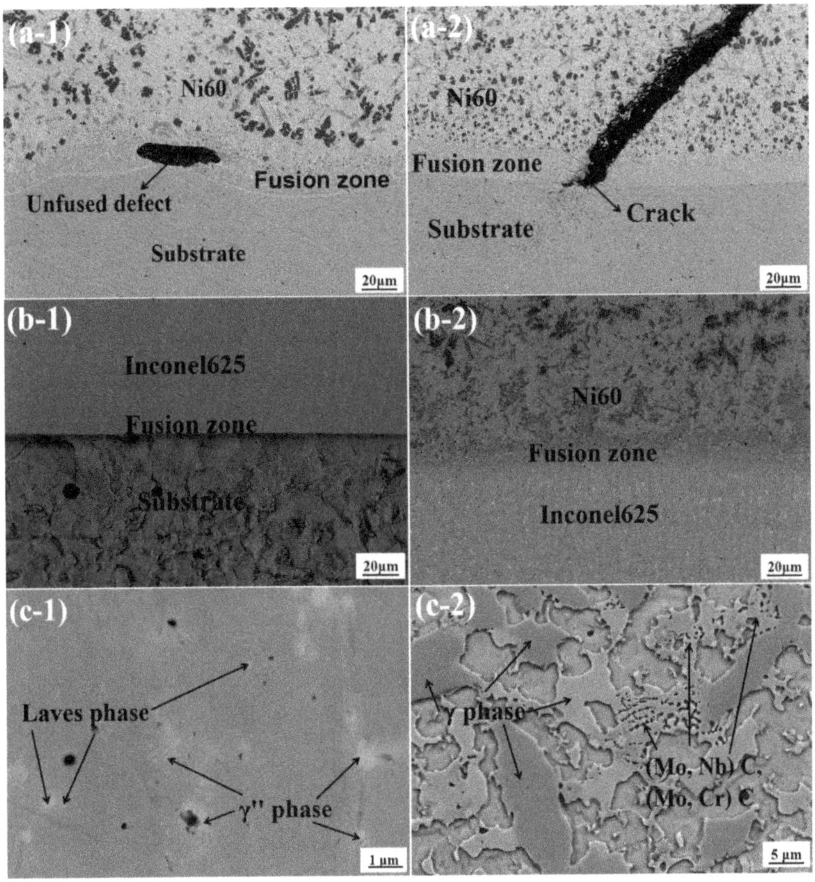

Figure 1. Interface micromorphology. (**a-1**) Unfused defect at the interface of a single cladding layer. (**a-2**) Crack at the interface of a single cladding layer. (**b-1**) Micromorphology of the FeMnNiSi-Inconel625 interface of the laminated structure. (**b-2**) Micromorphology of the Inconel625-Ni60 interface of the laminated structure. (**c-1**) micromorphology of Inconel625 transition layer in laminated structure (**c-2**) micromorphology of Ni60 cladding layer in laminated structure.

In Figure 1(b-1), no obvious cladding defects are observed at the interface of the laminated structure FeMnNiSi-Inconel625. Transition layer Inconel625 has excellent plasticity and a thermal expansion coefficient similar to matrix FeMnNiSi, effectively alleviating thermal stress concentration. As shown in Figure 1(b-2), there are also no obvious cladding defects at the interface of the laminated structure Inconel625-Ni60. The cladding layer Inconel625 has a face-centered cubic structure and solid solution-strengthening effect. Its high toughness forms a gradient transition with Ni60, inhibiting crack propagation [20].

In Figure 1(c-1), at a magnification of 20,000 times, the micromorphology of the Inconel625 transition layer of the laminated structure reveals the precipitation of multiple white bright spots in the interdendritic gaps. Analysis of the percentage of alloying elements at each point reveals that this phase is a nanoscale Laves phase, which can reduce the segregation of harmful elements (such as Cr and Mo) at grain boundaries, reducing grain boundary brittleness. At the same time, pinning the grain boundaries inhibits excessive grain growth, achieving grain refinement and effectively suppressing the generation of cladding defects. A γ' phase is also generated in the transition layer region, forming a coherent strain field with the matrix. The mismatch is only 0.08%, significantly improving the yield strength of the alloy. The uniform precipitation of the γ' phase can greatly increase the room temperature tensile strength, which also reflects the crack arrest effect of the FeMnNiSi-Inconel625-Ni60 laminated structure. As shown in Figure 1(c-2), at a magnification of 5000 times, a γ-Ni solid solution phase is observed in the Ni60 laminated surface cladding layer. This phase provides plastic deformation capability, preventing the spalling of the cladding layer due to the brittleness of carbides. Furthermore, the γ-Ni matrix absorbs impact energy through plastic deformation, slowing crack propagation. Furthermore, a high-hardness and wear-resistant (Mo,Nb)C or (Mo,Cr)C solid solution-strengthening phase is generated in the surface cladding layer [21]. The coherent interface (stellate precipitation structure) with the γ-Ni matrix reduces stress concentration and improves the bonding strength between the cladding layer and the substrate.

In summary, from the microstructural observation, it was found that the laminated structure FeMnNiSi-Inconel625-Ni60 effectively inhibited the initiation and propagation of cracks through the dual design of transition layer strengthening and cladding layer toughening, providing a stable foundation for matrix surface strengthening.

3.2. Interface Element Analysis

To investigate the crack-arresting mechanism of the FeMnNiSi-Inconel625-Ni60 laminated structure, we performed scanning electron microscopy (SEM) coupled with energy-dispersive X-ray spectroscopy (EDS) using an INCA OXFORD system. This enabled simultaneous microstructural characterization and elemental mapping at the substrate-cladding interface. Figure 2 shows the microstructure and element distribution curve of the FeMnNiSi-Ni60 single cladding layer interface; Figure 3 shows the microstructure and element distribution curve of the FeMnNiSi-Inconel625 interface in the FeMnNiSi-Inconel625-Ni60 laminated structure; and Figure 4 shows the microstructure and element distribution of the Inconel625-Ni60 interface in the FeMnNiSi-Inconel625-Ni60 laminated structure.

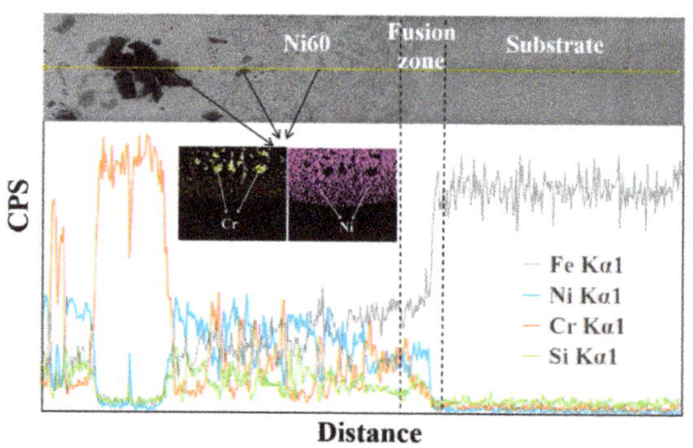

Figure 2. Element analysis of FeMnNiSi-Ni60 interface.

As shown in Figure 2, the black phase area of the online scanning shows the phenomenon of multiple-Cr-element aggregation and a significant reduction in Ni element. Cr-rich areas are prone to forming hard and brittle BCC phases, such as $Cr_{23}C_6$ carbides, which can lead to cladding defects. While Cr elements are gathering, the Ni element content is significantly reduced. The Ni element reduction area is mainly composed of FCC phases. The difference in thermal expansion coefficients between the BCC phase and the FCC phase leads to interfacial stress concentration, thereby increasing the risk of crack generation. On the other hand, Cr and Ni elements show opposite trends of change. The anti-correlated distribution of Cr-Ni promotes the precipitation of brittle phases such as $(Cr, Fe)_{23}C_6$ at the grain boundary. The interface between these hard phases and the substrate often becomes the area where cracks preferentially extend [22]. It is because of the above situation that cracks and obvious cladding defects were found in the fusion zone in the microstructure observation of a single cladding layer.

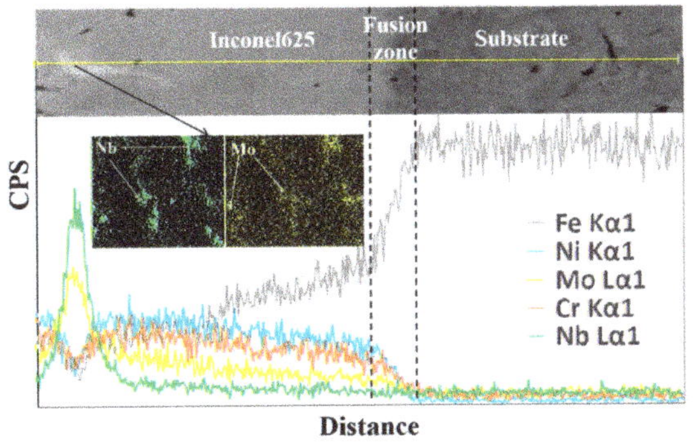

Figure 3. Element analysis of FeMnNiSi-Inconel625 interface.

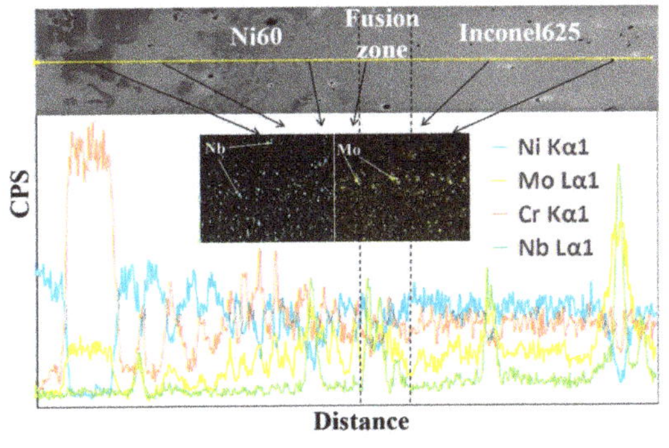

Figure 4. SEM and EDS of Inconel625-Ni60 interface.

Figure 3 shows the elemental analysis of the FeMnNiSi-Inconel625 interface of the laminated structure. It can be seen from the observation that in the line scanning area of the Inconel625 cladding layer, Nb and Mo elements are concentrated in the white phase, while Fe, Cr, and Ni elements are reduced. The enrichment of Nb and Mo elements will promote the precipitation of γ'' phase Ni_3Nb. Its coherent strain with the matrix can absorb residual stress and reduce the local stress concentration coefficient. In addition, the synergistic segregation of Nb-Mo can form a nanoscale Laves phase, which can achieve stress redistribution by pinning dislocation movement. The study of Rao G G and other phase field methods shows that Nb Mo co-segregation can increase the energy barrier of dislocation motion by 0.8 eV, which is consistent with the stress redistribution phenomenon achieved by pinning dislocation motion in this study [23]. The segregation of Nb can stabilize grain boundary carbides and reduce the initiation of microcracks caused by grain boundary sliding at high temperatures. The aggregation of Mo can increase the local yield strength through solid solution strengthening, so that crack propagation requires higher energy consumption, thereby reducing the occurrence of cracks [24]. The reduction in Fe and Cr can inhibit the formation of brittle σ phase Cr-Fe and avoid stress mutation caused by hard and brittle phase. The moderate reduction in Ni can reduce the difference in the thermal expansion coefficient between the Inconel625 cladding layer and FeMnNiSi substrate layer, as well as reduce interface thermal stress, and the reduction in Cr can avoid the continuous precipitation of carbides at grain boundaries and improve the grain boundary bonding state [25]. Through the above element analysis, we found that for the laminated structure with Inconel625 as the intermediate laminate, its bonding interface with the substrate presents a dual mechanism of reducing stress concentration and inhibiting cracks.

Figure 4 shows the elemental analysis of the Inconel625-Ni60 interface of the laminated structure. Via observation, it can be found that there are many fine white phases distributed throughout the interface. The positions of these phases all show the aggregation of Nb or Mo elements. In particular, there are two areas in the fusion line area where Nb and Mo elements are aggregated, while Cr and Ni elements are reduced. This phenomenon shows that Nb-Mo synergistically segregates in the fusion line area to form a nanoscale Laves phase, which can not only inhibit the formation and aggregation growth of carbides, thereby achieving the grain refinement effect, but also form pinning dislocations, causing the crack propagation path to deflect and reducing crack generation [26].

3.3. Phase Composition Analysis

To further verify the crack arrest effect of the FeMnNiSi-Inconel625-Ni60 laminated structure, diffraction pattern analysis of the laminated structure was carried out. The black broken line in Figure 5 is the XRD diffraction pattern of the Inconel625 cladding layer, and the red broken line is the XRD diffraction pattern of the Ni60 cladding layer. The two samples are the transition layer and the surface cladding layer of the laminated structure cladding layer, respectively. The characteristic peak positions of the FCC phase of the Inconel625 cladding layer are at 44.2°, 51.5° and 75.7°, the diffraction peaks of the Laves phase are at 38.3°and 56.2°, and the diffraction peaks of the γ'' phase are at 42.1° and 46.1°. The characteristic peak positions of the FCC phase of the Ni60 cladding layer are at 44.8°, 52°, and 76.2°.

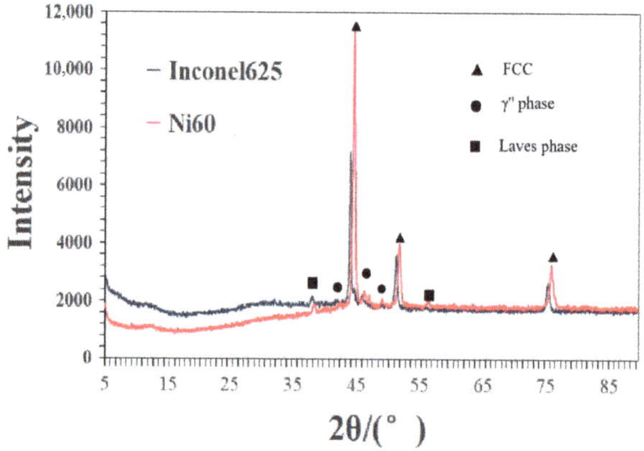

Figure 5. XRD diffraction pattern of the phase composition of the cladding layer.

Because this area is a surface cladding layer of a laminated structure, Nb and Mo elements are segregated from the Inconel625 cladding layer to this area, so the diffraction peak here may also correspond to the Laves phase or γ'' phase.

In comparing the specific data of the crystal plane diffraction peaks of the two cladding layers, it was found that the diffraction peaks of the FCC phase of the Inconel625 cladding layer shifted to the left. This is because the high content of Nb and Mo and other large-atomic-radius elements in Inconel625 replace the Ni lattice, resulting in the expansion of the γ-Ni lattice. According to the Bragg equation, $2d \cdot \sin\theta = n\lambda$, the increase in the interplanar spacing d will lead to a decrease in the diffraction angle θ, which is manifested as a left shift in the peak position. The presence of FCC phase in both cladding layers indicates that the cladding layers have good high-temperature stability, and the Cr element in the cladding layer can promote the formation of the passivation film and improve the corrosion resistance of the cladding layer in 3.5% NaCl solution [25]. The Laves phase acts as a pinning phase at the grain boundary to inhibit grain boundary sliding and maintain the stability of the structure under high-temperature conditions of laser cladding. Moreover, the preferential precipitation of the Laves phase can consume Nb, reduce the precipitation amount of NbC carbides, and reduce the peak strength of carbides [26]. The γ'' phase pins the grain boundary to hinder the continuous network precipitation of carbides, achieving a grain refinement effect and dominating the low-temperature strengthening. The synergistic effect of the two structures can improve the performance of the cladding layer and effectively inhibit the generation of cracks.

3.4. Microhardness

Regarding the microhardness measurement of the FeMnNiSi-Inconel625-Ni60 laminated structure cross-section, microhardness measurements were performed along a horizontal cross-section of the cladding layer, with test points spaced at 200 μm intervals from the cladding surface to the substrate. The average microhardness values ($HV_{0.3}$) reveal a distinct gradient: the substrate (FeMnNiSi) exhibits a hardness of 165.22 $HV_{0.3}$, while the intermediate Inconel 625 cladding layer reaches 340.71 $HV_{0.3}$ (2.06× the substrate). The surface Ni60 layer achieves the highest hardness at 641.31 $HV_{0.3}$ (3.88× the substrate), as shown in Figure 6. This progressive increase in hardness—from substrate to surface—suggests that the Inconel625 layer acts as an effective buffer, mitigating stress concentrations between the softer substrate and the harder Ni60 surface layer. Such a graded hardness profile is likely to enhance the mechanical integrity and wear resistance of the laminated structure. Sainath Krishna Mani Iyer found in the study of SS316L/Inconel625-based composite functional gradient coatings that the transition of the intermediate layer reduces interfacial stress concentration by more than 40%, providing theoretical support for this study [27]. During data recording, we also discovered that the hardness values of one area between the different cladding layers differed, as did the hardness of the two layers. This phenomenon is primarily due to the significant differences in hardness between the different alloy materials. While the cladding retains its original properties, the laser heat input causes elemental migration. During the laser cladding process, some elements also generate new species, all of which lead to changes in hardness. Overall hardness analysis shows that the cladding layer in the FeMnNiSi-Inconel625-Ni60 stack is significantly harder than the substrate, achieving a surface strengthening effect.

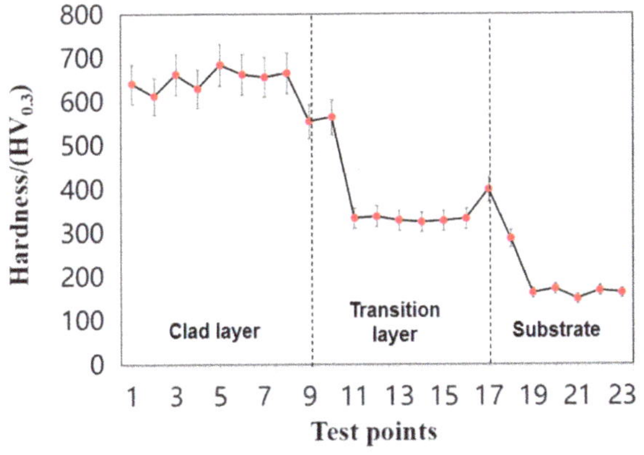

Figure 6. Microhardness distribution of the laminated structure interface.

3.5. Friction and Wear Properties

As shown in Figure 7a, the friction coefficient of both the substrate and the laminated cladding layer exhibits an initial transient period before stabilizing after approximately 5 min, indicating the transition into steady-state wear. The stabilized average friction coefficient (Figure 7b) reveals a 17.95% increase for the laminated structure (0.552) compared to the substrate (0.543). According to Archard's wear theory (Equation (1)), wear volume (V) is directly proportional to the sliding distance (S) and normal load (N), while inversely proportional to material hardness (H) [28].

$$V = kNS/H \tag{1}$$

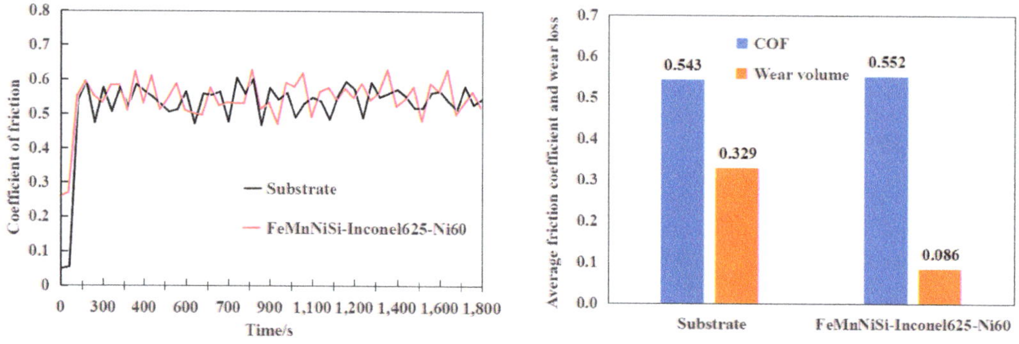

(**a**) (**b**)

Figure 7. Friction coefficient curve of FeMnNiSi-Inconel625-Ni60 laminated structure cladding layer and substrate. (**a**) Friction coefficient curve. (**b**) Average friction coefficient and wear volume.

Under identical testing conditions (20 N load, 5 mm sliding distance), the wear volume of the substrate (0.329 mm^3) significantly exceeds that of the laminated cladding layer (0.086 mm^3)—a reduction of 0.243 mm^3 (Figure 8b). The specific parameters are shown in Table 2. This pronounced improvement in wear resistance can be attributed to the graded hardness profile of the laminated structure, where the Inconel625 intermediate layer and Ni60 surface layer collectively enhance durability while maintaining a modest increase in friction coefficient.

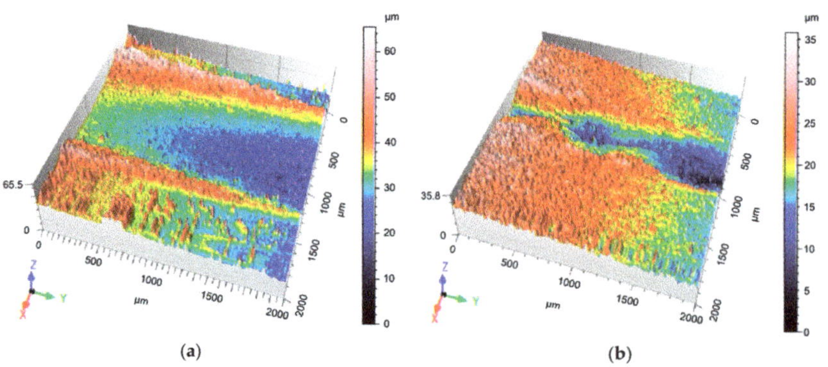

(**a**) (**b**)

Figure 8. Three-dimensional contours of wear scars on substrate and FeMnNiSi-Inconel625-Ni60 cladding layer. (**a**) Three-dimensional profile of wear scar on substrate. (**b**) Three-dimensional profile of wear scar on FeMnNiSi-Inconel625-Ni60 cladding layer.

Table 2. Wear volume calculation parameters.

Sample	Average Coefficient of Friction	Test Load /N	Sliding Distance /mm	Microhardness /$HV_{0.3}$	Wear Volume /mm^3
Substrate	0.543	20	5	165.22	0.329
Sandwich structure	0.552	20	5	641.31	0.086

To further evaluate the friction and wear, the wear scars of the cladding layer and the substrate were extracted and characterized using 3D contours. Figure 8a shows the 3D contour of the wear scar of the substrate, and Figure 8b shows the 3D contour of the wear scar of the FeMnNiSi-Inconel625-Ni60 laminated cladding layer. The two figures show that the wear scar of the laminated cladding layer is shallow and narrow, while the wear scar of the substrate is deep and wide, visually verifying that the wear resistance of the FeMnNiSi-Inconel625-Ni60 laminated cladding layer is significantly higher than that of the substrate.

The three-dimensional contour morphology of the scratch was characterized, yielding the comprehensive dataset presented in Table 3. In order to further verify the advantages of the laminated structure, the wear rate was calculated by substituting Formula (2) into the data obtained by scanning the three-dimensional profile [29]. The specific formula is as follows:

$$W = \frac{V}{LF} \tag{2}$$

Table 3. Comparison of friction and wear test data between FeMnNiSi-Inconel625-Ni60 laminated structure and substrate.

Sample	Wear Depth /μm	Wear Width/ μm	Wear Cross-Sectional Area/μm^2	Wear Volume/ mm^3	Wear Rate/ mm^3/(N·m)
Substrate	18.6	780	7806	0.329	3.29×10^{-2}
FeMnNiSi-Inconel625-Ni60	16.5	460	3883	0.086	0.86×10^{-2}

In Formula (2), V is the wear volume; L is the wear distance; F is the wear load; and W is the wear rate.

Wear scar morphology analysis (Table 3) reveals that the FeMnNiSi-Inconel625-Ni60 laminated cladding layer exhibits superior wear resistance compared to the substrate. Specifically, the maximum wear depth (16.5 μm vs. 18.6 μm) and cross-sectional area (3883 μm^2 vs. 7806 μm^2) are significantly reduced, leading to a 73.86% decrease in wear rate (0.86×10^{-2} $mm^3/(N{\cdot}m)$ vs. 3.29×10^{-2} $mm^3/(N{\cdot}m)$). These findings demonstrate that the laminated cladding structure effectively enhances the wear resistance of FeMnNiSi steel.

Scanning electron microscopy (SEM) examination of the wear scar morphology (Figure 9) reveals distinct wear mechanisms between the substrate and the laminated cladding layer. The substrate (Figure 9a) exhibits severe adhesive wear characteristics, including thick material peeling layers, deep grooves, and extensive material transfer, accompanied by abrasive wear due to large grinding debris accumulation. In contrast, the laminated cladding layer (Figure 9b) demonstrates primarily abrasive wear behavior, with fine grinding debris particles generated by the strengthening phase under grinding force, forming shallow grooves alongside the cutting action of silicon nitrideceramic balls. Minor adhesive wear features are observed, but no large-scale material peeling occurs [30].

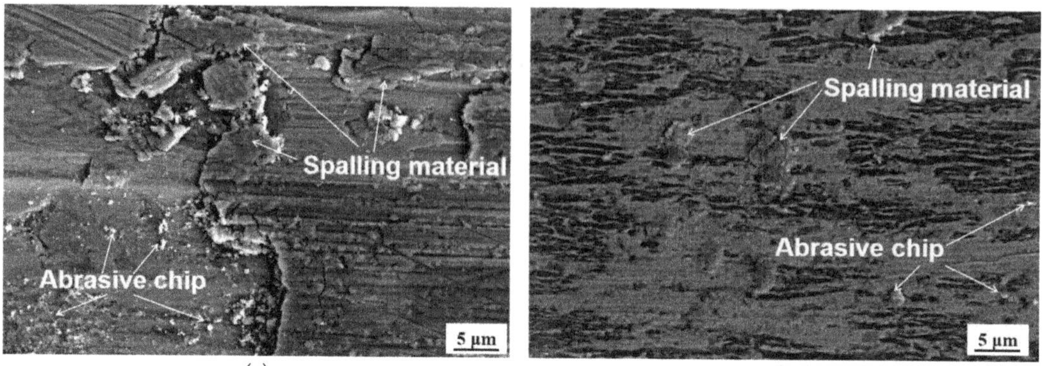

(a) (b)

Figure 9. Micromorphology of wear marks on substrate and laminated cladding layer. (**a**) Micromorphology of wear scar on substrate. (**b**) Micromorphology of wear scar on Ni60 cladding layer.

The superior wear resistance of the laminated cladding layer is attributed to two synergistic effects: (1) solid solution strengthening and grain refinement, which enhance hardness and deformation resistance; and (2) a strengthening phase that impedes dislocation motion, further suppressing wear. In comparison, the substrate's lower hardness renders it susceptible to failure under combined normal and shear stresses during abrasive contact [31].

3.6. Corrosion Resistance

Figure 10 presents the Tafel polarization curves of the FeMnNiSi-Inconel625-Ni60 laminated cladding layer and the substrate in 3.5% NaCl solution. A marked difference in electrochemical behavior is evident: the laminated cladding layer displays a well-defined smooth-curve region during corrosion, whereas the substrate lacks such a feature entirely. This smooth region represents the passivation region, indicating the good stability of the passivation layer. This demonstrates that the laminated cladding layer exhibits superior corrosion resistance compared to the substrate. Polarization resistance (Rp) directly reflects a material's ability to resist corrosion. The greater the polarization resistance, the greater the corrosion resistance. It is inversely proportional to the corrosion current density, and the proportional value can be obtained from the slope of the polarization curve. In electrochemical experiments, the polarization curve data measured in CS Studio software (SP5) was imported into CS (Version: V2.1.586) Analysis software for analysis. CS Analysis software automatically analyzes the polarization curve and calculates the polarization resistance Rp. The data show that the Rp values of the substrate and the laminated cladding are 2426.9 Ω and 8623.6 Ω, respectively. The laminated cladding exhibits markedly improved corrosion resistance relative to the substrate, featuring a 71.86% reduction in corrosion rate and elevated polarization resistance. Electrochemical analysis indicates that the cladding's corrosion potential (Ecorr) (−0.780 V) is more positive than that of the substrate (−0.844 V), while its corrosion current density (icorr) (8.32×10^{-6} A/cm^2) is approximately one order of magnitude lower than the substrate's value (6.546×10^{-5} A/cm^2). These findings collectively demonstrate that the laminated structure achieves slower surface degradation, diminished corrosion propensity, and enhanced protective performance under uniform environmental conditions [32].

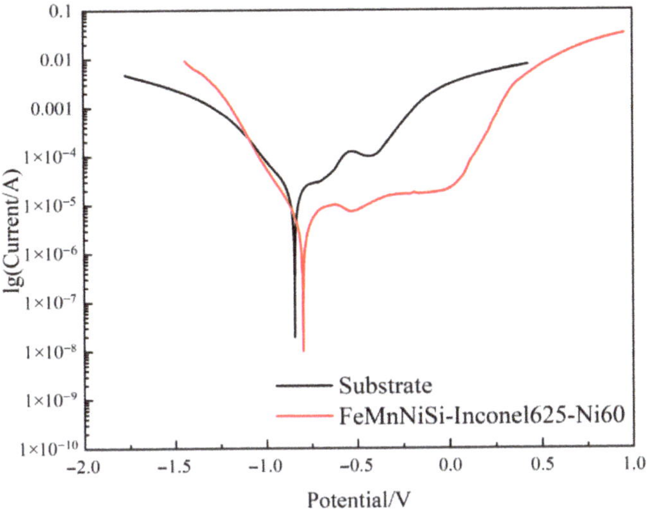

Figure 10. Tafel polarization curves of cladding layer and substrate.

To elucidate the electrochemical corrosion behavior of the FeMnNiSi-Inconel625-Ni60 laminated cladding layer and its substrate, electrochemical impedance spectroscopy (EIS) analysis was conducted, as illustrated in Figure 11. Figure 11a presents the Nyquist plots for both the laminated cladding and the substrate, while Figure 11b shows the corresponding Bode diagrams.

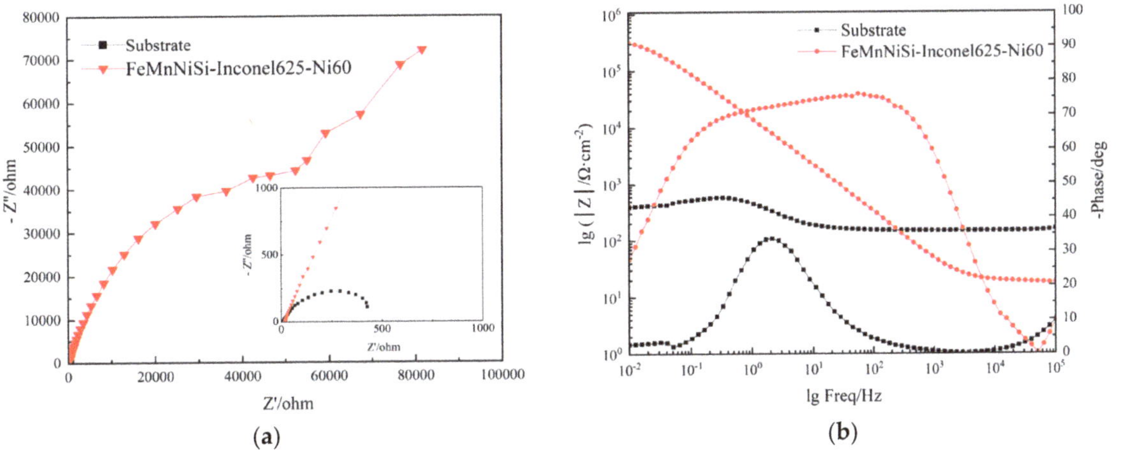

Figure 11. Electrochemical impedance spectroscopy of the laminated structure cladding layer and substrate. (**a**) Nyquist curve of the laminated structure cladding layer and substrate. (**b**) Bode diagram of the laminated structure cladding layer and substrate.

In Figure 11a, the laminated cladding exhibits a substantially larger capacitive arc radius than the substrate. A larger capacitive arc radius correlates with higher resistance to electrode reactions, slower corrosion rates, and superior corrosion resistance [33]. Figure 11b further reveals that the impedance modulus |Z| of the laminated cladding is significantly higher than that of the substrate. Notably, at the peak |Z| frequency [34], the laminated cladding demonstrates a broad frequency stability region, whereas this region is virtually absent in the substrate. These observations collectively confirm the enhanced corrosion resistance of the FeMnNiSi-Inconel625-Ni60 laminated structure.

Figure 12 shows the microscopic morphology of the substrate and cladding surface after electrochemical corrosion. As shown in Figure 12a, the corroded surface of the substrate has obvious pitting corrosion pits, and the substrate test surface is corroded as a whole, with basically no intact areas. Compared with Figure 12b, there are areas of uncorroded cladding in the laminated structure, and a small amount of clustered pitting corrosion clusters is present within the corroded grain boundaries. However, the corrosion marks are relatively shallow, indicating that the laminated structure cladding has better corrosion resistance. Therefore, in terms of corrosion resistance, the laminated structure cladding has better corrosion resistance than the substrate.

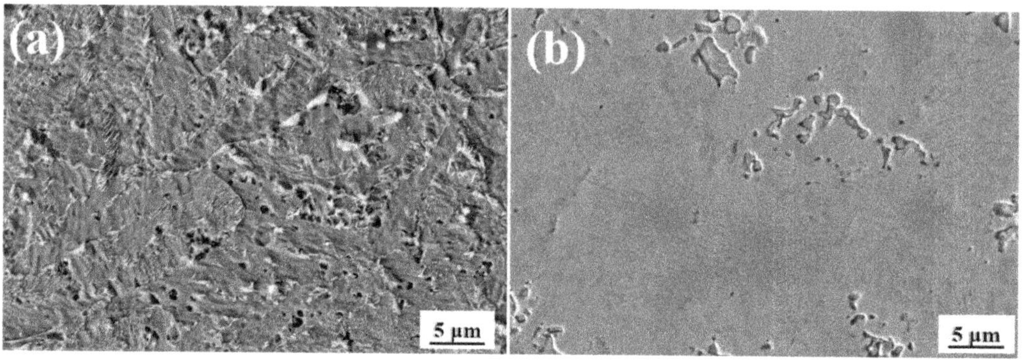

Figure 12. Micromorphology of substrate and laminated structure cladding layer after electrochemical corrosion. (**a**) Micromorphology of substrate. (**b**) Electrochemical corrosion micromorphology of laminated structure cladding layer.

4. Conclusions

This study successfully solved the cracking problem of Ni60 cladding layer on FeMnNiSi substrate through the design of a FeMnNiSi-Inconel625-Ni60 laminate structure, as well as revealed the intrinsic mechanism of interface microstructure regulation and performance enhancement and provided a theoretical basis for the additive manufacturing of heterogeneous materials. It also provided a basis for the surface modification of key components in industrial and mining applications such as aviation and marine equipment. The main conclusions are as follows:

(1) By observing the micromorphology of the interface area between the Ni60 single cladding layer and the FeMnNiSi-Inconel625-Ni60 laminated structure, we found unfused defects and obvious cracks in the fusion zone of the single cladding layer, while there were no obvious cladding defects in the two cladding layer areas of the laminated structure.

(2) Through EBSD micromorphology, interface elements, and phase composition analysis, it can be seen that the Cr element aggregates at the interface of Ni60 single coating and the Ni element decreases, thus forming a hard brittle phase and FCC phase. The difference in the thermal expansion coefficients of the two phases leads to interface stress concentration, which increases the risk of crack generation; the enrichment of Nb and Mo elements at the interface of the stacked structure FeMnNiSi-Inconel625 promotes the precipitation of γ'' phase and forms a nanoscale Laves phase under the synergistic segregation of Nb-Mo, which refines the grains while reducing local stress concentration; two Nb-Mo synergistic segregation phenomena appear in the fusion

line area of the stacked structure Inconel625-Ni60 interface, which can not only inhibit the formation and aggregation of carbides to achieve the effect of grain refinement but also form pinning dislocations, deflect the crack propagation path, and reduce crack generation. The FeMnNiSi-Inconel625-Ni60 stacked structure presents a dual mechanism of reducing stress concentration and inhibiting cracks.

(3) The hardness test results show that the average microhardness of the Ni60 surface cladding layer of the laminated structure is 641.31 $HV_{0.3}$, which is 3.88 times the average microhardness of the substrate (165.22 $HV_{0.3}$). The average microhardness of the base layer Inconel625 is 340.71 $HV_{0.3}$, which is 2.06 times the microhardness of the substrate. The wear resistance test results show that the wear volume of the laminated cladding layer is 0.086 mm^3, which is 0.243 mm^3 less than the wear volume of the substrate 0.329 mm^3; the wear rate of the laminated cladding layer is 0.86×10^{-2} $mm^3/(N{\cdot}m)$, which is 73.86% lower than the wear rate of the substrate 3.29×10^{-2} $mm^3/(N{\cdot}m)$; the substrate wear scar surface has a large area of adhesion and material peeling, resulting in severe wear, while the surface wear of the laminated cladding layer is significantly weakened, mainly due to abrasive wear, and has better wear resistance.

(4) The electrochemical corrosion test results indicate that under identical environmental conditions, the laminated structure cladding layer exhibits superior corrosion resistance compared to the substrate. Specifically, the cladding layer demonstrates a higher self-corrosion potential (−0.780 V vs. −0.844 V for the substrate), a lower corrosion current density (8.32×10^{-6} A/cm^2 vs. 6.546×10^{-5} A/cm^2), and a significantly higher polarization resistance (8623.6 Ω vs. 2426.9 Ω). Additionally, the frequency-stable region width at the highest impedance modulus $|Z|$ is notably wider for the cladding layer, and its corrosion rate is 71.86% slower than that of the substrate. These findings confirm that the laminated structure has a slow corrosion rate, minimal corrosion tendency, and excellent overall corrosion resistance.

Author Contributions: Validation, L.D., W.L. and J.C.; Writing—original draft, L.D. All authors have read and agreed to the published version of the manuscript.

Funding: This research was funded by [National Natural Science Foundation of China] grant number [51975264].

Institutional Review Board Statement: Not applicable.

Informed Consent Statement: Not applicable.

Data Availability Statement: The original contributions presented in this study are included in the article. Further inquiries can be directed to the corresponding author.

Conflicts of Interest: The authors declare no conflicts of interest.

References

1. Zhang, Q.; Li, Z.; Yu, Z. Optimization of Process Parameters for Laser Cladding FeNiMoCoCrTi High Entropy Alloy Coating on 45 # Steel. *Mech. Strength* **2021**, *2*, 308–312. [CrossRef]
2. Cao, P.; Gu, X.; Yan, S. The effect of shot blasting pretreatment on gas carburizing efficiency and layer characteristics of 45 steel. *Met. Heat Treat.* **2022**, *47*, 143–146. [CrossRef]
3. Chen, X.; Zhang, D.; Sun, W.; Wang, Y.; Zhang, J. Effect of scanning speed on deformation and microstructure of laser cladding thin high-speed steel plate. *Surf. Technol.* **2019**, *48*, 150–157. [CrossRef]
4. Dong, M.; Jin, K.; Huang, G.; Zhao, H.; Xuan, L.; Qiao, X.; Chen, P. Influence of scanning path on residual stress and tribological properties of multi-pass laser cladding layer. *China Surf. Eng.* **2025**, *38*, 75–86. [CrossRef]
5. Li, T.; Long, H.; Qiu, C.; Wang, M.; Li, D.; Dong, Z.; Gui, Y. Multi-Objective Optimization of Process Parameters of 45 Steel Laser Cladding Ni60PTA Alloy Powder. *Coatings* **2022**, *12*, 939. [CrossRef]
6. Luo, X.; Yao, Z.; Zhang, P.; Zhou, K. Corrosion resistance of Al-Cr composite reinforcement layer on 45 steel surface. *Rare Met. Mater. Eng.* **2018**, *47*, 7.

7. Liu, L.; Li, S.; Dou, W.; Han, F.; Lin, K. Process optimization and performance study of laser cladding Ni60 alloy coating on 316L stainless steel surface. *China Laser* **2024**, *51*, 1602207. [CrossRef]
8. Cao, J.; Yang, X.; Wang, S.; Zhang, H.; Yang, L.; Qiao, Y.; Li, C. Wear and Corrosion Resistance of Ni60 TiC Ceramic Coating Laser Cladded on 45 Steel Surface. *Rare Met. Mater. Eng.* **2020**, *49*, 611–617.
9. Wang, S.; Zheng, S.; Liu, W.; Zhu, X.; Zhou, M.; Jiang, H. Optimization of Process Parameters for Laser Cladding Fe35A Alloy on 45 Steel Surface. *Adv. Laser Optoelectron.* **2021**, *58*, 0914007. [CrossRef]
10. Wang, Y.; Zhou, H.; Wang, Z.; Xiang, N.; Cao, X. The influence of electromagnetic field on the microstructure and properties of laser cladding nickel based coatings. *Surf. Technol.* **2024**, *53*, 198–207. [CrossRef]
11. Lu, X.; Liu, X.; Yu, P.; Chen, Y. Effects of post heat-treatment on microstructure and tribological properties of Ni60/H-BN self-lubricating anti-wear composite coating on 304 stainless steel by laser cladding. *Tribology* **2016**, *36*, 49–54. [CrossRef]
12. Yang, W.; Li, S.; Xiao, Q.; Chen, D.; Wang, S.; Zhang, B. Research Status and Development Trends of Friction reducing and Wear resistant Laser Cladding Coatings. *Lubr. Seal.* **2023**, *48*, 171–182. [CrossRef]
13. Qian, S.; Qiu, Y. Research on the microstructure and properties of nickel-based alloy layer remanufactured by laser cladding. *Electroplat. Finish.* **2024**, *46*, 69–75. [CrossRef]
14. La, P.; Xue, Q.; Liu, W. Effects of boron doping on tribological properties of $Ni_3Al–Cr_7C_3$ coatings under dry sliding. *Wear* **2001**, *249*, 93–99. [CrossRef]
15. Shen, X.; Chen, L.; Wu, Y.; Chen, C.; Zhang, X.; Zhou, W.; Ren, X. Effect of Substrate Material Properties on Microstructure and Mechanical Properties of Ni60 Coatings Prepared by Extreme High-Speed Laser Cladding. *Steel Res. Int.* **2024**, *95*, 2300531. [CrossRef]
16. Wang, W.; Sun, W.; Zhou, H.; Xing, X.; Yang, K.; Yu, J. Research on Crack Formation Mechanism and Matching Method of Laser Cladding Nickel based Alloy Coating. *Hot Process. Technol.* **2022**, *51*, 83–88. [CrossRef]
17. Fang, Z.; Luo, F.; Yao, J.; Zhang, Q. Effect of transition alloy layer on microstructure and properties of high hardness iron-based multilayer laser cladding thick cladding. *Prog. Laser Optoelectron.* **2004**, *41*, 4.
18. Zhang, X.; Wang, A.; Zhang, X.; Qiao, X.; Huang, Z. Microstructure and properties of laser cladding Ni45-CaF2-WS2 self-lubricating coating. *J. Chin. Soc. Nonferrous Met.* **2008**, *18*, 215–220. [CrossRef]
19. Ou, J.; Zhang, G.; Peng, Y.; Chen, W.; Zheng, Z.; Zheng, K. Microstructure and abrasive wear behavior of $TiC–Cr_7C_3–Fe$ composites fabricated by spark plasma sintering. *Int. J. Refract. Met. Hard Mater.* **2025**, *131*, 107228. [CrossRef]
20. Singh, C.V.; Bagui, S.; Sahoo, B.K.; Umbare, O.S.; Tarafder, S.; Sivaprasad, S. Comparison of low cycle fatigue behaviour of additively manufactured and wrought Inconel625 alloys. *Mater. Sci. Eng. A* **2024**, *903*, 14. [CrossRef]
21. Li, J.; Wang, R.; Pan, L.; Zhou, X.; Yang, K. Effect of LaB6 on Microstructure and High-temperature Oxidation Resistance of Inconel625 Alloy Coating Fabricated by Plasma Transferred Arc Surfacing. *Surf. Technol.* **2025**, *54*, 98–107+142. [CrossRef]
22. Liu, Y.; Sun, Y. Effects of alloying elements Si, V and Ti on microstructure and properties of 40Cr13 martensitic stainless steel. *Jinshu Rechuli/Heat Treat. Met.* **2024**, *49*, 256–261. [CrossRef]
23. Rao, G.G.; Vasquez, V.R. Correction: Micromechanical model to study the deformation behavior and effect of misfits on mechanical properties in Ni-based superalloys—A phase-field approach. *J. Mater. Res.* **2025**, *40*, 1745. [CrossRef]
24. Wang, X.; Huang, Y.; Liu, Q.; Xie, L.; Xu, Q.; Huang, J. Effect of Molybdenum Content on the Microstructure and Wear Resistance of CoCrFeNiMo High Entropy Alloy Laser Cladding Coating on Carbon Steel Surface. *Electroplat. Coat.* **2020**, *39*, 1201–1208. [CrossRef]
25. Zhang, J.W.; Yu, L.M.; Liu, C.X.; Ding, R.; Liu, Y. Synergistic Strengthening of High-Cr Martensitic Heat-Resistant Steel and Application of Thermo-Mechanical Treatments. *Acta Metall. Sin.* **2024**, *60*, 713–730. [CrossRef]
26. Xiao, X.; Lu, S.; Hu, P.; Huang, M. Toughening Method of Laves Intermetallic Compound Cr_2Nb. *Aerosp. Mater. Technol.* **2007**, *37*, 5.
27. Iyer, S.K.M.; Ramasamy, K.; Subramaniam, P. Design and development of SS316L/Inconel625 based composite functional gradient materials for high-pressure pneumatic tools. *Rare Met. Mater. Eng.* **2025**, *54*, 363–376. [CrossRef]
28. Aramide, B.; Pityana, S.; Sadiku, R.; Jamiru, T.; Popoola, P. Improving the durability of tillage tools through surface modification—A review. *Int. J. Adv. Manuf. Technol.* **2021**, *116*, 83–98. [CrossRef]
29. Zhang, Z.; Cao, Z.; Wei, D.; Ge, P.; Zhou, W.; Chen, G.; Zu, Y.; Fu, X. The influence of laser shock surface micro pit texturing on the micro motion wear behavior of diamond-like carbon coatings. *China Laser* **2025**, *52*, 0402206. [CrossRef]
30. Xiao, B.; Jia, W.; Wang, J.; Zhou, L. Selective electron beam melting of WMoTaNbVFeCoCrNi refractory high-entropy alloy. *Mater. Charact.* **2022**, *193*, 112278. [CrossRef]
31. Walczak, M.; Nowak, W.J.; Szala, M.; Grądzka-Dahlke, M.; Maciaszek, N.; Vališ, D.; Pasierbiewicz, K. Effect of molybdenum addition on microstructure and behavior of AlCoCrFeNi high-entropy alloys in wet environments. *Arch. Civ. Mech. Eng.* **2025**, *25*, 205. [CrossRef]
32. Duan, S.; Ren, W.; Lei, W.; Wang, Y. Study on the microstructure and properties of rail cladding layer after laser quenching. *J. Manuf. Process.* **2023**, *108*, 14. [CrossRef]

33. Hu, P.; Song, R.; Li, X.J.; Deng, J.; Chen, Z.-Y.; Li, Q.-W.; Wang, K.-S.; Cao, W.-C.; Liu, D.-X.; Yu, H.-L. Influence of concentrations of chloride ions on electrochemical corrosion behavior of titanium-zirconium-molybdenum alloy. *J. Alloys Compd.* **2017**, *708*, 367–372. [CrossRef]
34. Feng, X.T.; Gu, H.; Zhou, S.F.; Lei, J. Microstructure and Electrochemical Corrosion Behavior of TC4 Titanium Alloy Cladding Layer Prepared with Powder Feeding Laser Additive Manufacturing. *Chin. J. Lasers* **2019**, *46*, 44–53. [CrossRef]

materials

MDPI

Article

Investigation on the Microstructure and Mechanical Properties of X70 Pipeline Steel Fabricated by Laser-Directed Energy Deposition

Zhandong Wang [1,2,*], Chunke Wang [1], Linzhong Wu [1] and Guifang Sun [2,*]

1 College of Mechanical and Electronic Engineering, Nanjing Forestry University, Nanjing 210037, China; ck_wang@njfu.edu.cn (C.W.); 3240300248@njfu.edu.cn (L.W.)
2 School of Mechanical Engineering, Southeast University, Nanjing 211189, China
* Correspondence: zdwang@njfu.edu.cn (Z.W.); gfsun@seu.edu.cn (G.S.)

Abstract

The laser-directed energy deposition (L-DED) technique, with its excellent environmental adaptability and superior repair capability, shows great potential for the repair of damaged X70 pipeline steel. In this work, the microstructure and mechanical properties of L-DED repaired X70 steel were systematically investigated. The deposited material exhibited inhomogeneity along the building direction. From the bottom to the top, the grains gradually coarsened, and the proportion of polygonal ferrite increased. This was mainly attributed to increasing thermal accumulation with deposition height, which reduced the cooling rate and promoted solid-state transformations at higher temperatures. Meanwhile, the heat accumulation and intrinsic heat treatment reduced the dislocation density and promoted Fe_3C precipitation within grains and along boundaries. Microhardness was highest in the bottom region and decreased along the building direction due to the gradual coarsening of microstructure and decreasing in dislocation density. The L-DED X70 showed lower yield strength (435 MPa) and ultimate tensile strength (513 MPa) compared to the base material and API 5L requirements. The elongation of the L-DED X70 was 42.9%, which was 58% higher than that of the base material, indicating excellent ductility. These results revealed a thermal history-dependent strength–ductility trade-off in the L-DED repaired X70 steel. Therefore, more efforts are needed to control the L-DED thermal process, tailor the microstructure, enhance strength, and meet the service requirements of harsh environments.

Keywords: pipeline steel; additive manufacturing; directed energy deposition; thermal process; microstructure; tensile strength

1. Introduction

X70 pipeline steel, characterized by excellent strength-toughness balance and favorable weldability, has become a critical material for long-distance transportation of energy sources such as oil and natural gas [1]. However, during prolonged service, pipelines are frequently exposed to harsh operating conditions involving complex stress states, corrosive environments, and coupled multiphase media, rendering their surfaces highly susceptible to damage such as cracking, pitting corrosion, and mechanical wear [2,3]. Leakage incidents not only endanger human safety and cause environmental pollution but also result in substantial economic losses. Therefore, the development of efficient and reliable on-site repair technologies is of significant engineering importance for ensuring the safe operation of pipeline networks and extending their service life.

Arc welding repair is currently the most widely adopted technique. However, its high heat input tends to induce microstructural coarsening and residual stress, making it difficult to meet the quality requirements of high-standard engineering applications. In recent years, laser-directed energy deposition (L-DED) technology has gained widespread application in aerospace, automotive, and marine industries [4,5]. L-DED exhibits many advantages, such as low heat input, narrow heat-affected zone, and minimal thermal distortion [6]. In addition, rapid laser melting and solidification effectively suppress grain growth, yielding a dense, fine-grained microstructure that enhances the overall mechanical performance of the repaired zone [7]. Thus, investigating the applicability of L-DED technology for repairing damaged X70 pipeline steel holds considerable scientific value and engineering potential.

The L-DED melt pool solidifies under non-equilibrium, rapid solidification conditions, resulting in a unique solidification microstructure within the repaired region. As the repaired zone is constructed via sequential track-by-track and layer-by-layer metal deposition, thermal cycling further modifies the characteristics of the solidified microstructure. Cheng et al. [8] demonstrated that the extremely high cooling rate inherent to L-DED significantly suppresses austenite grain growth and promotes the formation of fine-grained transformation products such as bainite and martensite, in contrast to the coarse columnar grains typically observed in conventional welding. Rashid et al. [9] investigated the microstructural features of additively manufactured 300 M ultra-high-strength steel. They demonstrated that the deposited zone consists primarily of untempered martensite, while overlapping scan tracks, subjected to additional thermal cycles, develop tempered martensite. Ning et al. [10] pointed out that the microstructure of X80 pipeline steel fabricated by the wire-fed additive manufacturing technique is dominated by lath bainite (LB). The grain size and phase composition are distinctly different from those produced by conventional methods. These results are also reported by Seede et al. [11]. Collectively, these studies indicate that the additive manufacturing process parameters and solidification characteristics jointly govern the microstructural constitution of the repaired zone, revealing fundamental differences compared to traditional techniques.

The unique microstructure of the L-DED repaired zone profoundly influences its mechanical properties. Sun et al. [12] applied laser cladding to repair AISI 4340 high-strength steel. They claimed that while the repaired zone exhibited markedly elevated microhardness, its tensile properties were significantly degraded. Zhan et al. [13] reported that the thermal cycling in the repair layer of AerMet100 ultra-high-strength steel induced grain refinement, leading to substantially higher microhardness and strength than the substrate, albeit with a pronounced reduction in ductility. Similar results are reported by Jing et al. [14]. Sun et al. [15] investigated the L-DED repair of HSLA-100 high-strength steel. They found that the complex thermal histories result in non-uniform microhardness distribution. These findings demonstrate that the heat input, composition, and cooling path modulate the microstructure, thereby profoundly affecting the mechanical response of the repaired zone. Despite abundant research on L-DED of high-strength steels, systematic investigations specifically targeting damage repair of X70 pipeline steel remain lacking, particularly concerning the mechanisms of microstructural evolution in the repaired zone and its quantitative correlation with mechanical properties.

To the best of our knowledge, powder-blown L-DED is firstly applied to repair the damaged X70 pipeline steel. The effects of thermal cycles and heat accumulation during L-DED on the microstructure, microhardness, and tensile properties of the repaired region are systematically investigated. Through comprehensive comparison with rolled base material (BM), the underlying mechanisms by which the L-DED process affects microstructural evolution and mechanical properties of the pipeline steel are thoroughly elucidated. This work aims to provide a scientific foundation for high-performance repair of X70 pipeline steel by L-DED.

2. Experimental Procedure

2.1. Materials

The BM is standard X70 pipeline steel supplied by Wuyang Iron and Steel Co., Ltd. (Pingdingshan, China). This steel plate is produced by the thermo-mechanical control process (TMCP). The dimensions of the prepared BM are 200 mm × 100 mm × 14 mm (length × width × height). The additively manufactured sample (AM) was fabricated using the X70 powder. The X70 powder was produced from the as-received X70 steel plates via the gas atomization technique. The particle size distribution of the X70 powder ranges from 45 to 105 μm. The chemical composition of the X70 plate and X70 powder is listed in Table 1. The chemical composition of the steel plate was analyzed using optical emission spectroscopy (SPECTRO LMF15; SPECTRO Analytical Instruments GmbH, Kleve, Germany). The chemical composition of the powder was measured using inductively coupled plasma optical emission spectroscopy (Spectro ICP-OES, SPECTRO Analytical Instruments GmbH, Kleve, Germany), while the contents of the non-metallic elements C and S were determined using an inert gas fusion analytical instrument (NCS CS-2800; Luoyang Shibida Precision Instruments Co., Ltd., Luoyang, China).

Table 1. Chemical composition (wt.%) of the X70 plate and X70 powder.

Element	C	Mn	Ni	Cr	Si	Mo	V	Cu	Nb	P	S	Fe
X70 plate	0.085	1.57	0.02	0.27	0.21	0.003	0.003	0.024	0.05	0.010	0.005	Bal.
X70 powder	0.057	1.47	0.08	0.36	0.20	0.013	0.002	0.02	0.04	0.012	0.008	Bal.

2.2. L-DED Process

The L-DED process was performed on a laser cladding forming system (AFS-C1280, Beijing Longyuan AFS Co., Ltd., Beijing, China). This system primarily consisted of the following modules: a fiber-coupled laser device (Trudisk 6002, Trumpf, Ditzingen, Germany), a three-axis numerically-controlled machine tool, and a coaxial powder-feeding laser cladding head (T003, Nanjing Huirui Photoelectric Technology Co., Ltd., Nanjing, China). The laser beam was delivered via a 600 μm core-diameter optical fiber to the laser cladding head and focused onto the substrate surface with a focused spot diameter of 2 mm. The laser cladding head was precisely controlled by the CNC machine tool to execute spatial trajectory movements, synchronously feeding X70 powder into the melt pool. The L-DED process is illustrated in Figure 1a.

As shown in Figure 1b, the L-DED scan path employed a layered reciprocating scanning strategy, with the dimensions of the formed sample being 40 mm × 40 mm × 4.5 mm (length × width × height). To prevent the oxidation of the melt pool during L-DED, high-purity nitrogen was used as the shielding gas at a flow rate of 18 L/min. Based on the results of preliminary process exploration, optimized process parameters were selected for the repair experiment, with specific parameters detailed in Table 2. This work utilized the energy volume density (EVD) to characterize the heat input resulting from the process parameters. It was calculated using Equation (1) [16]:

$$EVD = \frac{P}{v \cdot d \cdot h} \quad (1)$$

where P is the laser power, v is the scanning speed, d is the diameter of laser spot, h is the layer height. Based on the measurement result of the single track, the average height of the single track was 1.2 mm. Substituting the corresponding parameters into the equation yielded a EVD value of 50 J/mm^3.

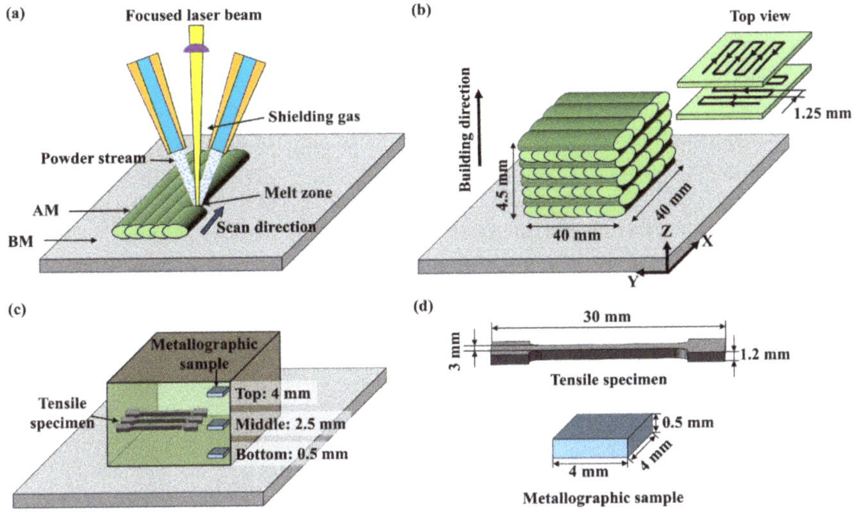

Figure 1. (**a**) L-DED process, (**b**) scan strategy, (**c**) sampling location, and (**d**) dimension of the tested samples.

Table 2. L-DED process parameters.

Parameter	Value
Laser power (W)	2000
Laser spot diameter (mm)	2
Scanning speed (mm/min)	1000
Powder feed rate (g/min)	10
Hatching distance (mm)	1.25
ΔZ increment (mm)	0.75
Layer number	8

2.3. Microstructure Characterization

The sampling locations for the metallographic specimens are shown in Figure 1c, with specific dimensions and shapes detailed in Figure 1d. Samples were taken from the top, middle, and bottom regions to comparatively analyze the microstructural characteristics at different deposition heights. The specimen surfaces were sequentially mechanically ground using SiC abrasive papers of 80, 240, 600, 1000, and 2000 grit, followed by fine polishing with 2.5 μm diamond polishing compound. A 4% nitric acid alcohol solution was used for chemical etching to reveal the microstructure. After etching, the microstructure was examined using optical microscopy (OM; Primotech, Carl Zeiss, Oberkochen, Germany) and scanning electron microscopy (SEM; Nova Nano, FEI Company, Hillsboro, OR, USA).

The internal grain structure was further examined using transmission electron microscopy (TEM; Tecnai G2, FEI Company, Hillsboro, OR, USA). The twin-jet electrolytic polishing method was employed, using an electrolyte composed of a mixture of 10 vol.% perchloric acid ($HClO_4$) and 90 vol.% ethanol (C_2H_5OH), to thin the samples until they became electron beam transparent. Thin areas of the TEM samples were further used for electron backscatter diffraction analysis (EBSD; Oxford Instruments, Abingdon, UK). The

EBSD scanning field of view was 250 × 190 μm^2, with a sampling step size of 0.36 μm. The AZtecCrystal V2.1 software was used for analyzing the obtained EBSD data.

2.4. Microhardness and Tensile Testing

Microhardness testing was performed using a microhardness tester (HXD-1000 TMSC/LCD, Shanghai Taiming Optical Instrument Co., Ltd., Shanghai, China). The applied load was 200 gf, with a dwell time of 15 s. For the AM specimen, microhardness measurements were initiated from the topmost surface and conducted at intervals of 0.5 mm along the depth direction. Total ten measurement points were used for the analysis. Each set of measurements was repeated for three times, and the average value was taken to minimize experimental error. For the BM specimen, ten measurement points were randomly selected at different locations. The average microhardness value was calculated for comparative analysis with the microhardness distribution of the AM specimen.

Tensile specimens were sampled as indicated in Figure 1c, with specific dimensions and geometry shown in Figure 1d. Tensile specimens were extracted from the middle region of the AM block to represent the characteristic mechanical behavior under stable deposition conditions. The initial layers had high cooling rates, while the last layers experienced severe heat accumulation. Both conditions led to non-uniform microstructures and localized property variations. In contrast, the middle region experienced relatively stable thermal cycles, resulting in a more representative and homogeneous microstructure that reflected the steady-state condition of the process.

The tensile specimens were mounted on an electronic universal testing machine (NKK-4050, Shenzhen Nanfang Jingke Instrument & Equipment Co., Ltd., Shenzhen, China) for quasi-static uniaxial tensile testing at a strain rate of 0.3 mm/min. Each test was repeated three times. The average values were used to ensure data reliability. The ultimate tensile strength (UTS) and post-fracture elongation of both AM and BM specimens were experimentally determined. Due to the absence of an extensometer, the yield strength (YS) could not be measured directly. Instead, it was estimated from the engineering stress–displacement curve using the deviation analysis method. The yield strength was determined as the stress corresponding to a 5% deviation from the linear elastic behavior predicted by Hooke's law [17]. In addition, the fracture surfaces of the tensile specimens were analyzed by SEM.

3. Results

3.1. Microstructure

Figure 2 shows the microstructural characteristics of the AM and BM X70 specimens. Figure 2a–c show that the microstructure of the AM specimen is primarily composed of polygonal ferrite (PF) and quasi-polygonal ferrite (QF). However, the grain size and phase fraction varied significantly across different regions. Specifically, the top region of the AM specimen exhibited coarse microstructure which was predominantly composed of large PF (Figure 2a). In the middle region, the grain size was refined. PF was more frequently observed than QF (Figure 2b). In contrast, the bottom region contained a higher proportion of QF compared to PF (Figure 2c). Figure 2d shows that the microstructure of the BM specimen is predominantly composed of QF. QF tended to exhibit a flattened morphology resulting from the rolling process during TMCP. In addition, the grain size of the BM specimen was smaller than that of the AM specimen.

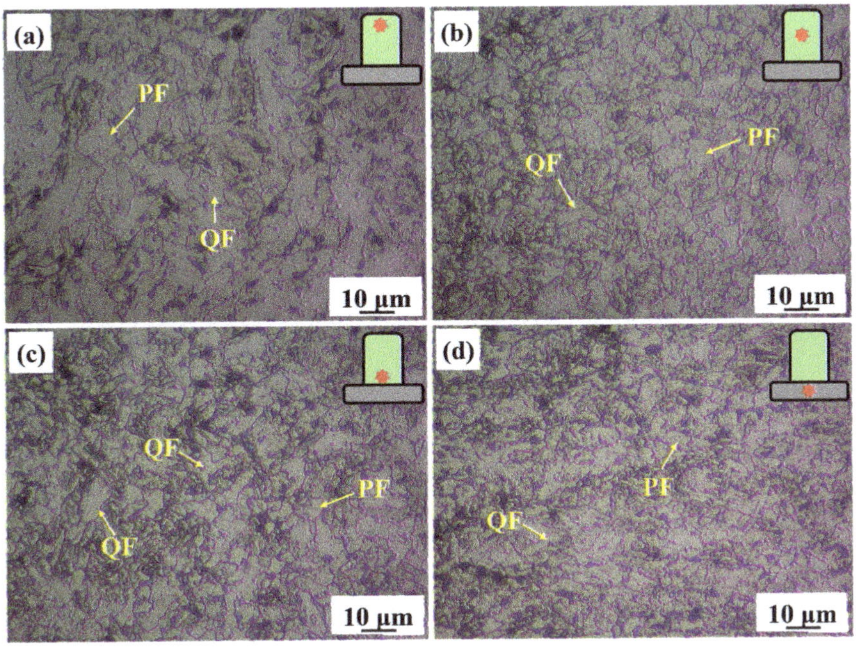

Figure 2. OM images showing the microstructural features of the X70 specimens. (**a**) Top region of AM specimen, (**b**) middle region of AM specimen, (**c**) bottom region of AM specimen, and (**d**) BM specimen.

For a concise and clear comparison between the AM and BM microstructures, the microstructure from the middle region of the AM specimen was selected for detailed analysis. The SEM images in Figure 3 showed the microstructures of the AM and BM specimens. Figure 3a reveals that the microstructure of the AM specimen is predominantly composed of regularly shaped PF, exhibiting equiaxed or near-equiaxed morphology. PF grain boundaries are clearly defined. Figure 3b clearly shows that the M/A island-like constituents are located along grain boundaries and distributed between PF grains. In addition, numerous spherical inclusions were dispersed within the grains. In contrast, Figure 3c shows that the microstructure of the BM specimen is primarily composed of QF, with smaller, irregularly shaped grains and wavy grain boundaries. Figure 3d exhibits the carbides and small M/A constituents along the grain boundaries. The M/A constituents were predominantly located at grain boundary regions.

Figure 4 presents the TEM analysis results of the middle region of the AM specimen. Figure 4a shows that dispersed inclusions are distributed within the PF grains. Grain boundaries between adjacent grains were clearly discernible. Figure 4b,c show the high-magnification images of the grain interiors, further revealing the distribution characteristics of carbides. Figure 4b shows that numerous rod-shaped carbides are interconnected and locally clustered within the grains. Selected area electron diffraction (SAED) analysis verified that the precipitated carbide was cementite (Fe_3C). Figure 4c shows that many near-spherical carbides are enriched along grain boundaries, exhibiting pronounced segregation behavior. Figure 4e,f compare the size distributions of the carbides with different shapes. The average diameter of the rod-shaped carbides (110 $\pm$ 25 nm) was much larger than that of the near-spherical carbides (32 $\pm$ 8 nm). Figure 4d shows that the dislocation lines in the AM X70 specimen are straight and sparsely distributed. This suggests minimal plastic strain accumulation during the deposition process, resulting in a significantly low dislocation density, which is consistent with the characteristics of PF as a diffusion-controlled transformation product.

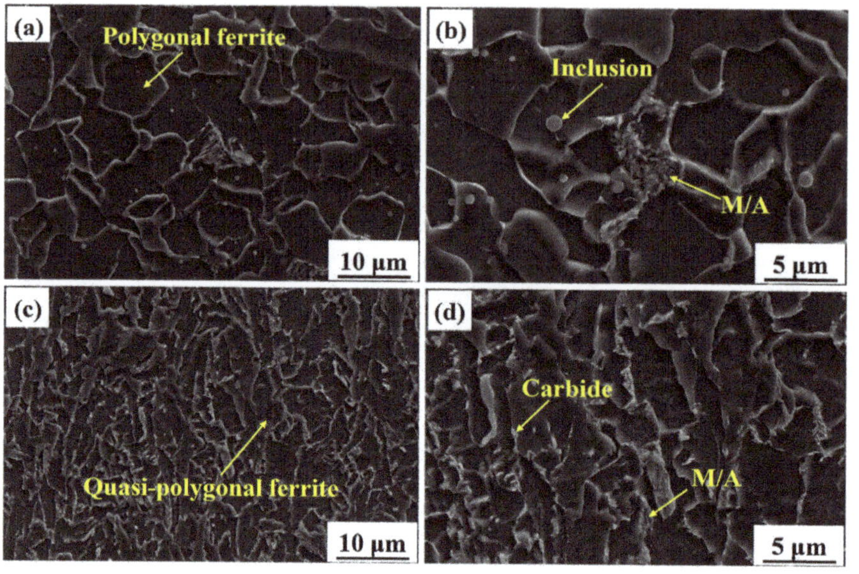

Figure 3. SEM images showing the typical microstructures of the X70 specimens. (**a**,**b**) Middle region of AM specimen and (**c**,**d**) BM specimen.

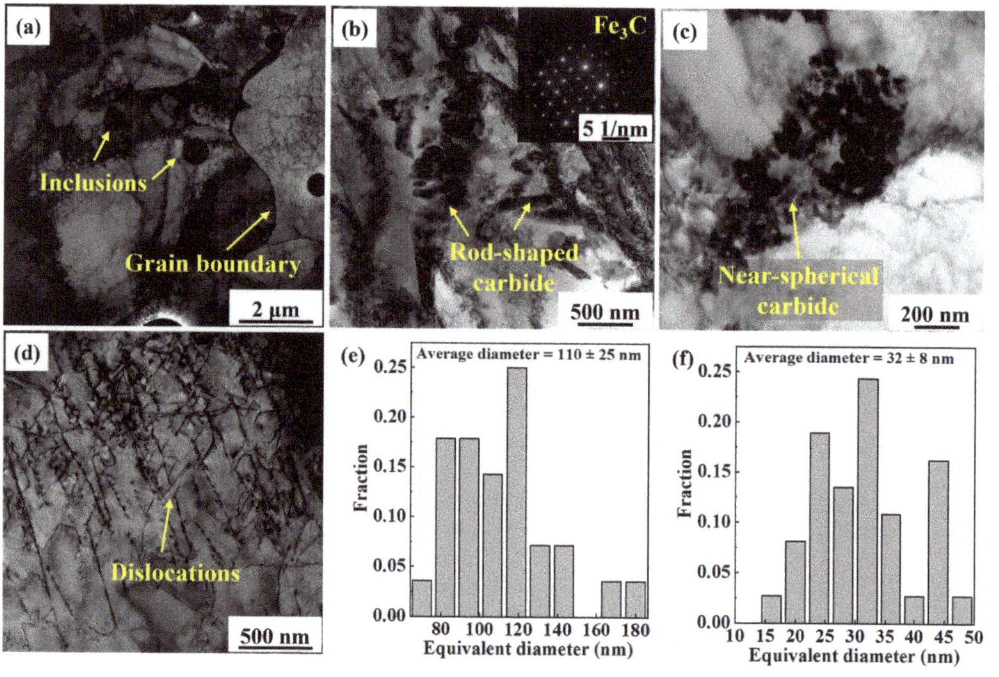

Figure 4. TEM bright-field images showing the microstructure in the middle region of AM X70 specimen. (**a**) Low-magnification of microstructure, (**b**) rod-shaped carbide aggregation, (**c**) near-spherical carbide aggregation, (**d**) sparse and straight dislocation lines, (**e**) size distribution of the rod-shaped carbides in (**b**), and (**f**) size distribution of the near-spherical carbides in (**c**).

Figure 5 presents the TEM analysis results of the BM specimen. Figure 5a shows a low-magnification image of the BM microstructure. Carbides precipitated at the trigeminal grain boundaries. In addition, some dislocation lines were distributed in the grains. Figure 5b shows that the carbides precipitate at the straight grain boundaries of two grains. The morphology and distribution of the carbides in the BM specimen were distinctly different from those of the AM specimen (Figure 4b,c). Figure 5c shows that the interior of the BM grains contains a dense network of entangled dislocations, with dislocation density notably higher than that in the AM specimen (Figure 4d). Dislocation density was even higher in grain boundary regions, forming typical dislocation pile-up structures, as shown in Figure 5a.

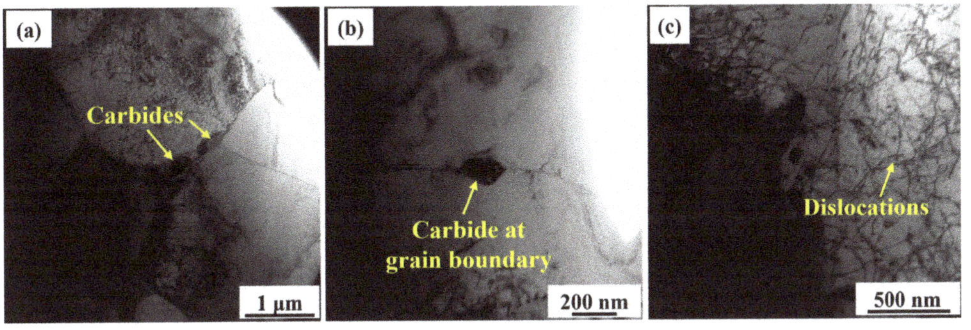

Figure 5. TEM bright-field images showing the microstructure of the BM X70 specimen. (**a**) Low-magnification of microstructure, (**b**) carbide distributed at the grain boundary, and (**c**) entangled and intersecting dislocations.

The EBSD analysis results of the middle region of the AM specimen are shown in Figure 6. The inverse pole figure (IPF) in Figure 6a revealed a random crystallographic orientation distribution. The grain size distribution in Figure 6d showed an average grain size of approximately 6.41 μm, with most grains smaller than 20 μm and a few grains reaching up to 40 μm. The grain boundary map in Figure 6b showed that the high-angle grain boundaries (HAGBs, ≥15°) accounted for 72.6% of total boundaries, while the low-angle grain boundaries (LAGBs, 2–15°) constituted 27.4%. The fractions of the HAGBs and LAGBs are further illustrated in Figure 6e. The HAGBs were predominantly located at PF boundaries, whereas the LAGBs were mainly distributed within PF grains. The kernel average misorientation (KAM) map in Figure 6c showed that the local misorientation was confined to a few grains, suggesting limited localized plastic strain accumulation. The distribution of the local misorientation agreed well with the distribution of LAGBs (Figure 6b). The KAM histogram in Figure 6f showed an average value of ~0.36°, with most values below 1.5°. The average density of the geometrically necessary dislocations (GND) can be calculated by the equation $\rho_{GND} = \frac{2KAM_{ave}}{\mu b}$, where μ is the step size of EBSD experiment, and b is the length of the Burgers vector. The ρ_{GND} of the AM specimen was $(1.89 \pm 1.64) \times 10^{14}\ m^{-2}$.

The EBSD analysis results of the BM specimen are shown in Figure 7. Figure 7a shows that no obvious preferred orientation is observed in the BM specimen. The grain size distribution in Figure 7d showed an average grain size of 4.91 μm. The grain size of BM specimen was smaller than that of the AM specimen. The fractions of HAGBs and LAGBs were 66% and 34%, respectively, as shown in Figure 7b. The fraction of HAGBs in the BM specimen was lower than that in the AM specimen (Figure 6b). Figure 7e shows that the misorientation angle distribution of LAGBs peaks around 2°, indicating that they are primarily formed by dislocation lines. The KAM map in Figure 7c showed that the distribution of local misorientation in the BM specimen was relatively uniform. The KAM histogram in Figure 7f showed an average value of ~0.53° which was higher than that in the AM specimen (0.36°), indicating that the local strain in the BM specimen was relatively high. The ρ_{GND} of the BM specimen was $(2.56 \pm 1.83) \times 10^{14}$ m^{-2} which was higher than that of the AM specimen.

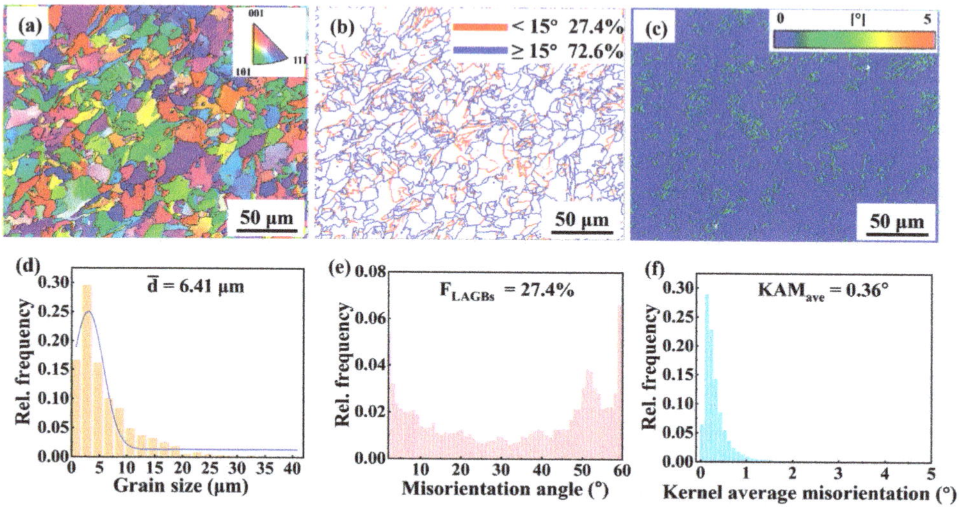

Figure 6. EBSD maps of the middle region of the AM specimen. (**a**) IPF map, (**b**) grain boundary distribution map, (**c**) KAM map, (**d**) distribution histogram of grain size, (**e**) distribution histogram of grain boundary misorientation angle, and (**f**) distribution histogram of KAM.

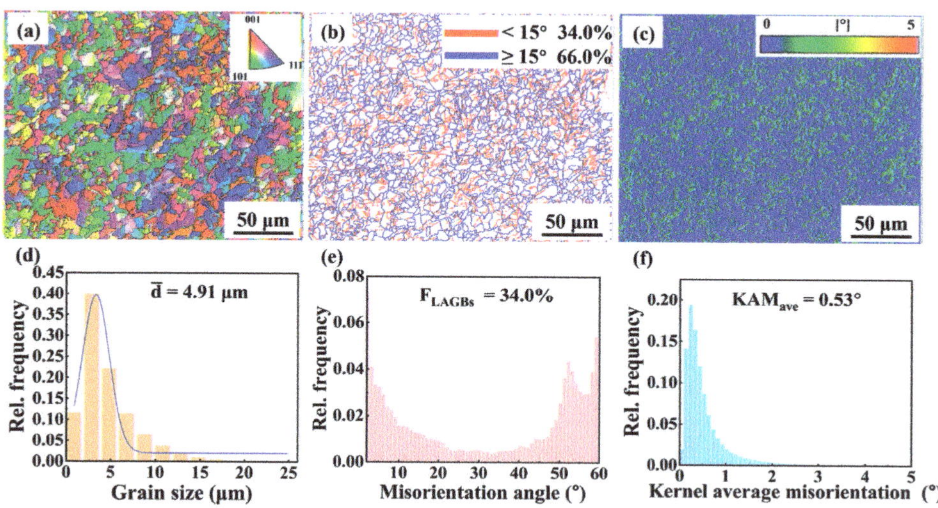

Figure 7. EBSD maps of the BM specimen. (**a**) IPF map, (**b**) grain boundary distribution map, (**c**) KAM map, (**d**) distribution histogram of grain size, (**e**) distribution histogram of grain boundary misorientation angle, and (**f**) distribution histogram of KAM.

3.2. Microhardness

Figure 8 shows the microhardness distribution along the depth direction of the AM specimen. The microhardness of the top region was the lowest, while that of the bottom region was the highest. The microhardness gradually increased along the depth direction, which was consistent with the microstructural heterogeneity through the building direction (Figure 2a–c). The average microhardness of the AM specimen was calculated as 179 ± 11 HV. The average microhardness of the BM specimen was 235 ± 7 HV. The microhardness of the AM specimen is 23.8% lower than that of the BM specimen, attributed to microstructural differences such as grain size and dislocation density.

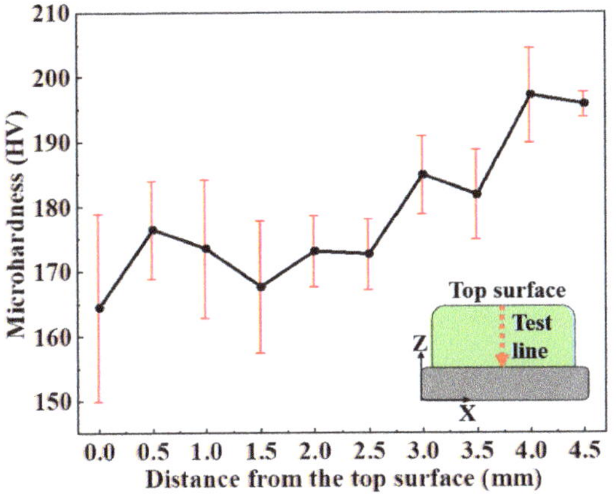

Figure 8. Microhardness distribution along the depth direction of the AM X70 specimen.

3.3. Tensile Properties

Figure 9 shows the engineering stress-displacement curves of the AM and BM specimens. The tensile specimens were taken from the middle region of the AM block, as shown in Figure 1c. The corresponding tensile test results are listed in Table 3. The AM specimen exhibited a clear yield plateau, indicating the presence of a yield point phenomenon. The BM specimen exhibited no distinct yield plateau. Due to the absence of an extensometer, the work hardening behavior beyond the yield point cannot be quantified. However, a qualitative comparison of the engineering stress-displacement curves of the AM and BM specimens revealed that the AM specimen exhibited a lower degree of work hardening than the BM specimen. Compared to the AM specimen, the BM specimen exhibited higher YS and UTS but lower elongation. According to the API 5L specification standard for X70 steel [18], the specified minimum YS is 485 MPa, and the minimum UTS is 570 MPa. However, the measured YS of the AM specimen was 434.8 MPa, which was quite below the required 485 MPa, indicating that the AM X70 did not meet the API 5L standard. This is mainly due to the formation of soft PF in the as-deposited microstructure.

Figure 10 shows the tensile fracture morphology of the AM specimen. Figure 10a reveals a significant necking area, with a measured reduction in area of 84.2% ± 2.7%. This indicated a significant plastic deformation during the fracture process of the AM specimen. Figure 10b reveals the presence of gas pores, inclusions and dimples on the fracture surface, demonstrating a ductile fracture of the AM specimen. The interface between the inclusions and the matrix was weak, making the interface prone to form micro-voids around them under tensile loading. Figure 10c,d show magnified views of the dimples. The uniform distribution and significant depth of the dimples were consistent with the microvoid coalescence (MVC) mechanism. The dimple sizes were concentrated in the 2–4 μm range.

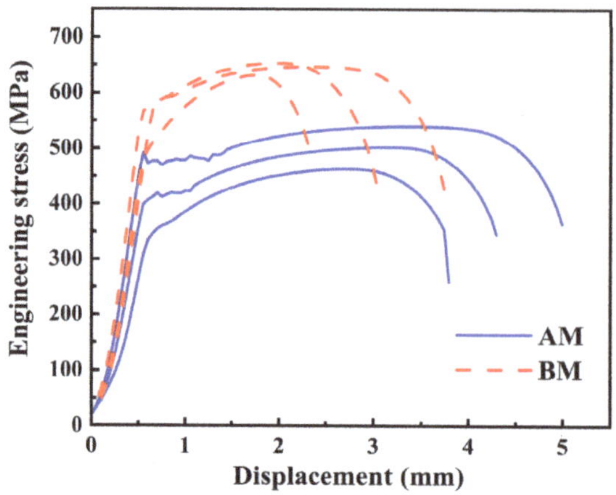

Figure 9. Engineering stress-displacement curves of the AM and BM specimens.

Table 3. Summary of the tensile test results of the AM and BM specimens.

Material	Yield Strength (MPa)	Ultimate Tensile Strength (MPa)	Elongation (%)
X70 AM	434.8 ± 36.4	512.6 ± 24.2	42.9 ± 3.5
X70 BM	549.7 ± 46.4	643.7 ± 10.9	27.1 ± 7.1
X70 API 5L	≥485	≥570	≥20

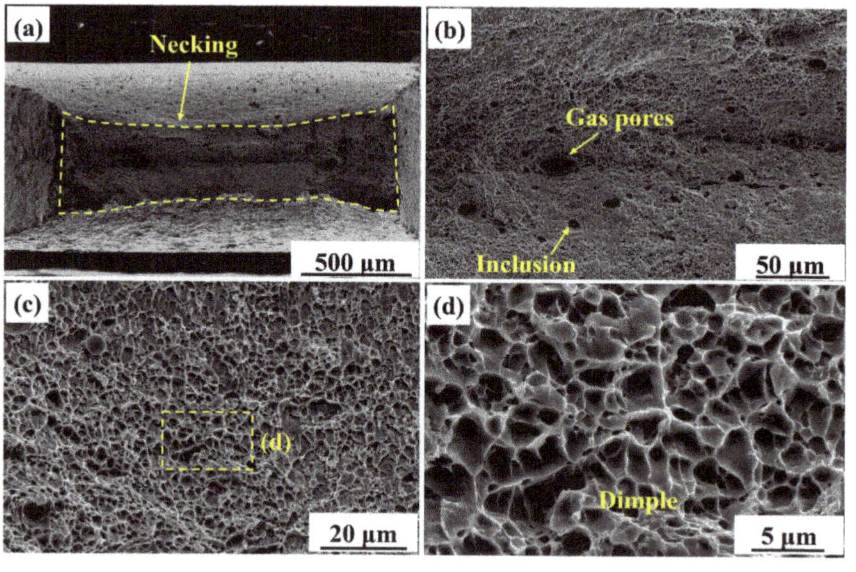

Figure 10. Fracture morphology of the AM tensile specimen. (**a**) Macroscopic morphology, (**b**) gas pores and inclusions, (**c**) uniformly distributed dimples, (**d**) magnified view of the dimples.

Figure 11 shows the tensile fracture morphology of the BM specimen. Figure 11a shows significant necking in the BM specimen. The measured reduction in area was 83.4% ± 3.5%, which was similar to that of the AM specimen. This indicated that both specimens exhibited comparable necking behavior. Figure 11b reveals the presence of micro-voids and dimples on the fracture surface. The formation, growth and coalescence of micro-voids were typical mechanisms of ductile fracture. Figure 11c,d show the magnified views of the dimples. The dimples in the BM specimen, which varied in size, were irregularly distributed. Figure 11d reveals shallower dimples compared to those in the AM specimen, indicating a lower degree of plastic deformation and inhibited micro-void expansion. The plastic deformation capacity of the BM specimen was lower than that of the AM specimen.

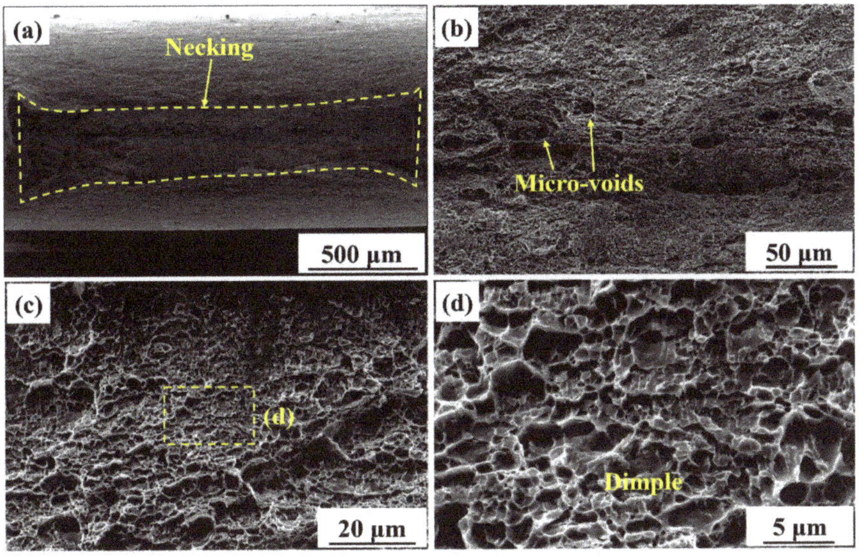

Figure 11. Fracture morphology of the BM tensile specimen. (**a**) Macroscopic morphology, (**b**) micro-voids on the surface, (**c**) uneven distribution of dimples, and (**d**) magnified view of dimples.

4. Discussion

4.1. Influence of L-DED Process on Microstructure Evolution

Experimental results revealed significant differences in the microstructures of the AM and BM specimens, particularly in grain morphology, size, and carbide precipitation. These microstructural variations directly influence the strength and ductility of the materials. The fundamental cause of these microstructural differences lies in the distinct thermal histories experienced during processing, with cooling rate and thermal cycling being key determinants of the microstructural evolution pathway [19].

4.1.1. Thermal Process During L-DED

In L-DED, a high-energy laser beam was used to generate a melt pool on the substrate or previously deposited layers, into which metal powder was simultaneously injected. Three-dimensional components were fabricated through successive layer-by-layer deposition. This layer-wise buildup imparted a complex thermal history to the material. Regions within the deposited layers underwent multiple thermal cycles, each involving rapid heating to a peak temperature followed by cooling [20].

During the early stages of L-DED, heat was primarily conducted into the substrate. The low initial substrate temperature created a steep thermal gradient between the melt pool and the substrate, facilitating efficient heat extraction. As the deposition height increased, heat accumulated in the deposited material. Subsequent heat must pass through the previously deposited layers before reaching the substrate. These layers were at elevated temperatures, which reduced the overall thermal gradient and impaired the heat dissipation efficiency, resulting in progressively lower cooling rates for subsequent layers [21]. The above-mentioned thermal process significantly influenced the formation and evolution of microstructures.

4.1.2. Phase Transformation of L-DED X70 Steel

It is well established that the cooling time from 800 °C to 500 °C ($t_{8/5}$) is a critical factor determining the microstructure formed during solid-state transformation in low-carbon steels [22]. The CCT curve of X70 steel was calculated by Jmatpro V7.0 software, as shown in Figure 12. During continuous deposition, the cooling rates range from approximately 2 to 20 °C/s [23]. According to the CCT curve of the X70 steel (Figure 12), these cooling rates favored the formation of PF [24,25]. However, different cooling rates and transformation temperature ranges lead to the formation of different types of ferrite. In the first deposited layer, the low substrate temperature resulted in a relatively high cooling rate, shifting austenite-to-ferrite transformation to a lower temperature range (600–680 °C). The solid-state transformation occurring within this temperature range was favorable for the formation of QF (Figure 2c). Figure 1b shows that the deposited block is relatively small. As the deposition height increased, heat accumulation within the build became increasingly significant. The temperature range for the transformation gradually increased. In the middle region, the moderate cooling rate promoted the formation of PF, resulting in an increased proportion of this phase (Figure 2b). At the top, the temperature range for the transformation was the highest (650–750 °C), coupled with the slowest cooling rate. This allowed sufficient time for the microstructure to fully grow, resulting in the formation of coarse PF and QF (Figure 2a).

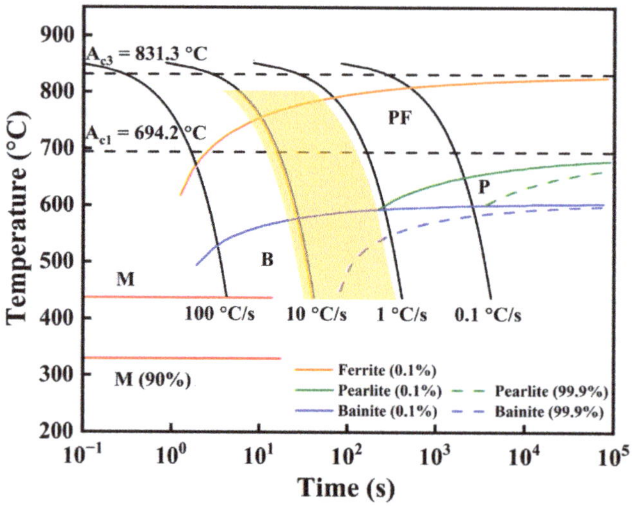

Figure 12. CCT curve of the X70 pipeline steel. The translucent region refers to the range of cooling curves in this work corresponding to the transformation from austenite to ferrite.

Related studies have also pointed out that the cooling rates during AM significantly affect the formation of the deposited microstructure. Ning et al. [10] demonstrated that the microstructure of the wire-fed AM X80 pipeline steel consists of LB due to the low heat input and rapid cooling rate. With increasing cooling rate, LB is preferentially formed. The formation of LB is accompanied by a shift in the transformation mechanism from a diffusion-controlled process to a mixed-mode mechanism involving both shear and diffusional characteristics. Wang et al. [26] pointed out that the cooling time $t_8/_5$ governs the continuous cooling transformation behavior of undercooled austenite. Compared to other heat affected zones in the X100 pipeline steel, the $t_8/_5$ of the laser-welded zone is the shortest (0.6 s), resulting in the formation of microstructure consisting of lath martensite and large bulk bainitic ferrite. As can be seen from the above comparison, the microstructure of the pipeline steel is primarily dependent on the cooling rate. PF tends to form under relatively slow cooling conditions.

EBSD results show that the grain size of the AM specimen is larger than that of the BM specimen. This difference arose from their distinct processing routes and thermal histories. During the layer-by-layer deposition, heat accumulation occurred progressively. This thermal buildup promoted repeated thermal cycling, leading to gradual coarsening of the microstructure into coarse PF. No external plastic deformation was applied to the as-deposited block, except for the mechanical constraints imposed by the substrate [27]. Therefore, the solid-state transformation and microstructural evolution were primarily governed by the high-temperature thermal cycles during L-DED. In contrast, the TMCP for X70 steel involved controlled rolling followed by accelerated cooling. As a result, the final microstructure was dominated by QF, characterized by a flattened morphology induced by plastic deformation and a refined grain structure due to fast cooling [28,29]. The differences in processing routes and thermal histories lead to significant variations in microstructure and mechanical properties.

4.1.3. Carbide Precipitation

TEM results showed that there were two types of carbides formed in the AM X70, as shown in Figure 4b,c. The rod-shaped carbides in Figure 4b were mainly distributed along the grain boundary. During the diffusional phase transformation from austenite to PF, carbon atoms were expelled to the austenite grain boundaries [30,31]. In addition, the quite large heat accumulation and slow cooling rates provided sufficient driving force for carbon atom diffusion and Fe_3C nucleation and growth [23]. The intrinsic heat treatment during L-DED is somewhat like the tempering treatment used in the production routes of steel [32]. The main difference between this intrinsic heat treatment and conventional tempering is that the heating and cooling processes of the deposited layers are relatively fast. The thermal cycles promoted the precipitation of Fe_3C. In contrast, conventional tempering treatments involve longer durations and higher temperatures, under which $M_{23}C_6$ generally precipitates from the microstructure and aggregates at grain boundaries [33]. The $M_{23}C_6$ ($(Cr, Mn, Fe)_{23}C_6$) exhibits higher stability than the M_3C ($(Cr, Mn, Fe)_3C$) [34].

The near-spherical carbides in Figure 4c were concentrated in a small region. These carbides were transformed from the M/A constituents formed during the cooling process. During the solid-state phase transformation process, the carbon atoms were ejected from the ferrite and partitioned into austenite. The carbon can improve the stability of the austenite. Further cooling led to the transformation of part of the austenite into martensite. As a result, a small amount of M/A constituents was retained in the microstructure [35]. The intrinsic heat treatment promoted the decomposition of the M/A constituents, resulting in the precipitation of theses carbides.

4.1.4. Grain Boundary and Dislocation Evolution

EBSD analysis revealed that the proportion of HAGBs in the AM specimen was 72.6%, significantly higher than that in the BM specimen (66.0%). This difference can be attributed to the distinct thermal cycles experienced during AM and TMCP. When the deposited layers were subjected to high temperatures (above A_{c3} or between A_{c3} and A_{c1}), grain growth and recrystallization occur, modifying the as-deposited microstructure. When the temperature was below A_{c1}, in situ tempering occurred [23]. In these processes, small grains merged into large grains, leading to the formation of HAGBs. Meanwhile, the diffusional phase transformation from austenite to PF were not accompanied by the formation of dislocations. The pre-existed dislocations underwent rearrangement and annihilation when they were exposed to the thermal cycles [36], leading to the reduction in dislocation density. As a result, the fraction of LAGBs was reduced in the AM specimen. In contrast, the rolling and controlled cooling processes during TMCP resulted in the formation of a large number of subgrain boundaries and dislocations in the microstructure. This significantly increased the fraction of the LAGBs in the BM specimen.

4.2. Influence of Microstructure on Microhardness

The AM specimen exhibited lower microhardness than the BM specimen. Compared with the BM specimen, the coarse PF grains and low dislocation density in the AM specimen contributed to its reduced microhardness. In addition, the AM specimen exhibited a pronounced microhardness gradient along the building direction. This microhardness gradient was determined by the spatially varying microstructural features. The microstructure in the bottom region was dominated by QF. QF formed through a mixed mechanism of diffusion and shear [24]. Its grain boundaries were slightly curved and uneven. It contained some dislocation tangles and subgrains inside. Therefore, the QF in the bottom region had higher microhardness. The middle region had a higher content of PF. Fe_3C precipitates were presented within the ferrite grains. The interstitial solid solution strengthening effect of carbon atoms was significantly reduced due to carbon partitioning into precipitated carbides. As a result, the microhardness of the middle region was decreased. The coarse microstructure in the top region contributed to the reduction in microhardness. In contrast, the BM specimen possessed a fine-grained microstructure and a high dislocation density, both of which significantly increased the resistance to dislocation motion, leading to higher deformation resistance [37,38].

Table 4 summarizes the microhardness of the X70 pipeline steel fabricated by different methods. The microhardness of the arc-welded X70 is lower than the standard, whereas that of the laser-welded X70 is higher [39–41]. The microhardness is influenced by the microstructure, which is determined by the cooling rate. The microhardness distribution of the AM specimen is well below the standard, reflecting the low strengthening capability of the PF-dominated microstructure.

Table 4. Summary of the microhardness of X70 pipeline steel fabricated by different methods.

Method	Microhardness (HV)	Refs.
Submerged arc welding	220–240	[39]
Submerged arc welding	206–220	[40]
Laser welding	370–380	[41]
API 5L	240–280	[18]
L-DED	165–197	This work

4.3. Influence of Microstructure on Tensile Properties

4.3.1. Tensile Strength and Ductility

The YS and UTS of the AM specimen were significantly lower than those of the BM specimen. According to the Hall-Petch relationship, the YS of the metallic material is inversely proportional to the square root of its grain size [42]. EBSD analysis revealed that the AM specimen was dominated by coarse PF and exhibits a low grain boundary density, as shown in Figure 6a,b. In addition, the TEM and EBSD results were consistent in showing that the AM specimen had a low dislocation density. The L-DED X70 exhibited limited strengthening capability due to its coarse microstructure and low dislocation density [43].

As shown in Figure 9, the AM specimen exhibited excellent ductility, which was closely related to its internal microstructure. The AM specimen contained a high proportion of HAGBs. These HAGBs can deflect the propagation direction of cracks within the material, causing them to consume more energy during propagation, thereby effectively inhibiting crack growth and enhancing the ductility [44]. Additionally, HAGBs facilitated dislocation transmission across grain boundaries, promoting a more uniform strain distribution throughout the polycrystalline structure and delaying localized necking.

Furthermore, the initial dislocation density in the AM specimen was relatively low, allowing dislocations to have higher mobility during the early stages of plastic deformation [45]. This not only alleviated the premature pile-up of dislocations at grain boundaries [46], which hindered dislocation movement, but also reduced stress concentrations that typically triggered micro-void nucleation, further enhancing the ductility.

4.3.2. Fracture Morphology

The fracture surface of the AM specimen exhibited typical ductile fracture characteristics, as shown in Figure 10. The coarse PF matrix, high fraction of HAGBs, and low dislocation density collectively contributed to a more homogeneous plastic strain distribution. Micro-void nucleation occurred predominantly at inclusions and second-phase particles. Subsequently, they underwent stable growth and coalescence under uniform stress conditions. As a result, deep dimples were formed through the MVC mechanism. In contrast, the dimple morphology in the BM specimen was shallow and non-uniform. The BM specimen consisted of fine grains and exhibited a high dislocation density. The BM specimen had high strength but limited ductility, leading to poorly developed dimples, as shown in Figure 11d.

4.4. Limitations

The L-DED process offers notable advantages for repairing X70 pipeline steel, including in situ repair capability, controllable heat input, adjustable powder composition, and fast repair speed. However, this study also reveals certain limitations. Significant heat accumulation and slow cooling rates associated with L-DED lead to the formation of PF. This deteriorates the mechanical properties of the as-deposited X70 steel. The tensile strength does not meet the mechanical property requirements specified in the API 5L standard [18]. It is widely reported that the ideal microstructure for X70 steel is AF, which exhibits superior mechanical properties, including high strength and toughness [47,48]. However, AF formation typically requires higher cooling rates. To optimize the mechanical properties of the deposited X70, future research could focus on controlling thermal cycles and cooling profiles to locally enhance the cooling rate, thereby promoting AF formation.

Potential technical solutions include: (i) Adjust the interlayer residence time. Heat accumulation can be reduced by extending the interlayer cooling time [49]. The next layer begins to deposit on a substrate at a lower temperature. (ii) Near-immersion active cooling. The build platform is lowered to bring the deposited layer into contact with a cooling medium, thereby enhancing heat dissipation [50]. (iii) Spray cooling. A spray cooling system can be integrated into the L-DED process for interlayer temperature control [51]. The cooling medium can be high-pressure air or liquid nitrogen. These strategies will be applied to L-DED X70 pipeline steel to improve its mechanical properties through controlled microstructure development.

5. Conclusions

This study employed the L-DED technique to repair the X70 pipeline steel. The microstructural characteristics and mechanical properties of the L-DED X70 steel were systematically investigated and compared with those of the BM. The influence mechanisms of the L-DED process on the comprehensive performance of the AM X70 were elucidated. The main conclusions are as follows:

(1) The microstructure exhibited inhomogeneity along the building direction. From the bottom to the top, the grains gradually coarsened, and the proportion of PF increased. From the bottom to the top, thermal accumulation increased, resulting in decreased cooling rates, which promoted grain coarsening and the formation of PF. The grain size of the AM X70 was larger than that of BM.

(2) AM X70 exhibited a significantly reduced dislocation density. The high thermal input and slow cooling rates led to substantial annihilation of dislocation structures. Fe_3C carbides were dispersed both within grains and along grain boundaries. Meanwhile, the intrinsic heat treatment provided favorable kinetic conditions for carbon diffusion, as well as for the nucleation and growth of Fe_3C.

(3) The microhardness gradually decreased along the building direction. The bottom region had a high content of QF, resulting in higher microhardness. In contrast, the top region was predominantly composed of coarse PF, leading to lower microhardness. The average microhardness of the AM specimen was 179 $\pm$ 11 HV which was 23.8% lower than that of the BM specimen (235 $\pm$ 7 HV).

(4) The AM X70 exhibited a YS of 435 MPa and an UTS of 513 MPa. Both the YS and UTS of the AM X70 were lower than those of the BM and API 5L standard. The elongation of AM X70 reached 42.9%, which was 58% higher than that of the BM. The reduction in strength stemmed from the grain coarsening and decreased dislocation density induced by the L-DED thermal cycles. Meanwhile, the plastic accommodation ability was enhanced, resulting in substantially improved ductility. In harsh service environments characterized by complex stress states, the strength of repaired pipeline steel becomes critically important. Given the current limitations in mechanical performance, future work will primarily focus on enhancing the strength of the deposited X70 steel.

Author Contributions: Conceptualization, Z.W. and C.W.; methodology, Z.W. and C.W.; data curation, L.W.; formal analysis, C.W.; writing—original draft preparation, Z.W.; writing—review and editing, Z.W. and G.S.; supervision, Z.W. and G.S. All authors have read and agreed to the published version of the manuscript.

Funding: This research was funded by the China Postdoctoral Science Foundation, Grant number 2024M751202.

Institutional Review Board Statement: Not applicable.

Informed Consent Statement: Not applicable.

Data Availability Statement: The original contributions presented in this study are included in the article. Further inquiries can be directed to the corresponding authors.

Conflicts of Interest: The authors declare no conflicts of interest.

References

1. Nguyen, T.T.; Park, J.; Kim, W.S.; Nahm, S.H.; Beak, U.B. Effect of low partial hydrogen in a mixture with methane on the mechanical properties of X70 pipeline steel. *Int. J. Hydrogen Energy* **2020**, *45*, 2368–2381. [CrossRef]
2. Beidokhti, B.; Dolati, A.; Koukabi, A.H. Effects of alloying elements and microstructure on the susceptibility of the welded HSLA steel to hydrogen-induced cracking and sulfide stress cracking. *Mater. Sci. Eng. A* **2009**, *507*, 167–173. [CrossRef]
3. Fekete, J.R.; Sowards, J.W.; Amaro, R.L. Economic impact of applying high strength steels in hydrogen gas pipelines. *Int. J. Hydrogen Energy* **2015**, *40*, 10547–10558. [CrossRef]
4. Zhang, S.B.; Hou, P.J.; Kang, J.Y.; Li, T.Y.; Mooraj, S.; Ren, Y.; Chen, C.H.; Hart, A.J.; Gerasimidis, S.; Chen, W. Laser additive manufacturing for infrastructure repair: A case study of a deteriorated steel bridge beam. *J. Mater. Sci. Technol.* **2023**, *154*, 149–158. [CrossRef]
5. Piscopo, G.; Iuliano, L. Current research and industrial application of laser powder directed energy deposition. *Int. J. Adv. Manuf. Technol.* **2022**, *119*, 6893–6917. [CrossRef]

6. Kang, L.; Song, P. Bending behaviour of surface corroded and perforated corroded steel tubes repaired by laser cladding additive manufacturing. *Thin-Walled Struct.* **2024**, *203*, 112213. [CrossRef]
7. Ning, J.; Wen, J.H.; Zhang, L.J.; Na, S.J. Assessment of the universality of duplex stainless steel powder in laser additive repair based on Schaeffler diagram. *Addit. Manuf.* **2022**, *55*, 102864. [CrossRef]
8. Cheng, H.; Lei, B.L.; Liu, X.; Deng, Z.R.; Chen, K.Q.; Wei, S.Y.; Luo, X.C.; Wu, X. Development of low transformation temperature alloys for laser additive manufacturing repairing the defects in HSLA-100 plates: Focusing on the microstructure, residual stress and performance. *J. Mater. Res. Technol.* **2025**, *38*, 1472–1486. [CrossRef]
9. Rashid, R.A.R.; Nazari, K.A.; Barr, C.; Palanisamy, S.; Orchowski, N.; Matthews, N.; Dargusch, M.S. Effect of laser reheat post-treatment on the microstructural characteristics of laser-cladded ultra-high strength steel. *Surf. Coat. Technol.* **2019**, *372*, 93–102. [CrossRef]
10. Ning, J.; Yu, Z.S.; Sun, K.; Hu, M.J.; Zhang, L.X.; Zhang, Y.B.; Zhang, L.J. Comparison of microstructures and properties of X80 pipeline steel additively manufactured based on laser welding with filler wire and cold metal transfer. *J. Mater. Res. Technol.* **2021**, *10*, 752–768. [CrossRef]
11. Seede, R.; Shoukr, D.; Zhang, B.; Whitt, A.; Gibbons, S.; Flater, P.; Elwany, A.; Arroyave, R.; Karaman, I. An ultra-high strength martensitic steel fabricated using selective laser melting additive manufacturing: Densification, microstructure, and mechanical properties. *Acta Mater.* **2020**, *186*, 199–214. [CrossRef]
12. Sun, S.D.; Liu, Q.C.; Brandt, M.; Luzin, V.; Cottam, R.; Janardhana, M.; Clark, G. Effect of laser clad repair on the fatigue behaviour of ultra-high strength AISI 4340 steel. *Mater. Sci. Eng. A* **2014**, *606*, 46–57. [CrossRef]
13. Zhan, X.H.; Fang, Y.C.; Gao, Q.Y.; Guan, X.H.; Ci, S.W.; Cai, X.K.; Guan, S.Z. Study on the microstructure evolution performance improvement of Invar alloy repaired by laser cladding Aermet100 ultra-high strength steel. *J. Mater. Process. Technol.* **2024**, *331*, 118517. [CrossRef]
14. Jing, G.Y.; Wang, Z.M. Influence of molten pool mode on microstructure and mechanical properties of heterogeneously tempered 300M steel by selective laser melting. *J. Mater. Process. Technol.* **2021**, *296*, 117188. [CrossRef]
15. Sun, G.F.; Yao, S.; Wang, Z.D.; Shen, X.T.; Yan, Y.; Zhou, R.; Ni, Z.H. Microstructure and mechanical properties of HSLA-100 steel repaired by laser metal deposition. *Surf. Coat. Technol.* **2018**, *351*, 198–211. [CrossRef]
16. Shang, F.M.; Chen, S.Y.; Zhou, L.; Jia, W.M.; Cui, T.; Liang, J.; Liu, C.S.; Wang, M. Effect of laser energy volume density on wear resistance and corrosion resistance of 30Cr15MoY alloy steel coating prepared by laser direct metal deposition. *Surf. Coat. Technol.* **2021**, *421*, 127382. [CrossRef]
17. Claeys, L.; Deconinck, L.; Verbeken, K.; Depover, T. Effect of additive manufacturing and subsequent heat and/or surface treatment on the hydrogen embrittlement sensitivity of 316L austenitic stainless steel. *Int. J. Hydrogen Energy* **2023**, *48*, 36142–36157. [CrossRef]
18. American Petroleum Institute. *API Specification 5L: Specification for Line Pipe*, 46th ed.; American Petroleum Institute: Washington, DC, USA, 2018.
19. DebRoy, T.; Wei, H.L.; Zuback, J.S.; Mukherjee, T.; Elmer, J.W.; Milewski, J.O.; Beese, A.M.; Wilson-Heid, A.; De, A.; Zhang, W. Additive manufacturing of metallic components-process, structure and properties. *Prog. Mater. Sci.* **2018**, *92*, 112–224. [CrossRef]
20. Müller, J.; Goulas, C.; Hensel, J. Analysis of thermal cycles during DED-Arc of high-strength low-alloy steel and microstructural evolution. *J. Mater. Res. Technol.* **2024**, *31*, 3661–3674. [CrossRef]
21. Zhang, X.T.; Wang, L.; Zhao, N.; Liu, R.C.; Zhang, L.; Wu, W.D.; Yang, D.Q.; Huang, Y.; Wang, K.H. The effect of thermal cycle on microstructure evolution and mechanical properties of Co-free maraging steel produced by wire arc additive manufacturing. *J. Mater. Process. Technol.* **2024**, *332*, 118582. [CrossRef]
22. Lazić, V.N.; Sedmak, A.S.; Živković, M.M.; Aleksandrović, S.M.; Čukić, R.D.; Jovičić, R.D.; Ivanović, I.B. Theoretical-experimental determining of cooling time (t8/5) in hard facing of steels for forging dies. *Therm. Sci.* **2010**, *14*, 235–246. [CrossRef]
23. Rodrigues, T.A.; Duarte, V.; Avila, J.A.; Santos, T.G.; Miranda, R.M.; Oliveira, J.P. Wire and arc additive manufacturing of HSLA steel: Effect of thermal cycles on microstructure and mechanical properties. *Addit. Manuf.* **2019**, *27*, 440–450. [CrossRef]
24. Liang, C.; Song, G.X.; Wang, W.L.; Sohn, I.L.; Zeng, J. In situ observation of the austenite to ferrite transformation in low-carbon steels from different initial phases at defined cooling rates. *J. Mater. Res. Technol.* **2024**, *28*, 2116–2126. [CrossRef]
25. Xie, H.; Du, L.X.; Hu, J.; Sun, G.S.; Wu, H.Y.; Misra, R.D.K. Effect of thermo-mechanical cycling on the microstructure and toughness in the weld CGHAZ of a novel high strength low carbon steel. *Mater. Sci. Eng. A* **2015**, *639*, 482–488. [CrossRef]
26. Wang, G.; Wang, J.; Yin, L.; Hu, H.; Yao, Z. Quantitative correlation between thermal cycling and the microstructures of X100 pipeline steel laser-welded joints. *Materials* **2020**, *13*, 121. [CrossRef]
27. Bertsch, K.M.; Meric de Bellefon, G.; Kuehl, B.; Thoma, D.J. Origin of dislocation structures in an additively manufactured austenitic stainless steel 316L. *Acta Mater.* **2020**, *199*, 19–33. [CrossRef]
28. Xiao, F.R.; Liao, B.; Shan, Y.Y.; Qiao, G.Y.; Zhong, Y.; Zhang, C.L.; Yang, K. Challenge of mechanical properties of an acicular ferrite pipeline steel. *Mater. Sci. Eng. A* **2006**, *431*, 41–52. [CrossRef]

29. Lobanov, M.L.; Zorina, M.A.; Karabanalov, M.S.; Urtsev, N.V.; Redikultsev, A.A. Phase transformation crystallography in pipeline HSLA steel after TMCP. *Metals* **2023**, *13*, 1121. [CrossRef]
30. Odqvist, J.; Hillert, M.; Ågren, J. Effect of alloying elements on the γ to α transformation in steel. I. *Acta Mater.* **2002**, *50*, 3213–3227. [CrossRef]
31. Danoix, F.; Sauvage, X.; Huin, D.; Germain, L.; Gouné, M. A direct evidence of solute interactions with a moving ferrite/austenite interface in a model Fe-C-Mn alloy. *Scr. Mater.* **2016**, *121*, 61–65. [CrossRef]
32. Biro, E.; McDermid, J.R.; Vignier, S.; Zhou, Y.N. Decoupling of the softening processes during rapid tempering of a martensitic steel. *Mater. Sci. Eng. A* **2014**, *615*, 395–404. [CrossRef]
33. Xue, H.Y.; Li, X.; Zhang, Z.H.; Xiao, S.J.; Yan, M.Z. The effect of tempering temperature on reheating cracks in the coarse-grained heat-affected zone of 15Cr1Mo1V steel pipe after repair welding. *Int. J. Press. Vessel. Pip.* **2025**, *218*, 105590. [CrossRef]
34. Zhao, L.Y.; Wang, Q.M.; Shi, G.H.; Hu, B.; Wang, S.B.; Qiao, M.L.; Wang, Q.F.; Liu, R.P. The impacts of M/A constituents decomposition and complex precipitation on mechanical properties of high-strength weathering steel subjected to tempering treatment. *J. Mater. Res. Technol.* **2023**, *23*, 2504–2526. [CrossRef]
35. Tang, Z.H.; Stumpf, W. The role of molybdenum additions and prior deformation on acicular ferrite formation in microalloyed Nb-Ti low-carbon line-pipe steels. *Mater. Charact.* **2008**, *59*, 717–728. [CrossRef]
36. Zhou, Y.; Chen, S.Y.; Chen, X.T.; Cui, T.; Liang, J.; Liu, C.S. The evolution of bainite and mechanical properties of direct laser deposition 12CrNi2 alloy steel at different laser power. *Mater. Sci. Eng. A* **2019**, *742*, 150–161. [CrossRef]
37. Forouzan, F.; Vuorinen, E.; Mücklich, F. Post weld-treatment of laser welded AHSS by application of quenching and partitioning technique. *Mater. Sci. Eng. A* **2017**, *698*, 174–182. [CrossRef]
38. Saha, D.C.; Biro, E.; Gerlich, A.P.; Zhou, Y. Martensite tempering kinetics: Effects of dislocation density and heating rates. *Mater. Charact.* **2020**, *168*, 110564. [CrossRef]
39. Ren, T.; Mohammadijoo, M.; Wiskel, J.B.; Lazor, R.; Willett, E.; Ivey, D.G.; Henein, H. Effect of cold wire addition in tandem submerged arc welding on weld geometry and micro-hardness of heavy gauge X70 steel. *Int. J. Adv. Manuf. Technol.* **2022**, *121*, 7607–7625. [CrossRef]
40. Sharma, L.; Chhibber, R. Mechanical properties and hydrogen induced cracking behaviour of API X70 SAW weldments. *Int. J. Press. Vessel. Pip.* **2018**, *165*, 193–207. [CrossRef]
41. Gordienko, A.I.; Derevyagina, L.S.; Malikov, A.G.; Orishich, A.M.; Surikova, N.S.; Volochaev, M.N. The effect of the initial microstructure of the X70 low-carbon microalloyed steel on the heat affected zone formation and the mechanical properties of laser welded joints. *Mater. Sci. Eng. A* **2020**, *797*, 140075. [CrossRef]
42. Chen, Y.B.; Feng, J.C.; Li, L.Q.; Chang, S.; Ma, G.L. Microstructure and mechanical properties of a thick-section high-strength steel welded joint by novel double-sided hybrid fibre laser-arc welding. *Mater. Sci. Eng. A* **2013**, *582*, 284–293. [CrossRef]
43. Hayes, B.J.; Martin, B.W.; Welk, B.; Kuhr, S.J.; Ales, T.K.; Brice, D.A.; Ghamarian, I.; Baker, A.H.; Haden, C.V.; Harlow, D.G.; et al. Predicting tensile properties of Ti-6Al-4V produced via directed energy deposition. *Acta Mater.* **2017**, *133*, 120–133. [CrossRef]
44. Qin, Y.Q.; Feng, Q.S.; Hu, Y.B.; Zhong, F.L.; Zhao, Z.F.; Yu, Q.C.; Yin, S.B.; Jia, S.J. Effect of Cr microalloying on microstructure evolution and toughness in SCGHAZ of X80 pipeline steel. *J. Iron Steel Res. Int.* **2025**, *32*, 3573–3586. [CrossRef]
45. Ju, Y.Z.; Li, S.; Ba, L.Z.; Yang, X.C.; Li, C.N.; Di, X.J. Effect of cyclic heat treatment on microstructure and mechanical properties of ultra-high strength steel fabricated by wire + arc additive manufacturing with cold metal transfer. *Mater. Sci. Eng. A* **2025**, *927*, 147986. [CrossRef]
46. Wang, W.; Shan, Y.Y.; Yang, K. Study of high strength pipeline steels with different microstructures. *Mater. Sci. Eng. A* **2009**, *502*, 38–44. [CrossRef]
47. Zhao, M.C.; Hanamura, T.; Qiu, H.; Yang, K. Lath boundary thin-film martensite in acicular ferrite ultralow carbon pipeline steels. *Mater. Sci. Eng. A* **2005**, *395*, 327–332. [CrossRef]
48. Xiao, F.R.; Liao, B.; Ren, D.L.; Shan, Y.Y.; Yang, K. Acicular ferritic microstructure of a low-carbon Mn-Mo-Nb microalloyed pipeline steel. *Mater. Charact.* **2005**, *54*, 305–314. [CrossRef]
49. Lee, H.J.; Narayana, P.L.; Kim, J.H.; Park, C.H.; Hong, J.-K.; Yeom, J.-T.; Lee, T.; Lee, S.W. Effect of interlayer cooling on the microstructure and mechanical properties of titanium alloys fabricated using directed energy deposition. *J. Alloys Compd.* **2023**, *953*, 170202. [CrossRef]
50. da Silva, L.J.; Souza, D.M.; de Araújo, D.B.; Reis, R.P.; Scotti, A. Concept and validation of an active cooling technique to mitigate heat accumulation in WAAM. *Int. J. Adv. Manuf. Technol.* **2020**, *107*, 2513–2523. [CrossRef]
51. Kuprienko, A.; Stickle, J.; Harwig, D.; Carney, M.; Fink, C. Effect of liquid nitrogen cryogenic cooling during wire arc directed energy deposition of nickel aluminum bronze. *Weld. World* **2024**, *68*, 515–524. [CrossRef]

materials

MDPI

Article

Effect of Laser Energy Density During LPBF on the Structure and Mechanical Properties of Al–15Sn–5Pb Alloy

Nikolay M. Rusin [1], Alexander L. Skorentsev [1,2], Kirill O. Akimov [1,3], Vadim E. Likharev [1,4], Dmitry P. Il'yashchenko [5] and Andrey I. Dmitriev [1,3,*]

1 Institute of Strength Physics and Materials Science SB RAS, 2/4 Akademicheskii Av., Tomsk 634055, Russia; rusinnm@mail.ru (N.M.R.); skoralexan@mail.ru (A.L.S.); akimov_ko@ispms.ru (K.O.A.); vel6@tpu.ru (V.E.L.)
2 School of Nuclear Science and Engineering, National Research Tomsk Polytechnic University, 30 Lenin Av., Tomsk 634050, Russia
3 Department of Metal Physics, National Research Tomsk State University, 36 Lenin Av., Tomsk 634050, Russia
4 School of Advanced Manufacturing Technologies, National Research Tomsk Polytechnic University, 30 Lenin Av., Tomsk 634050, Russia
5 Yurga Technological Institute, National Research Tomsk Polytechnic University, 30 Lenin Av., Tomsk 634050, Russia; mita8@tpu.ru
* Correspondence: dmitr@ispms.ru

Abstract

Al–15Sn–5Pb (vol.%) alloy was fabricated by the Laser Powder Bed Fusion (LPBF) method at laser scanning speeds of 0.8, 1.0, and 1.2 m/s and laser powers ranging from 70 to 130 W. The samples were synthesized from a mixture of elemental powders using an ONSINT AM150 3D printer under a flowing argon atmosphere. The structure and mechanical properties under compression tests of the produced material were investigated as a function of the volumetric laser energy density (E) during LPBF. It has been established that low laser energy density during LPBF results in incomplete melting of aluminum particles and a non-uniform distribution of soft inclusions within the material. Increasing the energy density ensures a significantly more uniform distribution of the phases, resulting in the formation of a fine-grained three-phase alloy. It was established that both the ductility and strength of the alloy improve with the increase in E until a critical value is reached. As a result, at $E \geq 48\ \mathrm{J \cdot mm^{-3}}$, the ultimate strength of the alloy reaches 100 ± 5 MPa, and its deformation before fracture is $15 \pm 1\%$. Substituting one quarter of the tin volume with lead results in a significant increase in the ductility of the LPBF-fabricated aluminum alloy.

Keywords: laser powder bed fusion (LPBF); volumetric laser energy density; critical energy density; aluminum matrix alloy; microstructure; ductility; compressive strength

1. Introduction

Alloying aluminum with tin imparts antifriction properties due to the increased seizure pressure under dry and boundary friction. Such alloys find applications as antifriction materials in plain bearings and as phase change materials in thermal energy applications. This effect is caused by the ability of tin to be extruded onto the friction surface and to form a protective tribofilm [1–10]. According to the Russian standard GOST 14113-78 [11], the tin concentration in aluminum should be limited to 20 wt.% (~10 vol.%), since higher contents lead to the formation of continuous tin interlayers along the aluminum grain boundaries, which prevents their bonding into a strong load-bearing framework [12–14]. During plastic deformation, tin does not undergo strain hardening, and its interlayers in the loaded material act as areas of localized plastic flow; once incorporated into a shear band, these interlayers rapidly thin and fracture [14,15].

Therefore, the stronger the aluminum framework in such an alloy—that is, the lower the tin content—the higher the stress required to initiate plastic flow localization. For example, samples of the Al-20Sn-1Cu wt.% alloy (containing ~10 vol.% Sn) can withstand up to 30% tensile strain at an ultimate strength of 100–110 MPa. However, as shown in Ref. [16], an alloy with twice the tin content exhibiting comparable strength and high ductility can also be produced by sintering elemental Al and Sn powders. This alloy has a high seizure pressure under dry sliding against steel [14].

Since tin is insoluble in aluminum and is located between its grains, it could be expected that an increase in the specific surface area due to grain refinement would promote the formation of a stronger aluminum framework. However, experimental observations have shown that even at very high solidification rates of the Al–Sn melt, the precipitated aluminum solid-phase particles become surrounded by thin shells of liquid tin [17,18]. These interlayers, owing to their small thickness, have a limited reserve of ductility and rapidly fail once the flow stress reaches a certain level, which causes the fine-grained Al–20Sn (vol.%) alloy to exhibit high strength but low ductility.

It has been shown [3,19] that continuous tin interlayers can be fragmented into isolated inclusions if Al–Sn castings are subjected to severe plastic deformation followed by annealing. However, the shape and dimensions of the castings change considerably during such treatment, while their structure becomes coarser, making these methods economically inefficient for producing fine-grained materials. To prepare materials with a finely dispersed microstructure, the Laser Powder Bed Fusion (LPBF) process can be employed, which is characterized by extremely high heating and cooling rates of the deposited powder layer [20–22]. Nevertheless, as reported in [22,23], in Al–20Sn (vol.%) alloys produced by LPBF, a developed tin net still persists despite the reduced thickness of individual tin inclusions.

It was established in [24,25] that intergranular tin layers can also be disrupted by metallurgical means, namely by alloying Al–Sn systems with elements forming a eutectic with tin. For instance, lead—similar to tin—is insoluble in solid aluminum and, during cooling of the Al–Sn–Pb alloy, segregates to the periphery of aluminum grains as part of a liquid Sn–Pb film. This intergranular liquid layer decomposes at 183 °C into distinct tin and lead phases. The morphology of these intergranular regions depends on the volumetric Pb/Sn ratio. In Ref. [26], this ratio was 1:3, i.e., the volume fraction of tin in the melt was three times higher than that of lead. This Al–15Sn–5Pb alloy produced by LPBF, in which the total volume fraction of the soft phase (Sn + Pb) was equivalent to that in the Al–20Sn (vol.%) alloy, exhibited a lower strength but a threefold increase in ductility, which largely determines the wear resistance of antifriction materials used in sliding bearings [23]. Thus, the approach of improving ductility through the fragmentation of continuous intergranular tin layers proved to be effective. It should be noted that lead is less scarce than tin, and its addition significantly reduces the cost of the material. The mechanisms for improving the ductility of ternary LPBF alloys are not sufficiently studied, and their investigation is of great scientific interest.

Unfortunately, in the fabrication of this material, a narrow range of LPBF parameters was employed, differing only in the laser scanning speed (v) at a constant laser power of $P = 70$ W. Consequently, the volumetric energy density (E) delivered to a unit volume of the powder bed was limited [26]. This parameter can be increased, while keeping other LPBF parameters constant, only by increasing the laser power, since it is defined by the following relationship:

$$E = \frac{P}{hsv},$$

where h is the powder layer thickness and s is the hatch spacing [27].

According to Ref. [23], at a laser power of P = 130 W and a scanning speed of v = 1.2 m/s, a volumetric energy density of approximately $E \approx 40$ J·mm^{-3} was sufficient to produce Al–20Sn samples with maximum strength and low porosity. Further increases in P resulted only in a decrease in ultimate strength (σ_U), while the material ductility remained low ($\varepsilon \leq 3.5\%$). In the Al–15Sn–5Pb alloy, in addition to tin, the presence of the much heavier lead requires additional time to achieve uniform concentration in the molten bath. During cooling of a homogeneous melt, it can be expected that solidification will yield a material with a more uniform phase distribution and improved mechanical properties. The time required for concentration equalization at a fixed P corresponds to LPBF regimes with lower laser scanning speeds. Therefore, the aim of this study is to investigate the structure and properties of the Al–15Sn–5Pb (vol.%) alloy produced over a wide range of laser scanning speeds at laser powers between 70 and 130 W.

2. Materials and Experimental Procedure

For the synthesis of the material by the LPBF method, pre-sieved elemental powders of aluminum ASD-1 grade [TU 48-8-226-87], tin PO-1 grade [28], and lead PS-1 grade [29] with particle sizes ranging from 25 to 50 μm were used. The powders were mixed in a Turbula-type mixer for 3 h at a container rotation speed of 30 rpm. A powder blend of the following composition (vol.%) was prepared: Al–15Sn–5Pb, hereafter referred to as P5 (Figure 1). The base composite of Al–20Sn (vol.%) composition is designated as P0. After mixing, the prepared powder mixtures were dried in a vacuum furnace at 120 °C for 3 h.

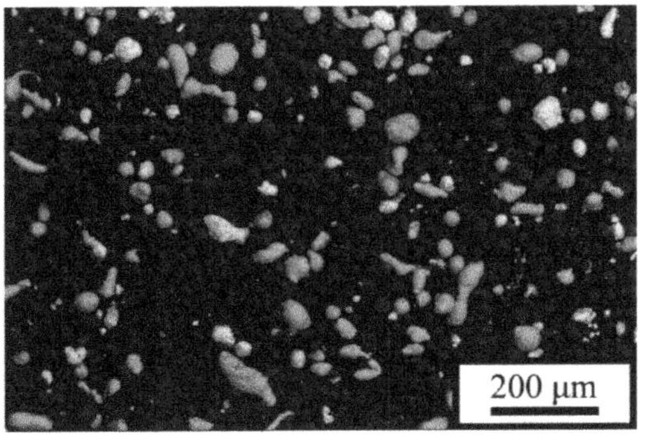

Figure 1. SEM image of the Al–15Sn–5Pb powder mixture. Bright particles correspond to Pb, gray particles to Sn, and dark particles to Al [26].

The samples were produced using an AM150 3D printer (ONSINT, Zelenograd, Russia). The LPBF process was carried out in a chamber filled with high-purity flowing argon, and the oxygen concentration in the chamber did not exceed 300 ppm. The thickness of the laser-irradiated powder layer was 30 μm, and this layer was applied to the surface of the synthesized sample with a squeegee. Each layer was deposited on top of the previous one onto a AMg6 alloy substrate preheated to 110 ± 5 °C using a rubber recoater blade. The laser scan strategy was linear with a 67° rotation between successive layers. The scanning speeds were 1.2, 1.0, and 0.8 m/s, the laser spot diameter was 75 ± 5 μm, and the hatch spacing was 90 μm. The laser power varied from 70 W to 130 W in 20 W increments.

The dimensions of the samples were 10 mm × 10 mm × 10 mm. The material porosity was determined by the hydrostatic weighing method in accordance with GOST 20018-74 [30]. Compression tests were performed using an Instron-1185 universal testing machine (Instron, Norwood, MA, USA) at a crosshead displacement rate of 0.5 mm/min. The specimens for compression were cut from the central region of the fabricated samples, with the loading axis oriented perpendicular to the LPBF build direction. Ends of the compressive specimens were lubricated with graphite prior to testing. For each LPBF processing condition, at least three specimens were tested.

Metallographic specimens were prepared using a standard procedure: grinding with SiC abrasive papers of progressively decreasing grit size, followed by polishing on a cloth pad using a diamond suspension containing particles smaller than 1 μm. The polished samples were then cleaned in an ultrasonic bath with alcohol.

The microstructure of the polished specimens was examined using a LEO EVO 50 scanning electron microscope (Carl Zeiss, Oberkochen, Germany) equipped with an energy-dispersive X-ray spectroscopy (EDS) attachment. X-ray diffraction (XRD) data were obtained using a DRON-8H diffractometer (Innovation Center Burevestnik, Saint Petersburg, Russia) with Cu Kα radiation (λ = 1.5406 Å) over a 2θ range of 10–140°. The diffraction patterns were processed using the PDWin V3.0 software package (Innovation Center Burevestnik, Saint Petersburg, Russia).

3. Results and Discussion

From the given image of the initial powder mixture, it is evident that its components are distributed relatively evenly, which indicates that the mixing conditions were chosen correctly (Figure 1). The samples synthesized by the LPBF method consisted of three phases—Al, α-Sn, and β-Pb (Figure 2)—in full agreement with the equilibrium phase diagram. The amount of energy transferred to a unit volume of the powder bed is presented in Table 1. As follows from the data, despite the different combinations of P and v, some LPBF regimes produced samples with similar values of E. This indicates that if E is the dominant LPBF parameter governing the structure and, consequently, the mechanical properties of the fabricated composites, then the deformation behavior of samples produced at comparable E values can be expected to be similar. A total of twelve types of samples were fabricated: two at $E \approx$ 26–28 J·mm^{-3}, three at $E \approx$ 32–34 J·mm^{-3}, three at $E \approx$ 40–42 J·mm^{-3}, and two at $E \approx$ 48–51 J·mm^{-3}. Additionally, single samples were produced at the minimum (E = 21.6 J·mm^{-3}) and maximum (E = 60 J·mm^{-3}) energy densities.

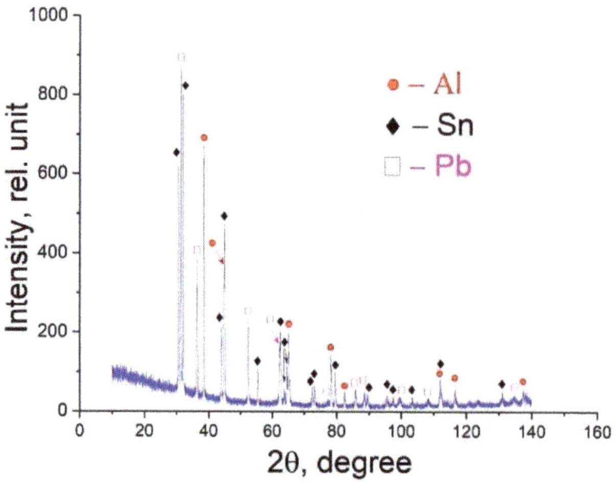

Figure 2. X-ray diffraction pattern of the P5 alloy produced at v = 0.8 m/s and P = 70 W.

Table 1. Dependence of the volumetric energy density *E* (J·mm^{-3}) on the LPBF processing parameters.

Scanning Speed v, m/s	Laser Power *P*, W			
	70	**90**	**110**	**130**
1.2	21.6	27.8	34.0	40.1
1.0	25.9	33.3	40.7	48.1
0.8	32.4	41.6	50.9	60.1

Figure 3 shows the corresponding compressive stress–strain curves for the fabricated samples. For comparison, the compression curves of LPBF-fabricated samples with composition P0 produced under the same processing conditions are also presented. As seen from the plots, the value σ_U of P0 alloy is in the range of 100–105 MPa and decreases slightly with increasing laser power. The ductility of the samples also improves marginally under these conditions (Figure 3a).

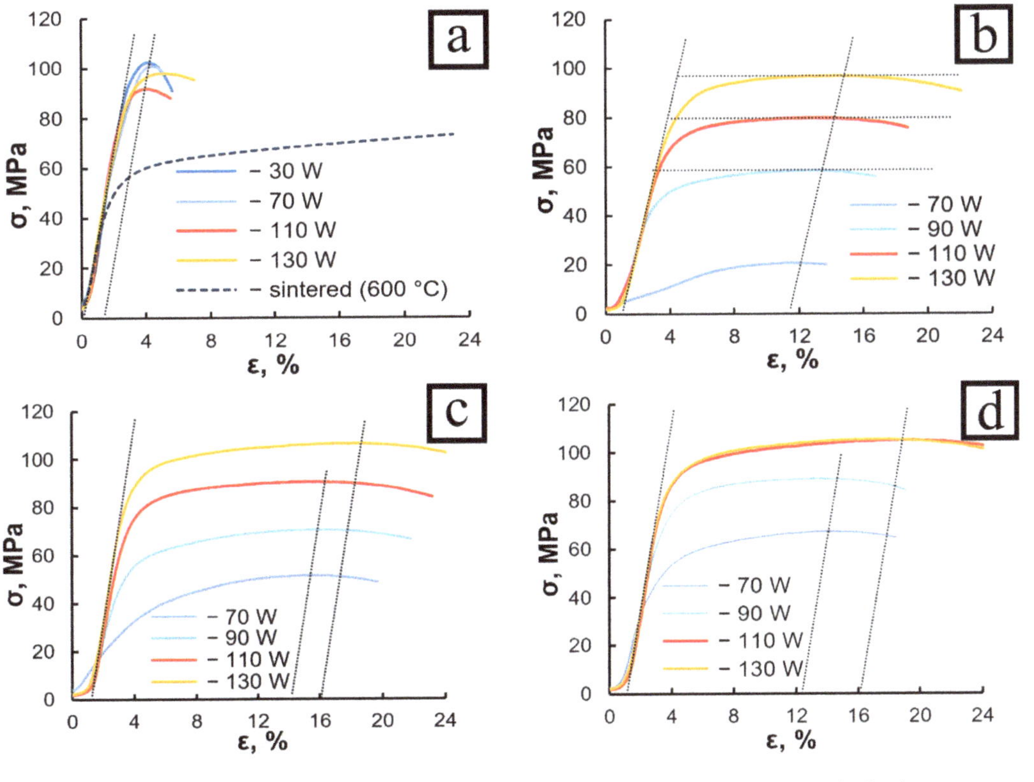

Figure 3. Compressive stress–strain curves of the P5 alloy fabricated by the LPBF method at laser scanning speeds (m/s): 1.2 (**a**,**b**), 1.0 (**c**), and 0.8 (**d**). Alloy compositions: P0 (**a**) and P5 (**b**–**d**).

A comparison of the plots in Figure 3a,b shows that under identical LPBF processing conditions and with the same volume fraction of the aluminum phase, the $\sigma(\varepsilon)$ curves of the P0 and P5 alloys differ significantly. For the P0 alloy, the segment of the curve following the elastic region is very short: the flow stress σ rapidly reaches its maximum value σ_U at a strain of approximately $\varepsilon \approx 1.5\%$ and then sharply decreases due to material fracture. Increasing the laser power results in a slight reduction in the strength of the P0 alloy and a moderate increase in ductility up to $\varepsilon \approx 3\%$. Thus, grain refinement in the P0 alloy leads to an increase in flow stress accompanied by a pronounced loss of ductility. The decrease in ductility of LPBF-fabricated alloys [23,26] is attributed to a sharp decrease in the ductility reserve of the intergranular tin layers as they become thinner. To support this conclusion, Figure 3a also shows the compression curve (dashed line) of a sample of the same composition produced by vacuum sintering. It can be seen that plastic flow of the sintered sample begins at relatively low stresses and continues even after $\varepsilon > 20\%$, while the material has not yet reached its maximum strength σ_U.

The shape of the compression curves for the P5 alloy differs from those of P0 by the greater length of the plastic flow region (Figure 3b). The maximum strength σ_U of the P5 samples is shifted toward higher strains, reaching ε = 11–12%. The strongest sample was prepared at $E = 40$ J·mm^{-3}, with its σ_U nearly matching that of the P0 alloy produced at the same energy density. The strength of P5 samples fabricated at lower E values was lower, although their maximum ductility remained high (ε = 11–12%).

A decrease in the laser scanning speed up to $v = 1.0$ m/s led to an increase in the volumetric energy density E. The samples produced at this scanning speed exhibited slightly higher compressive strength, with the maximum value shifted along the strain axis to approximately $\varepsilon \approx 14\%$ (Figure 3c). Thus, reducing v resulted in a slight improvement in ductility and an increase in σ_U, primarily due to a higher yield stress $\sigma_{0.2}$ and more pronounced strain hardening of the material. The strengthening effect was especially notable for samples fabricated at lower E values.

A further decrease in the scanning speed to v = 0.8 m/s led to an additional increase in the flow stress of the samples. In this case, the $\sigma(\varepsilon)$ curves of the specimens produced at laser powers of $P = 130$ and 110 W nearly coincided (Figure 3d). According to Table 1, these samples were fabricated at energy densities exceeding $E > 48$ J·mm^{-3}. Therefore, it can be assumed that this value of E represents a critical threshold for the P5 composition, and that its further increase—either by raising P or by reducing v—does not have a significant effect on the mechanical properties of the synthesized samples (Figure 3).

Thus, it can be concluded that substituting one quarter of the tin volume with lead in the LPBF-fabricated Al-20Sn alloy resulted in a significant improvement in its ductility. The alloy strength, however, decreased, especially at energy densities $E < 40$ J·mm^{-3}. Increasing E to approximately 48 J·mm^{-3}—either by raising P or by reducing v—provided samples with the highest combination of mechanical properties.

Since the influence of the energy density E on the mechanical properties of the P5 alloy is manifested through the characteristics of its resulting microstructure, the corresponding structural features were investigated. Figure 4 presents microstructural images of the P5 alloy produced at a laser power of 70 W and $v = 1.2$ m/s ($E = 21.6$ J·mm^{-3}) and $v = 0.8$ m/s ($E = 32.4$ J·mm^{-3}). That is, in this case, a decrease in v led to an increase in E by ≈50%.

The alloy produced at the lowest energy density E exhibits a heterogeneous phase distribution throughout the material volume. Moreover, in addition to the phases precipitated from the solidifying melt, regions of unmolten primary aluminum are present (dark particles in the micrographs, Figure 4a). These are presumably aluminum powder particles covered with a thick oxide film. Adjacent to them are gray regions with a eutectic structure, representing domains of Al–Sn alloy (Figure 4b). The samples also contained a small number of pores, the proportion of which was 1–2% and was practically independent of the laser energy at $E > 25$ J·mm^{-3}. The features of tin distribution indicate that the alloy formation during laser heating of the powder layer proceeded as follows: neighboring aluminum powder particles melted and coalesced into larger regions, while liquid tin and lead were located along their peripheries. Tin began to diffuse into the aluminum regions, enriching the upper layers of the molten aluminum with Sn, whereas the concentration decreased with depth. Thus, the upper areas of the molten aluminum regions consisted of a tin-enriched liquid, the intermediate layers contained less Sn, and the central regions were occupied by molten aluminum that had not yet mixed with tin. The undissolved tin was distributed between aluminum particles and mixed with lead, which precipitated as discrete particles upon cooling of the Sn–Pb melt. Hence, the relatively low temperature of molten aluminum and the presence of solid particles hindered the development of convective flow, so mixing with alloying elements occurred mainly through mutual diffusion.

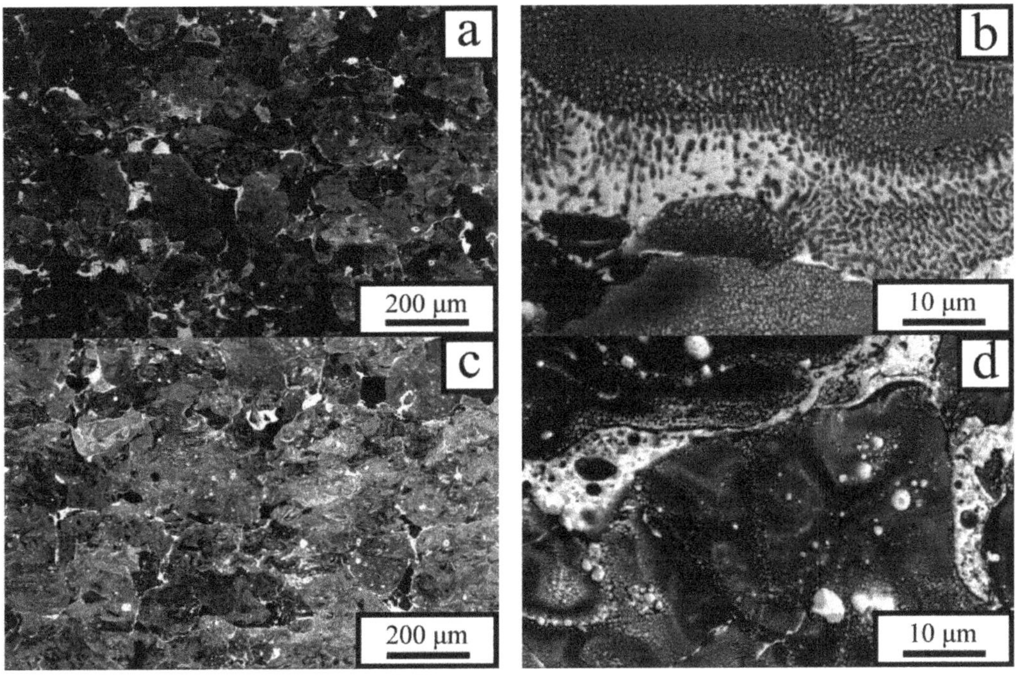

Figure 4. SEM images of the P5 alloy structure fabricated by LPBF at a laser power of $P = 70$ W and laser scanning speeds of $v = 1.2$ m/s (**a**,**b**) and $v = 0.8$ m/s (**c**,**d**).

The incident energy density E increases with decreasing laser scanning speed, which strongly affects the microstructure of the P5 alloy. The number of unmolten aluminum particles decreases markedly, while the fraction of gray regions—representing Al–Sn mixture areas—increases (Figure 4c). The size and number of Sn–Pb inclusions diminish as tin dissolves into aluminum, leading to a more uniform distribution of tin throughout the material. As seen in Figure 4d, convective flows begin to play an increasingly important role in the mixing of liquid tin with aluminum, alongside diffusion. These flows transport tin into the larger aluminum regions more efficiently than diffusion alone. The tin-enriched regions formed in this way are clearly visible in Figure 4d as light-contrast boundaries. A portion of the tin remains trapped inside the rapidly solidifying aluminum regions as isolated inclusions. Lead does not mix appreciably with aluminum and is located within the tin inclusions situated along the outer boundaries of the large aluminum regions.

This elemental distribution is confirmed by the image shown in Figure 5, where the intensity profiles of the elemental reflections are superimposed on the microstructural section along the scan line. The image reveals that there are regions of pure aluminum as well as eutectic regions with a constant Al/Sn ratio in the P5 alloy fabricated at low E. During solidification, all regions of the melt are divided into large fragments separated by thin tin interlayers. Lead, in contrast to tin, is practically insoluble in aluminum and is located in the intergranular regions together with tin; during solidification of the melt, it precipitates as discrete particles.

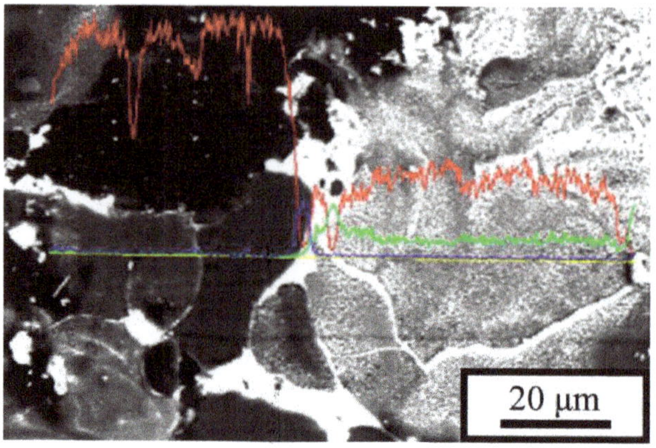

Figure 5. Microstructure of the P5 alloy and the corresponding elemental distribution. LPBF parameters: $P = 70$ W, $v = 1$ m/s. Elemental intensity profiles: aluminum (red), tin (green), and lead (blue). Electron beam scan line is marked by yellow.

The microstructure of the samples produced at the same laser scanning speeds but at a higher laser power of $P = 130$ W is shown in Figure 6. A comparison with the previous micrographs reveals that the structure of the alloy fabricated at higher laser power is more homogeneous than that of the alloy produced at $P = 70$ W. Even in the sample prepared at $v = 1.2$ m/s ($E \approx 40$ J·mm^{-3}), almost no unmolten or poorly mixed regions remain (Figure 6a), and the tin inclusions—both large and small—observed in Figures 4 and 5 as thin surface layers have nearly disappeared. Relatively fine intergranular inclusions corresponding to the Sn–Pb eutectic can be seen only at higher magnifications (Figure 6b).

Figure 6c,d show the microstructure of the samples exhibiting the best mechanical performance—high ductility combined with high strength. This is a consequence of the fact that the inclusions of soft phases are isolated and do not form a continuous net. Compared with the previous samples, this structure is even more homogeneous, with almost no tin segregations in the form of intergranular layers. No areas of unmelted aluminum were detected. The main part of the material volume is occupied by regions (outlined in yellow in Figure 6d) where aluminum is uniformly mixed with tin. However, some elongated regions with a compositional gradient of tin are still present; during solidification, tin was expelled toward their peripheries. Such regions are much more numerous and larger in Figure 6b than in Figure 6d, which explains the lower mechanical properties of the samples (Figure 3d).

A structure similar to the relatively homogeneous one observed in Figure 6d is also formed at $P = 110$ W, provided that the incident energy density on the sample surface exceeds 48 J·mm^{-3}. In this case, due to the high melt temperature, mixing of the components occurs predominantly through multiple convective flows. During solidification, a fine-grained structure is formed in which tin does not have sufficient time to produce flat inclusions by diffusion (Figure 7b) that would otherwise promote localized plastic flow. As a result, the strain distribution within the loaded sample remains relatively uniform, contributing to the simultaneous increase in both ductility and strength of the material (Figure 3d).

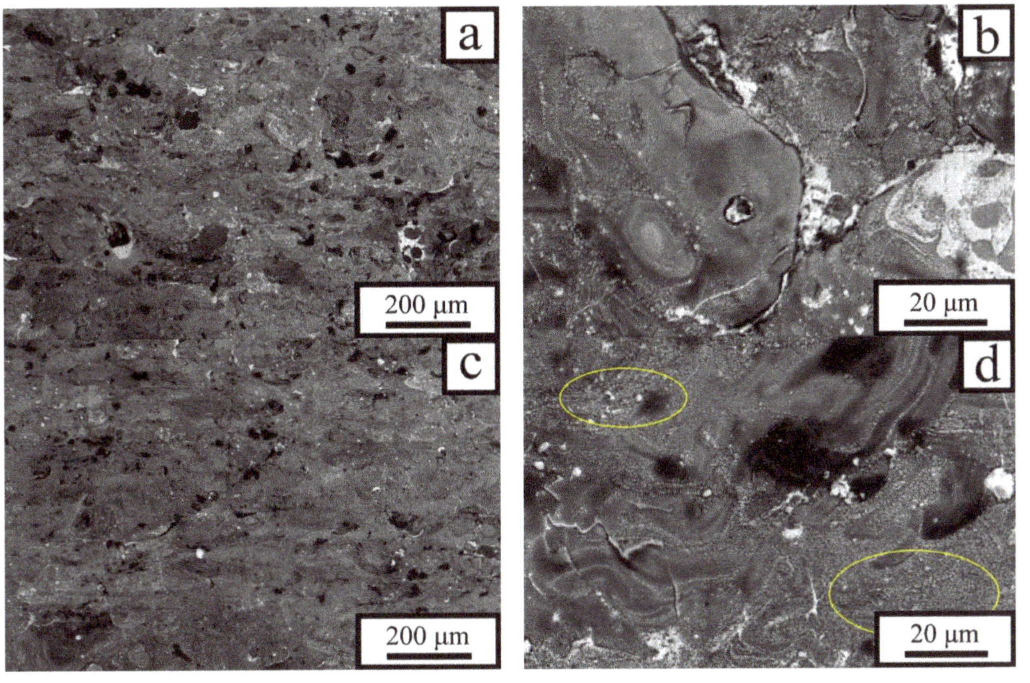

Figure 6. Microstructure of the P5 alloy. Laser power P = 130 W; laser scanning speeds v = 1.2 m/s (**a**,**b**) and 0.8 m/s (**c**,**d**).

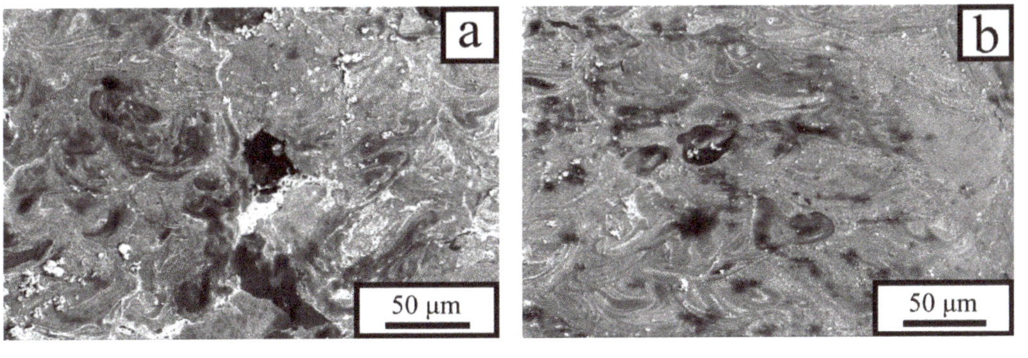

Figure 7. Microstructure of the P5 alloy. Laser power P = 110 W; laser scanning speeds v = 1.2 m/s (**a**) and 0.8 m/s (**b**).

Thus, the energy density E has a strong influence on the microstructure of the P5 alloy and, consequently, on its mechanical properties. When E exceeds 48 $J \cdot mm^{-3}$, numerous regions with a homogeneous distribution of tin in aluminum, which prevents deformation localization in the loaded material, are formed during the solidification. In this case, both the strength and ductility of the alloy are governed primarily by the mechanical behavior of the aluminum matrix. However, when E is insufficient to ensure intensive and uniform mixing of tin and aluminum, solidification results in the formation of regions containing planar tin inclusions (Figure 7a), which reduce the flow stress of the samples (Figure 3). In this case, lead acts as a beneficial additive, as its particles within the tin interlayers hinder the formation of extended bands of localized shear.

4. Conclusions

Based on the obtained results from the study of the structure and mechanical properties of the Al–15Sn–5Pb alloy fabricated by the LPBF method under various laser irradiation regimes, the following conclusions can be drawn:

- Substituting one quarter of the tin volume with lead results in a significant increase in the ductility of the LPBF-fabricated Al–15Sn–5Pb (vol.%) alloy. Therefore, a promising direction for further research is the study of Al-20(Sn + Pb) (vol.%) alloys with higher lead content.
- The strength of the Al–15Sn–5Pb (vol.%) alloy depends on the degree of compositional homogeneity, which is determined by the intensity of melt mixing. The higher the volumetric laser energy density E delivered to the powder bed, the more intensive the mixing of components and the higher the strength of the LPBF-fabricated ternary alloy.
- Increasing E beyond 48 $J \cdot mm^{-3}$ has practically no significant effect on the mechanical properties of the Al–15Sn–5Pb (vol.%) alloy.

Author Contributions: Conceptualization, N.M.R.; Methodology, N.M.R.; Formal analysis, A.L.S. and D.P.I.; Investigation, A.L.S., K.O.A. and V.E.L.; Data curation, A.L.S., K.O.A. and V.E.L.; Writing—original draft, N.M.R.; Writing—review & editing, K.O.A. and A.I.D.; Visualization, A.L.S., K.O.A. and D.P.I.; Supervision, A.I.D.; Project administration, A.I.D. All authors have read and agreed to the published version of the manuscript.

Funding: The work was performed within the framework of Assignment of the Russian Science Foundation (Project No. 24-79-10099).

Data Availability Statement: The original contributions presented in this study are included in the article. Further inquiries can be directed to the corresponding author.

Conflicts of Interest: The authors declare no conflict of interest.

References

1. Zeng, M.Q.; Hu, R.Z.; Song, K.Q.; Dai, L.Y.; Lu, Z.C. Enhancement of wear properties of ultrafine-structured Al–Sn alloy-embedded Sn nanoparticles through in situ synthesis. *Tribol. Lett.* **2019**, *67*, 84. [CrossRef]
2. Mironov, A.E.; Gershman, I.S.; Gershman, E.I. Influence of tin on the tribotechnical properties of complex antifriction aluminum alloys. *J. Frict. Wear* **2018**, *39*, 394–399. [CrossRef]
3. Heydari, V.; Alizadeh, M.; Pashangeh, S. Investigation into microstructural and mechanical properties of Al–Sn alloy fabricated using accumulative roll bonding combined with heat treatment. *J. Mater. Res. Technol.* **2023**, *26*, 3481–3495. [CrossRef]
4. Seo, N.; Lee, S.-H.; Lee, J.; Park, S.; Kim, M.-S.; Lee, S.-J.; Jung, J.-G. Microstructural evolution and aging behavior of multicomponent Al–Sn–Zn–Mg–Li alloy. *J. Alloys Compd.* **2025**, *1011*, 178381. [CrossRef]
5. Bertelli, F.; Freitas, E.S.; Cheung, N.; Arenas, M.; Conde, A.; Damborenea, J.; Garcia, A. Microstructure, tensile properties and wear resistance correlations on directionally solidified Al–Sn–(Cu;Si) alloys. *J. Alloys Compd.* **2017**, *695*, 3621–3631. [CrossRef]
6. Xu, M.; Yin, Y.G.; Li, C.M.; Zhang, G.T.; Duan, C.C. Effects of macrosegregation on mechanical and tribological properties of squeeze casting immiscible bearing alloys. *China Foundry* **2023**, *20*, 443–451. [CrossRef]
7. Huang, S.; Zhu, B.; Zhang, Y.; Liu, H.; Wu, S.; Xie, H. Microstructure comparison for AlSn20Cu antifriction alloys prepared by semi-continuous casting, semi-solid die casting, and spray forming. *Metals* **2022**, *12*, 1552. [CrossRef]
8. Sadawy, M.; Metwally, H.; Abd El-Aziz, H.; Adbelkarim, A.; Mohrez, W.; Mashaal, H.; Kandil, A. The role of Sn on microstructure, wear and corrosion properties of Al–5Zn–2.5Mg–1.6Cu–xSn alloy. *Mater. Res. Express* **2022**, *9*, 096507. [CrossRef]
9. Li, S.; Li, Y.; Guo, H.; Wen, Z.; Zhu, Z. Effect of electromagnetic stirring frequency on tribological performance and corrosion resistance of Al–Sn bearing alloy. *Mater. Today Commun.* **2022**, *32*, 103898. [CrossRef]
10. Barros, A.; Cruz, C.; Botelho, T.; Silva, A.; Casteletti, L.; Garcia, A.; Cheung, N. Dry sliding wear features of an Al–20Sn–5Zn alloy affected by microstructural length scales. *Lubricants* **2022**, *10*, 352. [CrossRef]
11. *GOST 14113-78;* Splavy Alyuminiyevye Antifriktsionnye. Marki. Gosstandart SSSR: Moscow, Russia, 1980.
12. Chikova, O.A.; Shishkina, E.V.; Konstantinov, A.N. Measurement of Young's modulus and hardness of Al–50 wt.% Sn alloy phases using nanoindentation. *Phys. Met. Metallogr.* **2013**, *114*, 616–622. [CrossRef]

13. Abed, E.J. Study of solidification and mechanical properties of Al–Sn casting alloys. *Asian Trans. Eng.* **2012**, *2*, 89–98.
14. Bushe, N.A.; Dvoskina, V.A.; Rakov, K.M. *Bearings Made of Aluminum Alloys*; Transport: Moscow, Russia, 1974.
15. Zeng, Z.; Li, Y.; Wang, H.; Lai, R.; Teng, J.; Li, Y. Tribological behavior of AlSn20Cu alloy manufactured by additive friction stir deposition. *Wear* **2025**, *560–561*, 205606. [CrossRef]
16. Rusin, N.M.; Skorentsev, A.L.; Kolubaev, E.A. Effect of equal channel angular pressing on mechanical and tribological properties of sintered Al–Sn composites. *J. Mater. Eng. Perform.* **2020**, *29*, 1955–1963. [CrossRef]
17. Valizadeh, A.R.; Kiani-Rashid, A.R.; Avazkonandeh-Gharavol, M.H.; Karimi, E.Z. Influence of cooling rate on the microstructure and microsegregation in Al–30Sn binary alloy. *Metallogr. Microstruct. Anal.* **2013**, *2*, 107–112. [CrossRef]
18. Ning, X.-J.; Kim, J.-H.; Kim, H.-J.; Lee, C. Characteristics and heat treatment of cold-sprayed Al–Sn binary alloy coatings. *Appl. Surf. Sci.* **2009**, *255*, 3933–3939. [CrossRef]
19. Hernandez, O.; Gonzalez, G. Microstructural and mechanical behavior of highly deformed Al–Sn alloys. *Mater. Charact.* **2008**, *59*, 534–541. [CrossRef]
20. Sercombe, T.B.; Li, X. Selective laser melting of aluminium and aluminium metal matrix composites: Review. *Mater. Technol.* **2016**, *31*, 77–85. [CrossRef]
21. Rosenthal, I.; Stern, A.; Frage, N. Strain rate sensitivity and fracture mechanism of AlSi10Mg parts produced by selective laser melting. *Mater. Sci. Eng. A* **2017**, *682*, 509–517. [CrossRef]
22. Confalonieri, C.; Gariboldi, E. Al–Sn miscibility gap alloy produced by powder bed laser melting for application as phase change material. *J. Alloys Compd.* **2021**, *881*, 160596. [CrossRef]
23. Rusin, N.M.; Akimov, K.O.; Skorentsev, A.L.; Dmitriev, A.I. Influence of laser power during laser powder bed fusion on the structure and properties of Al–40Sn alloy prepared from a mixture of elemental powders. *J. Alloys Compd.* **2025**, *1013*, 178648. [CrossRef]
24. Billur, A.C.; Gerçekcioglu, E.; Bozoklu, M.; Saatçi, B.; Ari, M.; Nair, F. Electrical, thermal conductivity, microstructure and mechanical properties of Al–Sn–Pb ternary alloys. *Solid State Sci.* **2015**, *46*, 107–115. [CrossRef]
25. Mosleh, A.O.; Kotova, E.G.; Kotov, A.D.; Gershman, I.S.; Mironov, A.E. Bearing aluminum-based alloys: Microstructure, mechanical characterizations, and experiment-based modeling approach. *Materials* **2022**, *15*, 8394. [CrossRef] [PubMed]
26. Skorentsev, A.L.; Akimov, K.O.; Rusin, N.M.; Likharev, V.E.; Dmitriev, A.I. Effect of laser beam scanning speed during SLM on the structure and physical–mechanical properties of Al–15Sn–5Pb alloy. *Russ. Phys. J.* **2025**, *68*, 221–228. [CrossRef]
27. Confalonieri, C.; Casati, R.; Gariboldi, E. Effect of process parameters on laser powder bed fusion of Al–Sn miscibility gap alloy. *Quantum Beam Sci.* **2022**, *6*, 17. [CrossRef]
28. *GOST 9723-73*; Poroshok Olovyannyy. Tekhnicheskie Usloviya. Gosstandart SSSR: Moscow, Russia, 1973.
29. *GOST 16138-78*; Poroshok Svintsovyy. Tekhnicheskie Usloviya. Izdatel'stvo Standartov: Moscow, Russia, 1978.
30. *GOST 20018-74*; Splavy Tverdyye Spechennye. Metod Opredeleniya Plotnosti. Izdatel'stvo Standartov: Moscow, Russia, 1974.

MDPI AG
Grosspeteranlage 5
4052 Basel
Switzerland
Tel.: +41 61 683 77 34

Materials Editorial Office
E-mail: materials@mdpi.com
www.mdpi.com/journal/materials

www.ingramcontent.com/pod-product-compliance
Lightning Source LLC
Chambersburg PA
CBHW061546210926
51850CB00009B/2016
* 9 7 8 3 7 2 5 8 7 1 6 6 7 *